Mark D. Hoover

Radiation Instruments

Health Physics Society
2001 Summer School

Edited by
Herman Cember

Medical Physics Publishing
Madison, Wisconsin

ISBN: 1-930524-03-X
Library of Congress Catalog Card Number: 2001087768

Published By
Medical Physics Publishing
4513 Vernon Boulevard
Madison, WI 53705
(608) 262-4021

On Behalf of
Health Physics Society
1313 Dolley Madison Blvd.
Suite 402
McLean, VA 22101
Phone: (703) 790-1745
FAX: (703) 790-2672
E-mail: hps@burkinc.com
Home Page: http://www.hps.org

Printed in the United States of America

PREFACE

One of the main characteristics that distinguishes radiation from most other environmentally-based health threats is the fact that humans have no sensory detectors for radiation to serve as warnings of the existence of a radiation field that may be great enough to be potentially harmful. As a result, we must depend on instruments. The purpose of using health physics instruments is to keep us apprised of the radioactivity and radiation levels in our environment in order to maintain radiation health and safety. To achieve this end the health physicist uses a wide variety of radiation measuring instruments for a number of different specific purposes. Some of these uses include, but are not limited to:

- Sensitive gamma ray detectors for searching for a possible leak in shielding
- Sensitive alpha or beta detectors for finding contamination
- Measurement of background environmental dose rates
- Measurement of dose rates in the workplace
- Identifying environmental radionuclides by spectroscopic methods
- Radiometric analysis for invitro bioassay
- In vivo (whole body counting) bioassay
- Personal dosimeters
- Personal (lapel type) air samplers
- Air sampling for particles
- Air sampling for vapors and gases

Although many different types of instruments are used in the practice of health physics, the operating principles are relatively few. The basic requirement for health physics instruments is that the detector interacts with the radiation in such a manner that the magnitude of the instrument's response is proportional to the radiation property or radiation effect. The response of a health physics dosimeter should be similar to that of human tissue, but must be much more sensitive than human tissue. That is, the instrument must respond to a level of radiation that is very much less than that which might be expected to cause an observable effect in humans. In this sense, a dose measuring instrument may be thought of as a surrogate for a person.

Although new detectors, such as thermoluminescent and optically stimulated luminescent detectors for personnel monitoring, and scintillation counters and semiconductor detectors for sensitive measurements and for spectroscopy have been introduced and are in use, photographic film and gas filled chambers are still widely used in the practice of health physics. However, there have been significant advances in the electronic processing of the output signals from the detectors and the treatment and analysis of the information carried by these signals. These advances have rendered the older instruments obsolete. The new health physics instruments are smaller and lighter than their predecessors. They also are "smart" instruments that can log, plot, and analyze data, and store the information in a form that can be downloaded into a computer for further analysis or for record keeping.

The purpose of this Health Physics Summer School is to update these students' knowledge of the attributes, limitations, and proper application of the respective health physics instruments in order to allow radiation sources to be safely used for the benefit of individuals and the community. To this end, we have recruited a group of eminent health physicists who are specialists in the topics that they present. We all expect to benefit from their knowledge and experience.

Herman Cember, Ph.D., PE, CHP
Academic Dean, 2001 Health Physics Summer School
Professor Emeritus, Northwestern University
Visiting Professor, Purdue University

CONTRIBUTORS

Joseph L. Alvarez
Auxier & Associates, Inc.
9821 Cogdill Rd.
Knoxville, TN 37932

Shane Brightwell
Shepard Miller Inc.
4192 Lookout Drive
Loveland, CO 80537

Frazier Bronson
Canberra Industries
800 Research Parkway
Meriden, CT 06450

Bernard L. Cohen
University of Pittsburgh
Pittsburgh, PA

F. Morgan Cox
Lovelace Respiratory Research Institute
Kirtland AFB Bldg 92000, Area Y
Albuquerque, NM 87115

David J. Derenzo
University of Illinois at Chicago
Radiation Safety Section, M/C 932
820 S. Wood St.
Chicago, IL 60612-7314

Douglas Draper
7400 Willey Road
Fernald, OH 45030

R.S. Eby
CH2M Hill
Oak Ridge, TN 37830

Shawn W. Googins
NIH
21 Wilson Drive
Bethesda, MD 20892-6780

N.E. Hertel
Neely Nuclear Research Center
Georgia Institute of Technology
900 Atlantic Drive NW, room 114
Atlanta, GA 30332

Mark D. Hoover
Lovelace Respiratory Research Institute
Kirtland AFB Bldg 92000, Area Y
Albuquerque, NM 87115

R.D. Ice
Neely Nuclear Research Center
Georgia Institute of Technology
900 Atlantic Drive NW, room 114
Atlanta, GA 30332

E. Jawdeh
Neely Nuclear Research Center
Georgia Institute of Technology
900 Atlantic Drive NW, room 114
Atlanta, GA 30332

Raymond H. Johnson, Jr.
CSI Radiation Safety Academy
481 North Frederick Ave., Suite 302
Gaithersburg, MD 20877

Alan Justus
Argonne National Laboratory
9700 S Cass Ave.
Argonne, IL 60439

Ronald Kathren
137 Spring Street
Richland, WA 99352-1651

Charles Kent
1748 Colonial Shores Drive
Hixson, TN 37343

Michael W. Lantz
Arizona Public Service Company
Palo Verde Nuclear Generating Station
5801 S. Wintersburg Road
Mail Station 7902
Tonopah, AZ 85354-7529

John A. Leonowich
Pacific Northwest National Laboratory
Richland, WA 99352

L.A. Lundberg
CH2M Hill
Oak Ridge, TN 37830

S.G. Marske
CH2M Hill
Oak Ridge, TN 37830

David W. Miller
Clinton Power Station
Illinois Power Company
P.O. Box 678
Clinton, IL 61727

Carson A. Riland
8221 Hollister Ave
Las Vegas, NV 89131

Joseph J. Shonka
Shonka Research Associates
4939 Lower Roswell Road
Suite 106
Marietta GA 30068

R. Craig Yoder
Landauer, Inc.
2 Science Road
Glenwood, IL 60425

TABLE OF CONTENTS

Chapter 1

ORIGINS OF MODERN HEALTH PHYSICS INSTRUMENTATION

Ronald L. Kathren

1.1 IONIZING RADIATION DETECTION AND MEASUREMENT: THE BEGINNING

1.1.1 The Discovery of X-rays and Radioactivity

During the slightly more than the full century that has passed since the discovery of x-rays and radioactivity, human understanding of the hazards associated with exposure to ionizing radiations has grown enormously as have the sophisticated instruments essential to the modern practice of health physics. The pivotal twin discoveries of x-rays and radioactivity in the closing years of the nineteenth century were both made with the simplest of passive devices that the modern day health physicist would classify as solid state devices. The discovery of x-rays was made late in the afternoon of Friday, November 8, 1895, when a screen of crystalline barium platinocyanide, exposed to the unseen radiation emanating from energized Crookes tube in the laboratory of German physicist Wilhelm Roentgen, emitted a pale green fluorescence attesting to the presence of a new and potentially utilitarian form of radiation. A few months later, in neighboring France, physicist Antoine Henri Becquerel discovered another new and curious energy source, later named radioactivity, when he observed blackening in a developed photographic plate on which some uranium ore had been placed. Thus, one discovery was made with a primitive scintillation detector and the other with a photographic emulsion -- the first record use of instruments for the detection of ionizing radiation. Incredible as it may seem, these same detection devices are still used today for quantification of radiation and radioactivity.

1.1.2 First Measurements of Ionizing Radiation

Initially, measurement progress was slow, with efforts directed more towards understanding just what the newly discovered phenomena were all about, and, in the case of x-rays, application to medical uses. Perhaps the earliest attempt at quantitation of energy deposition -- a forerunner of the modern measurement of absorbed dose -- was made by in 1897 by German physicist Frederich Ernst Dorn, better remembered as the discoverer of radon in 1900 and for his other researches on radioactivity. Dorn used a simple air thermometer to measure the heat produced in thin metal sheets exposed to x-rays, thus carrying out what were apparently the first calorimetry measurements of ionizing radiation (Dorn 1897). Within the next two years, the technique was greatly expanded and improved upon by Ernest Rutherford and R. K. McClung (1900) who used a sensitive bolometer to determine the change in the electrical resistance in metal strips heated from exposure to both x-rays and to radioactive materials and in the process made one of the earliest measurements of the work function W, the amount of energy required to produce one ion pair in air. Despite these initial early successes, calorimetry was quickly rejected as impractical (Hudson 1933) and even today is little used in health physics.

The first radiation protection measurements appear to have been made with photographic emulsions. In 1902, the legendary American x-ray safety pioneer William Herbert Rollins described a technique for using photographic emulsions to test the efficacy of "non-radiable" housings that he himself had designed for x-ray tubes (Rollins 1902). And the following year, Philadelphia dermatologist Samuel Stern suggested placing strips of photographic film on the skin of patients to verify the dose that they had been administered during x-ray therapy (Stern 1903).

1.1.3 Electroscopes, Electrometers and Chemical Pastilles

Almost from the outset it seemed clear that ionization based measurements held the greatest promise for radiation measurements. The simple gold leaf electroscope, which required no external power source and was relatively rugged and sensitive, proved highly useful. Total ionization could be determined in terms of the total deflection of the leaves an electroscope discharged by radiation, and rate measurements could be simply obtained by timing the movement of the gold leaves as they were discharged. More delicate ionization chambers and electrometers became a mainstay of the physics laboratory and in his now classic book *Radio-activity* published in1904, Rutherford observed that electrometers were the ". . . most generally convenient apparatus" (Rutherford 1904). He also described a crude parallel plate electrometer and a cylindrical one, much like the modern Geiger-Mueller tube in configuration. The signal from the electrometer was fed to a quadrant electrometer, a relatively simple and sensitive device analogous in some ways to a Wheatstone bridge that was used to determine the rate of charge in the ionization chamber. Readout was direct; there was no signal amplification or gain and the power supply was a simply battery. These instruments were used in the research laboratory where high sensitivity was needed to measure the emanations from the small quantities of radioactive materials under study.

In medicine, the need was for routine measurement of relatively high doses from x-rays used for therapy. Electrometers and ionization chambers were too delicate and sensitive for this purpose, and the needs of the radiotherapist were far better met by solid mixtures of chemicals which changed color in accordance with the dose received. The first of the chemical dosimeters was a secret mixture of solid fused hydrochloric acid and sodium carbonate developed by the prominent Viennese radiologist Guido Holtzknecht, and quickly commercialized under the trade name *Chromoradimeter.* This simple device gave rise to the first unit of dose for it was calibrated in terms of H (for Holtzknecht) units, which related the colorimetric change in the dosimeter with biological response, specifically the induction of skin erythema (Holtzknecht 1902).

1.1.4 The Spinthariscope

As the first decade after the twin discoveries of x-rays and radioactivity drew to a close, Rutherford (1906) identified three basic categories or methods of measuring radiations from radioactive substances: 1) photographic; 2) phosphorescence (e.g. scintillation); 3) gas ionization. To these three basic laboratory methods must be added the chemical pastilles used in medicine, which were based on radiation induced color changes; at least two chemical pastilles were available commercially, each with its own calibration and dosage unit. There was also a commercial instrument for the detection of radiation in widespread use both in the scientific laboratory and by the general public as a parlor curiosity. This was the spinthariscope of Sir William Crookes. The spinthariscope consisted of a small brass tube fitted with an eyepiece at one end and a thin base covered with a layer of ZnS scintillation material at the other (Crookes 1903; Soddy 1912). The method of use was simplicity itself; a radioactive ore or similar material was placed against the end containing the ZnS, and the resulting scintillations or light flashes observed through the eyepiece at the opposite end. Counting the number of flashes, an admittedly tedious job, permitted a quantitative measurement to be obtained. This simple instrument, invented in 1903, enjoyed great initial popularity, and was used for decades in high school and college physics and chemistry laboratories, enjoying a renaissance of renewed interest in the late 1940's and 1950's as both an educational tool to demonstrate radioactivity, and by prospectors looking for uranium.

1.2 LABORATORY COUNTING INSTRUMENTS AND ASSOCIATED ELECTRONICS

1.2.1 Pioneer Days: The Electrometer

Radiological measurements began in the laboratory with simple electroscopes and electrometers, instruments that go back to the earliest days of radioactivity and were the basic laboratory measuring devices of the radioactivity pioneers. Almost immediately after discovering the blackening of a photographic plate by uranium salts, Becquerel verified his discovery of radioactivity with an electroscope. The Curie quartz electrometer, originally developed to study piezoelectricity by the Curie brothers, Pierre and Jacques, was quickly adapted to radiation measurement by Marie Curie who fitted the device with a small ionization chamber. A similar device was used by Mme. Curie in 1913 to develop the first radium calibration standard. The early electrometers were essentially passive ionization chambers in which the total deflection or rate of deflection of a charged needle or foil was used to measure the radiation output from a radioactive source, or, somewhat later, to measure cosmic ray intensity. The quadrant electrometer, basically a high resistance voltmeter, was a popular early device used by such luminaries as Rutherford, A. S. Eve, B. B. Boltwood, and the Curies. The simplicity and ease of use of electrometers made them popular and numerous types were developed; some of the early mechanical types such as the Hoffman (1912) and Lindemann (1924) were so serviceable that they were used well into the 1940's and beyond. One might say they really knew how to build them in those days!

However, active electronic devices, specifically the vacuum tube electrometer, held greater promise and offered important advantages over mechanical electrometers, which were difficult to set up and operate and in general quite temperamental and easily influenced by seemingly trivial outside sources. Laboratory measurements of radioactivity were revolutionized by application of vacuum tubes electrometers in the 1920's to measure the very small ionization currents. The use of these electrometers was initiated by astronomers trying to measure the minute currents in photocells coupled to their telescopes. The method worked well, and was soon adapted to measurement of low intensity ionizing radiation from cosmic rays. Early electrometer tubes were unstable and relatively insensitive but improvements in vacuum tube technology, coupled with operation at 4 to 6 volts, well below the ionization potential of most gases, quickly made them the standard for low level measurements both in the laboratory and in medical practice. By 1930, several manufacturers including Victoreen, Westinghouse, Western Electric and General Electric were producing commercial electrometers (Metcalf and Thompson 1930). Perhaps the most popular was the General Electric Type FT 54 which used a single thermionic tube with a high grid resistance and low grid current to compensate for small changes in filament and plate supply voltages. By 1933, compensating circuitry was developed to minimize disturbances from slight changes in battery voltage, greatly improving sensitivity and stability and enabling measurements to be made in the attoampere (10^{-18} amp) region (DuBridge and Brown 1933). The vacuum tube also made possible the line amplifier, which, through the use of multiple triodes connected in plate-to-grid series, was able to achieve gains of 10^5 by the middle 1930's (Kathren 1980).

1.2.2 Early Counters and Scalers

Although the electronic pulse produced in a Geiger tube or similar electronic detector is of sufficient magnitude to produce a measurable deflection in an electrometer, electronic counting required special circuitry, particularly if measurements of any magnitude were to be made. The basic circuitry that led to the development of the binary scaler was put forth in 1918 in the form of the flip-flop circuit, more properly known as Eccles-Jordan trigger (Eccles and Jordan 1919). The Eccles-Jordan flip-flop circuit served as the heart of most early pulse counting systems and in 1932 was adapted to a practical binary scaler circuit by British physicist C. E. Wynn-Williams (Wynn-Williams 1932). By using a number of flip-flops or Eccles-Jordan triggers in series, a scale of 2^n counts could be obtained, in which the power n was simply the number of trigger circuits in series. The number of triggers that could be strung together was limited only by the cost and patience of the experimenter, for n was theoretically unlimited. In practice, six to eight flip-flop circuits were strung together, giving a scaling factor as it was called of 64 (2^6),128 (2^7), or 256 (2^8).

Readout was obtained by having a string of lights, one tied to each flip-flop, which were lighted or not depending on the counting state, and read out in binary fashion. The first light in the string had a value of $2^0 = 1$, the second 2^1 or two, the third 2^2 or 4, and so on. The first count to enter the system would drive the first flip-flop into its alternate position and cause the light associated with that particular flip-flop (actually the first in the string) to light; this would be read as one count. The second count would drive the flip-flop back to its original position, causing the first light in the string to turn off and sending a signal to drive the second flip-flop into its alternate position lighting the second light in the string; this light would be read as two counts. The next count to arrive would again drive the first flip-flop, and both lights would be lit hence registering 1 + 2 =3 counts. The fourth count would cause both the first and second flip-flops to reverse, thereby turning the first two lights in the string off and causing the third light in the string to be lit, and so on.

An significant improvement to the Wynn-Williams readout circuit was the addition of a mechanical register or counter driven by the last flip-flop which greatly extended the counting or scaling range of the instrument. Thus, in a scale of 64 or 2^6 counter, there would be a string of six lights; when 63 counts had been accumulated, all six lights in the string would be lighted. The 64th count would restore the flip-flop circuitry to its original state, extinguishing all the lights and causing a count to register on the mechanical counter. The electromechanical registers of the 1930's were notoriously slow and required high power inputs and thus were not particularly conducive to being driven by a vacuum tube. Despite their oft times balky nature, and resolving times of 0.1 seconds, mechanical registers offered the significant advantage of enormously extending the upper counting range of a binary scaler with only a modest expenditure of funds, and so found widespread application.

Extension of the lower end of the counting range was spurred by the needs of the cosmic ray physicists, who were at the time leaders in developing sensitive radiation counting instrumentation. This was accomplished through the development of coincidence counting circuitry and techniques in the late 1920's. Among those prominent in the development of coincidence and anticoincidence counting techniques were physicists Hans Bethe, Bruno Rossi, and American physicist Merle Tuve, an early associate of Ernest Lawrence.

The decade of the 1930's saw the birth and growth of the count rate meter. One of the earliest practical count rate meter circuits was described by F. V. Hunt in 1930 (Andrews 1988). The basic design of the early to middle 1930's fed equalized size pulses into a large capacitor shunted by a resistor. The resultant leakage current through the shunt resistor was proportional to the count rate and could be read directly with a milliammeter or amplified through a vacuum tube volt meter (Kathren 1980). In 1936, a major step advance was made through the introduction of the Massachusetts Institute of Technology (MIT) counting rate meter. This direct reading random pulse counter was basically an electronic amplifier and computing circuit with dc current or voltage output proportional to the number of pulses fed into the circuit by a Geiger-Mueller tube (Gingrich, Evans and Edgerton 1936). One of its earliest and certainly highly important applications was by one of its developers, Robley D. Evans, who took the device into the field and applied it to his studies of the radium dial painters (Evans 1980). Evans became a towering figure in both health and medical physics, receiving numerous honors and awards and serving as President of both the Health Physics Society and the American Association of Physicists in Medicine. Edgerton would achieve renown in the commercial sector as one of the three founders of the well-known company, EG & G, in whose name the 'E' stands for Edgerton.

1.2.3 Enter the Decade Scaler

One of more significant of the numerous important instrumentation contributions of the Manhattan District in World War II was the development of the decade scaler. The basic scale-of-ten or decade counting circuit was developed by H. V. Regener in 1945, and was basically a four stage binary scaler or scale of sixteen device plus an additional flip-flop which was designed to electronically cut off the last six counts and to drive either a second decade or a mechanical register (Regener 1946; Elmore 1949). The decade scaler offered the convenience of a direct readout of counts in the base 10 numerical system. Thus, there was no need to convert from the binary readout, a process that required the computation of the counts accrued by multiplying the register reading by the scaling factor and adding in the counts that were indicated by the by the lighted lights or interpolation in the binary string, a time consuming annoyance as well as a frequent source of error.

Commercial decade scalers made their appearance in the late 1940's, in large measure an enterprenurial postwar spin off from the various Manhattan District instrument development laboratories. Application of the Schmidt trigger, a much improved trigger or flip-flop circuits that had been developed in the 1930's, coupled with other electronics advances of the Manhattan District such as the thyratron, univibrators and multivibrators, made possible fast counting or scaler circuits as well as other improved instruments including survey meters (Elmore 1949, Kathren 1988). A commercial decade scaler was offered by the Potter Instrument Company of Flushing, New York, at least as early as 1947, and featured three plug-in type decades and a pulse resolution time of only 5 μs, very fast for the time (*Nucleonics* staff 1948). Also among the first was the Radiation Instrument Development Laboratory (RIDL) Model 1200, a stout 50 pound vacuum tube model which featured three decades driving a Veeder-Root register, a high voltage supply continuously variable from 500 to 5000 volts with regulation of 0.01% per 1% change in line voltage, and plug in interchangeability of decade modules with its competitors Berkeley Instruments and Delectron. There was a built in timer and options of preset time or preset count. Counting losses from hot samples were minimized by the fast electronics which produced a one microsecond resolution time for the first decade. Other early manufacturers of decade scalers were El-Tronics, Tracerlab, and Nuclear-Chicago. The early scalers all had a number of features in common -- they were typically large, bulky and heavy, with numerous vacuum tubes which in turn required considerable power. The heat produced by the operation of these early scalers was prodigious, and if not removed could heat a small laboratory to uncomfortable levels and damage the instrument itself.

Despite their obvious advantages and convenience of use, the switchover from binary to decade scalers was relatively slow. Perhaps some of the reason lay in a general human characteristic known as resistance to change, but a more cogent reason was the high initial cost of decade scalers relative to their binary counterparts, which for a much lower initial cost pretty much accomplished what was needed. Decade scalers too were more complex, and likely required or were thought to require, greater maintenance.

However, it was the transistor that enabled huge advances to be made not only with scaling circuits but in virtually all areas of radiological instrumentation. Transistorized scalers and ratemeters appeared during the 1950's, and were smaller, lighter, and far more stable and rugged than their vacuum tube counterparts. Power consumption was also far less, making the transistorized instruments truly portable and utilitarian in the field. Power supplies were far more rugged and stable, facilitating or making practical proportional counting, which had been demonstrated and made feasible by John Simpson at Argonne National Laboratory in the 1940's. Certainly by the 1960's, transistorized scalers became dominant, and by 1970 it was virtually impossible to purchase a new vacuum tube scaler. Modern day scalers are miniaturized solid state circuitry versions of their vacuum tube ancestors with such improved features as digital readout, temperature compensation, and ultrastable power supplies and fast electronics.

1.2.4 The Multichannel Analyzer

Multichannel analyzers (MCA's) are in essence a special and advanced form of the scaler, and indeed are sometimes referred to as multiscalers and operated in a multiscaling mode. The MCA or spectrometer utilizes sorting circuitry to route pulses according to size to different portions of the memory or, in earlier times, to different counting circuitry. An early forerunner of the MCA is the single channel analyzer (SCA) in which a window is created by a high and low energy discriminator to the basic linear amplifier. Only pulses falling within the pulse height range of the window, which could be adjusted to be very narrow or very broad, would be counted, thus enabling a crude pulse height spectrum to be obtained. If the pulse height from the detector were proportional to the energy of the incident particle or photon, and energy analysis or pulse height analysis (PHA) could be performed; viola, spectroscopy!

Although the roots of the PHA or MCA go back to the 1930's, what might well be considered the first true MCA was developed by Otto Frisch at Cambridge University in 1944 -- the same Otto Frisch, who escape from Nazi Germany who together with his aunt Lise Meitner put forth the liquid drop theory of nuclear fission while in exile in Sweden in 1939. This MCA had a 12 individual channels or bins each of which was a single channel analyzer in its own right. The MCA's of the 1950's were huge devices utilizing dozens or more likely hundreds of vacuum tubes, enormously complex, costly to build, tempermental and difficult to operate. Their enormous size and cost and limited application initially made them unaffordable or impractical for all but a few well funded governmental entities, and exerted a practical limitation on the number of channels to which pulses could be routed. An early 128 channel device might usurp much of the counter space in a standard size laboratory, and larger units required special rooms equipped with heavy duty cooling systems to remove the heat generated by the hundreds of vacuum tubes in the apparatus. Typical of the few large early units was a 1024 channel vacuum tube MCA built at the now defunct U.S. Naval Radiological Defense Laboratory in San Francisco in the 1950's. The unit was shoehorned into a secured room of perhaps 250 square feet, equipped with its own air conditioning system, and with access limited to specified personnel. The standing joke was that the MCA should have been located in the basement and used in lieu of a furnace to warm the entire laboratory on those chilly San Francisco mornings. However, what was to be done during those balmy summer afternoons was not specified by the jokesters.

Solid state electronics and a growing interest in computers put the commercial development of the MCA on a fast track, and resulted in truly extraordinary changes in capabilities in just a very few years. The first commercial MCA was marketed in 1952 by Atomic Instruments (later Baird-Atomic), a company that had been formed in 1947 to produce health physics instruments (Terry 1980). This large and bulky rack mounted system was about the size of a large home refrigerator, standing about six feet tall. Based on an Oak Ridge National Laboratory design, it boasted 20 channels and had full vacuum tube technology. Within a few years, RIDL and Tullamore (later Victoreen) were marketing much larger and more advanced systems, with up to 2048 channels and featuring such advances as magnetic core storage memory, advanced analogue to digital converters, and more efficient vacuum tubes which reduced the size and heat output considerably. Prices were, in terms of today's dollars, frightful; a large MCA of the day would sell for the equivalent of hundreds of thousands of year 2000 dollars.

In only a few years however, the emergent solid state electronics technology would radically change the MCA, reducing its size and cost and greatly increasing its capabilities. The first commercial fully transistorized MCAs appeared in 1959 as off the shelf products of Nuclear Data Corporation and Technical Measurements Corporation, which maintained a lively and aggressive competition for several years. Packed into these small desk top size boxes that became commonplace during the 1960's were affordable 100, 128, or 400 channel systems, with fast electronics, low power demands, and such extraordinary features as CRT readout along with the choice of paper tape or typewriter output. These early units were, by the standards of today, primitive; features taken for granted in the instrumentation of 2000 such as built in spectrum identification and isotope libraries, computerized peak identification, spectrum stripping, spectrum splitting and the ability to expand or otherwise manipulate the spectrum or portions thereof were simply not available. More recent advances brought out in the 1990's such digital signal processing, and the specific software applications such as Canberra's mathematical sourceless efficiency calibration were not even dreamt of a decade or so earlier.

The early MCA's typically had digital outputs which were used to drive a paper punched tape apparatus or an ordinary electric typewriter used as a teletype. The typical typewriter printout was a sheet of rows of numbers, virtually incomprehensible to the untutored. The unlabelled rows of numbers required decoding and tedious plotting of the spectrum by hand. Paper tapes were punched in digital format, and were almost impossible to decode manually. The noisy punches produced a lengthy strip of tape with the spectrum coded in binary format and were in themselves highly temperamental. Tapes were prone to tearing and breakage, but were a considerable improvement over the very first computer card outputs, which produced a large stack of punched cards readable only by a primitive mainframe computer.

1.3 THE GEIGER COUNTER

1.3.1 The Townsend Avalanche

Soon after the discovery of radioactivity and x-rays, a growing number of scientists recognized that the ionization produced by radiation in air provided a superior method of obtaining quantitative measurements of radiation. As early as 1905, New York radiotherapist, Milton Friedman proposed that radiation be quantified in terms of the ionization it produced in air, an idea that quickly spread internationally, promulgated by a number of prominent radiation scientists of the day including the British physicist Charles E. S. Phillips and the discoverer of gamma radiation, French physicist Paul Villard. In 1908, a dose quantity based on ionization was more or less formally defined, and although not an officially accepted quantity, was used primarily in x-ray measurements in medicine.

Measurement of ionization was relatively simple and direct. Ion chambers were simple and inexpensive devices that provided a direct measurement of the quantity under consideration, in this case electric charge, which of course was a measure of the radiation field strength or of the source strength depending on whether the device was used as a current averaging device or as a pulse counter. But the early ion chambers, although effective, had important limitations as well: they were sometimes bulky, and more significantly had very limited sensitivity. One means to the solution of the sensitivity problem was through application of a discovery in 1900 by James H. Townsend, a physicist working in J. J. Thomson's laboratory. Townsend observed that if the voltage across the plates or electrodes of an ionization chamber were increased above a certain critical level, the current produced by an ionizing event would be increased manyfold, producing a pulse of ionization hundreds or even tens of thousands of times greater than the initial ionizing event (Townsend 1901).

This phenomenon became known as gas multiplication or the Townsend avalanche and initially was little more than a laboratory curiosity. In 1908 Rutherford, working in conjunction with the German physicist Hans Geiger at Manchester, put it to practical use for the important purpose of determining the charge on the alpha particle, as well as to measurement of the emanation from a number of naturally radioactive substances. The two Manchester experimenters made cylindrical brass tubes measuring from 15 to 25 cm (approximately 6 to 10 inches) in length with an inside diameter of 1.7 cm (0.67 inches), with a central collecting axial wire 0.45 mm (0.018 inches) in diameter passing through ebonite insulating corks at each end of the tube. In size and shape, the Rutherford-Geiger tube was very much like a modern Geiger-Mueller tube. Rutherford and Geiger pressurized their tube to a few atmospheres and applied a potential in the vicinity of 1300 volts. Alpha particles were introduced directly into the detector through a special window in one of the insulting corks, and the resultant current output was very much larger and, in their own words, could be measured with ". . . an electrometer of moderate sensibility" (Rutherford and Geiger 1908). Rutherford and Geiger measured the current pulses produced by the alpha particles with an ordinary quadrant electrometer, which permitted a counting rate of up to perhaps as many as a few tens of counts per minute.

1.3.2 The Geiger-Mueller Tube

In actuality, the Rutherford and Geiger were operating their brass tubes in the proportional region of the gas amplification spectrum and hence doing alpha proportional counting rather than true Geiger counting. In the proportional region, the size of current pulse is proportional to the applied voltage, or potential across the electrodes of the detector. At still higher operating voltages, a plateau is reached in which all pulses are of the same size irrespective of changes in voltage. This is known as the Geiger region or Geiger plateau. At atmospheric pressure, potentials of a few thousand volts are required for operation in the Geiger region, and stable power supplies are required as well. It was not until after Geiger returned to his native Germany in 1912 that he produced a tube capable of operation in the Geiger region of gas amplification spectrum. These tubes found immediate application for laboratory counting, initially with electrometers. However, when coupled to the already mentioned Eccles-Jordan trigger circuit and the stable amplifiers devised in 1924 by the Swiss physicist Henri Greinacher, who first utilized vacuum tube triodes to amplifying signals from ionizing particles, a relatively satisfactory counting system was achieved that was vastly superior to ion chamber pulse counters. The weak link, however, was the Geiger tube itself; it was fragile, temperamental, subject to inconsistent operation and leakage, photosensitive, very dependent on voltage and had a limited counting range. Important improvements were yet to come. Geiger teamed with his countryman Walther Mueller to systematically study and improve his counting tubes, and in 1928, they published their classic paper, only three pages in length, on the workings and preparation of the detectors that are now known as Geiger-Mueller (G-M) tubes (Geiger and Mueller 1928). Among the significant advances they made was the use of a very fine wire, only 0.1 mm (0.004") in diameter as the center electrode, which permitted a lower operating voltage and stabilized operation of the tube.

Initially, all Geiger tubes were made by hand, and their fabrication was an art that required considerable skill. To minimize absorption and thereby allow more particles or photons to enter the sensitive volume when used for external counting, metal tubes were constructed with thinner walls. As these proved too delicate even for laboratory applications, and could not be pressurized adequately to improve sensitivity, skillfully blown glass tubes coated with an opaque conductive material were made. Although the basic geometry remain cylindrical with a conducting central wire that served as the anode, various other geometries and configurations were tried. Tubes were made with thin mica windows, pressurized to make available in their small volume more gas molecules for interaction, and the inert gas argon replaced air as the fill gas. Each tube, no matter how carefully constructed had its own unique characteristics, which had to be determined prior to use and frequently during the operational lifetime of the tube. Operating voltage, resolution time, and efficiency differed from tube to tube, and were not even constant from day to day within a single tube. These crucial operating parameters had to be determined prior to use, and on a frequent basis. It was not uncommon for the operating voltage to change by 50 or more volts, particularly as the tube aged, and until well into the 1970's, it was common practice many counting laboratories to run daily voltage plateaus on their G-M counting systems, even after manufacturing techniques and more exacting tolerances made production of consistent tubes commonplace.

1.3.3 The Problem of Quenching

The Geiger-Mueller counting tubes were an immediate success, quickly adopted by the cosmic rays physicists for whom ionization chambers had insufficient sensitivity and proportional counters were too unstable. A troublesome feature of the early G-M tubes was that a single ionizing event sometimes produced a repetitive series of pulses. This effect was attributable to the release of electrons from the wall of the detector which served as the cathode by the positive ions produced in the avalanche. In extreme cases, the effect could even result in detector breakdown and continuous discharge and permanent damage to the counting tube - a most serious and potentially costly limitation. The tendency towards continuous discharge could be mitigated by adding a high resistance, but this had the drawback of lengthening the dead time of the tube between pulses to as much as 1/100 of a second, unacceptably long for many counting applications and imposing a severe limit on the upper counting range of the tube. Furthermore, since a second ionizing event was likely to occur in the tube before the discharge produced by the first was complete, the pulse from this second event could not be distinguished or resolved from the pulse produced by the first event.

During the 1930's, special circuits were devised to surmount the response time problem. One approach was the Neher-Harper circuit, which short circuited the G-M tube voltage immediately after the counting event (Kathren 1980). A second approach, the Neher-Pickering circuit, temporarily disconnected the G-M from the high voltage power supply after the count (Neher and Pickering 1938), and resulted in much faster quenching times. Various other quenching circuits were also devised but none was fully satisfactory; electronic quenching was costly, cumbersome, not fully reliable even under the best of circumstances, and still left an undesirably lengthy (albeit much improved) quenching time. A different and far more satisfactory solution was that of Dutch physicist A. Trost, who found that addition of a tiny quantity of an organic vapor such as ethanol or amyl acetate to the G-M fill gas produced quenching right inside the tube. Later ethyl formate was used. Because of the ionization potential of the ethanol or other organic was lower than that of positive ions of the fill gas (argon was typically used; its ionization potential is 15.7 volts as compared with 11.3 volts for ethanol), the charge from the fill gas ions was transferred to the ethanol molecule. Hence only the ions of the organic compound would reach the cathode to be neutralized, and the energy released in this process would go into dissociation of the organic molecule rather than into releasing an electron from the cathode to start another Townsend avalanche. Although the organic compound was thus consumed in the quenching process, organic quenched G-M tubes had sufficiently long lifetimes -- 10^{10} counts -- to make them practical.

The halogen quenched G-M tube was developed about a decade after the introduction of the organic quenched tube. Although the mechanism of quenching was similar, the dissociated halogen recombined into diatomic gas molecules and so were not consumed in the process. Theoretically, then, a halogen quenched tube had an infinite lifetime and could also be operated at higher voltages, producing output pulses of 10 volts or even larger. Offsetting these apparent advantages were a more limited counting plateau, steeper and shorter than the plateau of their organic counterparts, which necessitated far better high voltage regulation. By the 1950's, virtually all G-M tubes were self quenching, and a wide variety of geometries, sizes, and wall materials were commercially available, both with organic or halogen quenching. But by the 1970's, improved voltage regulation, a standard G-M operating voltage of 900 volts, and other improvements gave halogen quench tubes the upper hand, and today organic quench tubes, although still used, are relatively rare.

1.3.4 Evolution of the Modern G-M Survey Meter

The term "Geiger counter" is synonymous in the mind of the public with measurement of radiation, and conjures up visions of a suitably garbed individual with an electronics package in one hand and a detecting probe in the other, perhaps wearing earphones, making measurements in the field. The first commercial Geiger-Mueller survey meter made its appearance in the 1930's, the commercial product of the Victoreen Instrument Company of Cleveland, Ohio. This early instrument featured a glass wall G-M tube as detector, earphones for aural detection, and an instrument package which, with batteries, weighed in at 20 pounds. Victoreen also produced a portable ion chamber survey meter in the 1930's that utilized a miniature electrometer tube that had been developed for use in hearing aids and an ion chamber hand crafted from cylindrical cardboard tubes coated with Aquadag (carbon in alcohol) to make them conductive (Terry 1980). At the start of World War II, these two instruments were the sum total of commercial portable survey meters available for use in the Manhattan District.

Accordingly, the Manhattan District embarked on an aggressive program of instrument development. The G-M Sets, as they were called, included both the commercial item available from Victoreen as well as improved and lighter weight survey meters of Manhattan District design. The G-M survey meter was a particular favorite, even though as a pulse counter it did not measure dose. However, it was simple to operate, very sensitive (and hence good for monitoring surface contamination and for finding radiation leaks), fast responding, and, unlike some of the ion chamber survey instruments, did not suffer from instability problems. Nevertheless, the early G-M survey meters had a number of annoying and potentially dangerous problems. Most significant was the problem of 'paralysis' or saturation which produced a down scale reading when the instrument was in a field that exceed its upper range. Since the upper range was relatively low, this would occur frequently, and the user did not know whether he was in a lower level field or one that might produce a high or even life threatening dose in a short time.

The problem of G-M saturation or paralysis was attacked and solved in several ways. One technique was simply to use a small size G-M tube; tube manufacturing improvements also helped by reducing dead times to 100 μs and even less in some cases. Perhaps most significant was the invention of the Integrator tube, which, when used with appropriate circuitry, switched from G-M pulse mode to current mode at high intensities (Anton 1956). Current output from these tubes, which were typically quite small measuring only a few centimeters in length, produces output currents on the order of several μa, sufficient to be read with a portable survey meter. Integrator tubes were incorporated into military G-M's, and into high range G-M survey meters with extendable long probes.

G-M survey meter developments in the Manhattan District were many and varied. They included a special unit for low level monitoring that was given the name 'Walkie-Talkie' which was derived from the fact that the user wore earphones and carried the 10 pound instrument package via a shoulder strap, just like the genuine World War II communications device. Indeed, because of their weight, which frequently exceeded 10 pounds, most early G-M survey meters were equipped with removable shoulder straps. After World War II, the G-M survey meter attained great popularity. In addition to being used for operational health physics purposes, smaller and inexpensive models were made specifically for uranium prospecting and for Civil Defense purposes; Sears and Roebuck and Montgomery Ward as well as general scientific supply houses retailed simple and inexpensive but fully serviceable G-M survey meters as early as the late 1940's, and thousands were purchased by ordinary citizens -- prospectors, hobbyists, civil defense devotees, and by teachers for classroom demonstrations. Large numbers of specially designed G-M survey meters were made for the military, which required considerably greater ruggedness and reliability.

The typical postwar G-M counter was in it basic respects and operation very similar to the small modern transistorized units in use today, but was much larger. Its vacuum tube electronics which required a large power supply that was provided by large, heavy 'B' 45 volt batteries, sometimes as many as four, to obtain the high voltage, and smaller but still large and heavy 22.5 volt batteries were used to produce the necessary tube filament heating. Operating voltages were not yet standardized, and the G-M tubes in some early units required 1500 or more volts for operation. The voltage regulator tube made its appearance in the 1950's, and operating voltages were standardized at 900 volts. No longer were large battery packs required; adequate power for the transistorized circuitry which became commonplace in the 1960's could be achieved with a pair of ordinary D cells.

Early G-M survey meters were typically equipped with analog meter readouts and cylindrical glass wall detector tubes. These were temperature dependent, subject to breakage, and, worse still, if the painted interior of the tube were less than perfect, the tube was photosensitive. Even a tiny scratch or chip in the coating could render a tube unusable. Sensitivity of the tubes was not constant over their volume, and it was common practice to determine the point of maximum sensitivity by sliding a small source along the tube, and marking with a dab of paint or nail polish the point of maximum response. Metal wall tubes made their appearance about 1950, and because of their ruggedness, consistency, and other desirable features, quickly began to replace glass walled tubes; within a decade, there were few if any commercial G-M survey meters manufactured with glass wall tubes.

Advances in tube manufacturing technology led to the demise of the glass wall tube in the 1950's and to the production of G-M survey meters equipped with metal wall tubes with a thin mica window on one end. Although a bit ungainly to use, and small in area, these gained wide acceptance as contamination monitors. the development of the 'pancake' tube in the 1960's was a major step forward, not only increasing the sensitive area of the tube, but providing thinner window which allowed the G-M to be used for alpha contamination surveys as well as for beta-contamination. The G-M pancake probe is, some four decades later, still in wide use.

Because G-M instruments were pulse counters rather than current measuring devices, and because of the materials of tube construction were high Z relative to tissue or air led to their energy dependence, they were considered detectors rather than dose rate meters. Nonetheless, it was far more common than not for commercial instruments to be provided with a scale 'calibrated' in mr/h, with no distinction made between beta and photon radiations. The meter face usually featured a scale calibrated in both mR/h and counts per minute, but commercial instruments with only the mR/h markings were not unknown. Most G-M survey meters were equipped with a small radioactive check source to verify operability, a practice that has continued through today.

With the exception of the instruments designed for military use, early G-M survey meters, although simple and easy to use, were not particularly rugged. Perhaps the weakest link was the meter, which was highly sensitive to shock. In the 1960's, taut band meters were introduced and when coupled with replacement of hard wired circuits and vacuum tubes with printed circuit boards and transistors resulted in truly rugged instruments that could stand up to harsh use in the field. Digital readouts also first appeared in the 1960's, but were slow to catch on; cost being no small factor. It was not until the 1980's that digital readout G-M's gained some measure of popularity, particularly for high end commercial survey meters.

1.4 ION CHAMBER APPLICATIONS

1.4.1 The First Ion Chamber Survey Meter

How and why the world's first and for some years the only ionization chamber survey meter is an interesting sidelight on the history of radiation instrumentation as well as exemplification of Benjamin Franklin's maxim that necessity is the mother of invention. In 1928, Lauriston Sale Taylor, then a young physicist at the U.S. National Bureau of Standards (now National Institutes for Standards and Technology) was experimenting with x-rays, and inadvertently received a whole body dose of unknown but expectedly large magnitude. Using dental film packs and an integrating condenser ion chamber of the type used in medicine, Taylor determined that his exposure was 250 $\pm$ 50 R (about 2.5 Gy in modern terms). Concerned with the possibility of recurrences and with the need for an effective and practical real time means of monitoring x-ray sources, Taylor designed and constructed a portable ionization chamber survey meter with three overlapping ranges, achieved through the use of different size chambers, an idea that would be resurrected decades later and adapted to commercial survey meters (Taylor 1967).

Taylor, now in his 99th year, went on to enjoy a long and distinguished career in health physics, making numerous important contributions through his long years as a member and first president of the National Council of Radiation Protection and Measurements and as a member of the International Commission on Radiological Protection as well. He apparently never suffered any ill effects from his large acute accidental whole body exposure to x-rays, and was wont to say that he never got sick or even suffered from nausea, likely because he didn't know that he was supposed to get sick.

1.4.2 John Victoreen and the Condenser R-meter

The application of x-rays to treatment of malignancies and other diseases necessitated accurate measurements of the doses delivered to the patient. Chemical methods and photographic measurements were too imprecise, and subject to large variations in result because of energy dependence which made their response per unit tissue dose vary considerable with only a slight change in the energy characteristics of the exposing spectrum. During the decade of the 1920's, the talented German born physicist Otto Glasser, then at the Cleveland Clinic, and Robert E. Fricke, a radiotherapist at the Mayo Clinic, combined their talent to produce the Fricke-Glasser X-ray Dosimeter which made its commercial appearance in 1928 as a product of the Victoreen Instrument Company and was to quickly into the Victoreen R-meter which for at least the next half century was the measurement standard for x-rays in medicine (Kathren and Brodsky 1996). The condenser R-meter was made possible by the availability of precision wire wound megohm resistors which were built into the stem of a chrome polished cylinder atop which was perched a precisely made ion chamber with bakelite walls. The chamber was charged and read with a separate device. Originally and for more than thirty years, the charger was a direct current device and the chamber was charged by mechanical friction on a piece of leather. In the 1960's, an alternating current model was introduced, and over the years, a number of important dosimeters devolved from the R-meter, including a rate meter version known as the Radocon which itself went through several iterations.

John Austin Victoreen began the instrument company that bears his name in 1927, making it the first and oldest radiation instrument manufacturing company. Victoreen himself was a colorful and innovative person, who like so many of his contemporaries, was an amateur (ham) radio operator. Like many others, Victoreen found his way into a career in radiation instrumentation or electrical engineering by his youthful interest in radio. Unlike most of his peers, however, Victoreen's formal education had concluded before he finished high school, the result of a disagreement with his chemistry teacher. Disturbed by some scientifically erroneous statements that the teacher had made in class, the youthful Victoreen attempted to get the teacher to correct his errors. The teacher stubbornly refused. Discouraged, Victoreen sought counsel from his father, himself an engineer. The young student related what had happened to his incredulous father, and together they concluded that the best course would be for Jack to withdraw from high school rather than be taught scientific misinformation.

Despite his lack of a formal education, Victoreen, who would later be awarded an honorary doctorate by a college in his native Cleveland, made numerous important contributions not only to radiological measurements and instrumentation, but in other areas as well. He published several original and near classical papers on empirical measurement of photon cross sections and interactions with matter in the *Physical Review*, and was the developer of the precision high megohm resistor. In his later years, he performed pioneering work towards developing hearing aids that were individually fitted to the spectral hearing loss of the person, much as corrective lenses are for vision. Additional insight into Victoreen's contributions and personality can be gained from the fine and most interesting recapitulation of a conversation he had with Richard Terry, who compared Victoreen to ". . . a talking history book" (Terry 1980).

1.4.3 The Cutie Pie

Generically, the portable ion chamber survey meter is typically referred to as a "cutie pie," a name that goes back to its code name in the Manhattan District, but whose origin is uncertain. At least two plausible explanations have been offered for the origin of the term (Kathren 1975). One is that the operational health physics group gave to the instrument development specialists the specifications for what they desired in a dose rate survey meter. After a few weeks, the instrument developers returned with a prototype in the pistol grip shape of the classical cutie pie. When demonstrated to the health physics group, one member said, in the vernacular of the day, "Isn't that a little cutie pie?" A second, and perhaps more convincing explanation is that the name is derived from the mathematical expression that was used to calculate the response of the instrument which included the term $Q2\pi$ (Q for charge and 2π for the geometry factor), which, when pronounced aloud, sounded very much like "cutie pie."

The cutie pie was only one of several ionization chamber survey meters developed during the Manhattan District and which found their way into commercial manufacture after the war. Other notable ion chamber instruments were the boxlike Juno. It was available in both a high and low range model, and it featured a large surface area downward looking window making it ideal for monitoring bench tops and other flat surfaces for contamination in addition to serving as a dose rate meter. The unit was fitted with two sliding shields, one to admit betas and gammas, and one for admitting only gammas.

Although instruments featuring the same basic design and external physical appearance are still marketed, portable ionization type survey meters have undergone significant changes. Early models were slow responding, particularly on the most sensitive range, and relatively unstable. One major advance was replacement of the electrometer tube with a metal oxide silicon field effect transistor (MOS-FET) which greatly increased sensitivity and stability. Modern instruments are equipped with automatic temperature compensation circuitry, and fast digital readouts, and are available over a dose rate range of several orders of magnitude, with remote probes, integrating capability, and flat energy response down to a few kilovolts.

1.4.4 Personal Monitoring With Ion Chambers

In 1940, the Victoreen Instrument Company introduced the Minometer, an instrument system patterned after its highly successful R-meter (Terry 1980). The Minometer consisted of a small fountain pen sized condenser (capacitor) type integrating ionization chamber designed to be worn on the person along with an a.c. powered charger-reader. The Minometer was used for measuring low doses of radiation, 0-200 mR (0-2 mSv) and was designed for personnel monitoring. As such, it was the first in a long line of devices known as pocket ionization chamber, variably known as PICs or pencils. Such devices found widespread application in the Manhattan District, in which work with radioactive materials was being carried out on a scale manyfold greater than in any previous human endeavor. Dosimeters were needed by the tens of thousands, and the Minometer and its early descendants filled that need.

The Minometer chambers were indirect reading in that they had to be read out on a separate and specific charger, and then recharged and used again. During World War II a direct reading type (similar to a gold leaf electroscope) was developed which had a self contained reticule and could be read at any time simply by holding one end up to the light and looking through the internal microscope eyepiece at the other end. The early model pocket ionization chambers were quite energy dependent and so were only really useful for photons with energies above about 100 keV. More significantly, they were often times unreliable, having a tendency to discharge due to insulator leakage, dust, or static charge buildup, particularly from rubbing against clothing. For this reason, two dosimeters were typically worn in the field.

Pocket dosimeters were widely used in the Manhattan District and thereafter; literally millions were made for use in the nuclear weapons testing program and for Civil Defense and the military. The typical range was 0-200 mR for normal radiation work, with higher ranges, including a 0-200 R pencil being made for Civil Defense purposes.

1.5 SOME RANDOM THOUGHTS IN CONCLUSION

To a significant extent, the origins of modern health physics instruments are rooted in the Manhattan District, born out of the necessity to protect the thousands of personnel working towards the development of an atomic bomb from the hazards of excessive exposure to ionizing radiation. The accelerated instrument development of that era will likely never be equaled, and the Manhattan District days might well be the Golden Age of radiological instrument development. To William P. Jesse and his dedicated group at the University of Chicago goes much of the credit for developing portable survey instruments. No less than 16 different survey meters for various types and applications related to radiation safety came from his group during the four WWII years.

The final words in this opening chapter are an apology to the reader, for it is simply not possible to cover in a few printed pages the fascinating and lengthy story of the development of modern of health physics instrumentation. One therefore is forced to pick and choose, and in so doing running the risk of including selected items deemed to be of interest and importance at the expense of others. Since it is likely that most readers will have at least nodding familiarity with modern radiological instrumentation, but more importantly to complement the oral presentation of my friend and colleague, Frazier Bronson, the decision was made, rightfully or otherwise, to emphasize instrument developments in the first half century or so after the discovery of radioactivity and to try to show how these paved the way and led to our modern instruments.

Regrettably, however, much more seems to have been left out than has been included, and perhaps future articles or book chapters will cover some of these topics including standards and standardization, neutron monitoring instrumentation, and the some of the special application devices such as portal monitors and *in vivo* counters.

In writing this chapter, the author has drawn upon primary sources as much as possible, and also liberally and often without attribution cribbed from an earlier work of his own published in the Proceedings of the Twenty-second Midyear Topical Meeting of the Health Physics Society on Instrumentation held in San Antonio December 4-8, 1988, which was graciously and competently hosted by the South Texas Chapter. Since there is no question of copyright infringement and plagerization of one's own work is not possible, the author hopes that reader will excuse any excessive transgressions or liberties, which may have been inadvertently taken.

1.6 REFERENCES

Andrews, H. L. Laboratory measuring instruments. In: Stannard, J. N. Radioactivity and Health. U.S. Department of Energy Publication DOE/RL/01830-T59, 1988, 1509-1535.

Anton, N. In: Proceedings of the International Conference on the Peaceful Uses of Atomic Energy, Volume 14, New York: United Nations; 1956: 279-281.

Crookes, W. Proc. Roy. Soc. 81:405; 1903: Cited in Rutherford, 1906.

Dorn, E. Uber der erwarmende Wirkung der Rongenstrahlen. Wied. Annelen 63:160-163; 1897.

DuBridge, L. A.; Brown, H. An improved d.c. amplifying circuit. Rev. Sci. Inst. 4:532-536; 1933.

Eccles, W. H.; Jordan, F. W. A trigger relay utilizing three electrode thermionic vacuum tubes. Radiat. Res. Rev. 1:143-146; 1919.

Elmore, W.C.; Sands, M. Electronics. National Nuclear Energy Series Div. 5, Vol. 1. New York: Mcgraw-Hill; 1949.

Evans, R.D. Origin of standards for internal emitters. In: Kathren, R. L.; Ziemer, P. L., eds. Health physics: A backward glance. New York:-Oxford: Pergamon Press, 1980, 141-157.

Geiger, H.; Muller,W. Electro counting tube for the measure of weak activities. Naturssenschaften. 16:617-618; 1928.

Geiger, H.; Müller, W. Das Elektronzählrohr I. Wirkungsweise und Herstellung eines Zählrohrs. Phys. Zeit. 29:839-841; 1929.

Gingrich, N.S.; Evans, R.D.; Edgerton, W.E.. A direct reading counting rate meter for radon pulses. Rev Sci. Instruments 7:441-445;1936.

Holtzknecht, G. Das Chromoradimeter. Cong. Intern. d'Electroloqie et de Rad. Med. 2:377-386; 1902.

Hudson, J. C. Roentgen-ray dosimetry. In: Glasser, O., ed. The science of radiology. Springfield:, IL: Charles C. Thomas; 1933: 120-138.

Kathren, R.L. On the etymology of "Cutie Pie". Health Phys 28:487; 1975.

Kathren, R. L. Before transistors, IC's, and all those other good things: The first 50 years of radiation monitoring instrumentation. In: Kathren, R. L.; Ziemer, P. L., eds. health physics: A backward glance; New York: Pergamon Press; 1980: 73-93.

Kathren, R. L. Instrumentation for monitoring and field use. In: Stannard, J. N. Radioactivity and health; U.S. Department of Energy Publication DOE/RL/01830-T59, 1988, 1537-1574.

Kathren, R. L.; Brodsky, A. Radiation protection. In: Gagliardi, R. A.; Almond, P. R., eds. A history of the radiological sciences: Radiation physics. Reston, VA: The Radiology Centennial, Inc.; 1996: 187-221.

Metcalf, G. F.; Thompson, B. J. A low grid-current vacuum tube. Phys. Rev. 36:1489-1494; 1930.

Neher, H. V.; Pickering, W. H. Modified high speed Geiger counter circuit. Phys. Rev. 53:316-318; 1938.

Nucleonics Staff. Decade scaler. Nucleonics 2(1):84; 1948.

Regener, H. V. Decade Counting Circuits. Rev. Sci Instrum. 17:185-189; 1946.

Rollins, W. H. Non-radiable cases for X-light tubes. Electrical Rev. 40:795-799; 1902.

Rutherford, E. Radio-activity. Cambridge: At the University Press; 1904: 67-89.

Rutherford, E. Radioactive Transformations. London: Archibald Constable and Company; 1906: 23.

Rutherford, E.; Geiger, H. An electrical method of counting the number of αparticles from radioactive substances. Proc. Roy. Soc. A, 81:141-161; 1908.

Rutherford, E.; McClung, R. K. Energy of Roentgen and Becquerel rays and the energy required to produce an ion in gases. Proc. Roy. Soc. 67:245-249; 1900.

Soddy, F. The interpretation of radium. New York: G. Putnam's Sons; 1912: 59-61.

Stern, S.A. method for measuring the quantity of x-rays. J. Cut. Dis. 21:568-572; 1903.

Taylor, L.S. An early portable survey meter. Health Phys. 13:1347-58; 1967.

Terry, R. D. Historical development of commercial health physics instrumentation. In: Kathren, R. L.; Ziemer, P. L., eds. Health physics: A backward glance. New York: Pergamon Press; 1980: 159-165.

Trost, A. Uber Zahlrohre mit Dampfzusatz. Zeit. Phys. 105:399-444; 1937.

Townsend, H. Philosophical Mag. February 1901; cited in Rutherford, 1903; 36.

Wynn-Williams, C. E. A thyratron 'scale of two' automatic counter. Proc. Roy. Soc. (London) A136:312-316; 1932.

Chapter 2

IONIZING RADIATION QUANTITIES AND UNITS

Ronald L. Kathren

2.1 INTRODUCTION

2.1.1 Quantities, Units and Measurements

A *quantity* is a physical phenomenon or characteristic that lends itself to numerical specification or quantification -- that is, it has the property of being able to be measured or numbered. A *unit* is a specified or defined sample of a quantity and as such can be used to specify the magnitude of the quantity. Quantities include such basic physical characteristics as length (or distance), mass, and time as well as more complex ones such as density and acceleration

A logical and orderly system of quantities and units is fundamental to obtaining meaningful measurements of any kind. Such a system must, of necessity have highly rigorous, exact and unambiguous definitions, often mathematical, to accomplish its purpose. Thus the quantity length, which can be simply defined as the distance between two points, is specified in terms of meters. The meter is simply an agreed to sample of specification of distance with a rigorous physical definition, currently and commonly agreed to by international convention as the path length or distance travelled by light a vacuum by light during a period of 1/299792458 seconds. Prior to 1983, when the current definition was adopted, the meter was defined as equal to 1,650,763.73 wavelengths in vacuum of the radiation from the transition between the $2p_{10}$ and $5d_5$ levels of ^{86}Kr. An earlier definition, which served until 1960, was much less rigorous and was referenced to the distance between two lines scratched into the surface of a platinum-iridium bar maintained by the International Bureau of Weights and Measures (CGPM) in France under rigidly controlled conditions of temperature (0°C) and pressure (760mmHg). It is interesting to note the U.S. Bureau of Standards (now NIST) defined the foot as 1200/3937 meter.

The above definition of the meter illustrates the highly precise and rigorous nature of definitions of various quantities and units, which is essential to the establishment of an unambiguous and exact and system of quantities and units. Of necessity, similar exactness and rigor is used for the definition of other physical units. In practice, however, most measurements do not require such a high degree of rigor, as hopefully will be exemplified by this chapter, which will utilize what might be considered a semirigorous treatment that is fully consistent with the correct and proper usage of quantities and units albeit not as rigorous in the language used to characterize them.

2.1.2 Towards a Coherent and Unambiguous System of Units: The SI

Although initially slow to develop, radiological quantities and units have undergone a significant evolution over the years, culminating in the current system of radiological quantities and units based on the *Systême International d'Unités*, more commonly known as the SI units (NBS 1981; NCRP 1985; Taylor 1991, BIPM 1998). The SI is the product of more than a decade of effort by International Committee for Weights and Measures of the CGPM, an international diplomatic scientific conference to develop a practical system of units suitable for international use. The new system of units was adopted by the CGPM in 1960 and subsequently, in 1977, accepted by all parties to the Meter Convention.

For simplicity and accuracy and to obtain coherence, an arbitrary hierarchical system of base, supplementary and derived units was defined within the SI. The SI is thus built on seven *base units*: meter, kilogram, second, ampere, kelvin, mole, and candela, characterizing the quantities length, mass, time, electric current, thermodynamic temperature, amount of a substance, and luminous intensity, respectively (Table 2.1). Two *supplementary units*, the unit of plane angle or radian, and the unit of solid angle, the steradian, were initially also defined. Although listed in Table 2.1 as supplementary units, these are now usually classified as dimensionless derived units. That these units are in fact dimensionless stems from their definitions which characterize the plane angle in terms of length over length, and the solid angle in terms of area over area. The *derived units* are the products or quotients of the base units, and hence can be expressed in terms of the seven base units. Thus the units for such commonly used quantities as area (m^2), volume (m^3), velocity or speed ($m\text{-}s^{-1}$), acceleration ($m\text{-}s^{-2}$), and mass density ($kg\text{-}m^{-3}$) are all derived units. Including the two supplementary units, twenty-two units with special names and symbols have been defined in the SI and are given in Table 2.1 along with their expression in terms of the SI base units. The commonly used SI radiological units are

Table 2.1: Hierarchical Tabulation of SI Quantities and Units

Base Quantities and Units

Physical Quantity	Unit	Symbol	Base Unit Representation
Length	meter	m	
Mass	kilogram	kg	
Time	second	s	
Electricurrent	ampere	a	
Thermodynamic temperature	kelvin	k	
Amount of substance	mole	mol	
Luminous intensity	candela	cd	

Supplementary Quantities and Units

Physical Quantity	Unit	Symbol	Base Unit Representation
Plane angle	radian	rad	$m\text{-}m^{-1}$
Solid angle	steradian	sr	$m^2\text{-}m^{-2}$

Derived SI Quantities and Units with Special Names and Symbols

Physical Quantity	Unit	Symbol	Base Unit Representation
Frequency	hertz	Hz	s^{-1}
Force	newton	N	$kg\text{-}m\text{-}s^{-2}$
Pressure, Stress	pascal	Pa	$N\text{-}m^{-2} = kg\text{-}m^{-1}\text{-}s^{-2}$
Energy, Work, Quantity of Heat	joule	J	$N\text{-}m = m^2\text{-}kg\text{-}s^{-2}$
Power, radiant Flux	watt	W	$J\text{-}s^{-1} = m^2\text{-}kg\text{-}s^{-3}$
Electric charge	coulomb	C	A-s
Electric potential difference	volt	V	$W\text{-}A^{-1} = m^2\text{-}kg\text{-}A^{-1}\text{-}s^{-3}$
Capacitance	farad	fd	C/V =
Electric resistance	ohm	Ω	$V\text{-}A^{-1} = m^2 \text{–}kg\text{-}A^{-2}\text{-}s^{-3}$
Electric conductance	siemens	S	$A\text{-}V^{-1} = A^2\text{-}s^3\text{-}m^{-2}\text{-}kg^{-1}$
Magnetic flux	weber	Wb	$V\text{-}s = m^2\text{-}kg\text{-}A^{-1}\text{-}s^{-2}$

Magnetic flux density	tesla	T	$Wb\text{-}m^{-2} = kg\text{-}A^{-1}\text{-}s^{-2}$
Inductance	henry	H	$Wb/A = m^{2}\text{-}kg\text{-}A^{-2}\text{-}s^{-2}$
Temperature	celsius	^{o}C	K
Luminous flux	lumen	lm	cd-sr = cd
Illuminance	lux	lx	$lm\text{-}m^{-2} = cd\text{-}m^{-2}$
Catalytic activity	katal	kat	$m\text{-}s^{-1}$

Derived SI Quantities and Units with Special Names and Symbols for Purposes of Safeguarding Human Health

Physical Quantity	Unit	Symbol	Base Unit Representation
Activity	becquerel	Bq	s^{-1}
Absorbed dose, kerma	gray	Gy	$J\text{-}kg^{-1} = m^{2}\text{-}s^{-2}$
Dose equivalent, Effective dose equivalent quantities	sievert	Sv	$J\text{-}kg^{-1} = m^{2}\text{-}s^{-2}$

included among the 22 units with specials names and symbols, specially defined SI units (Table 2.1) because of their need and application to the practice of radiation safety (NBS 1981; Taylor 1991, BIPM 1998).

The SI is what is known as a *coherent* system of units, i.e. no conversion factors other than unity or powers of ten are needed to construct units derived from the base and supplementary units, and, as is illustrated in Table 2.1, all physical quantities in the SI can be expressed in terms of the base and supplementary units exclusively. The SI also uses a system of prefixes to denote multiples or submultiples of units; these are shown in Table 2.2. With the exception of multiples of 10^3 or less, capital letters are used to denote multiples, and lower case letters for all submultiples. Another convention is to capitalize the symbols for units named after people (e.g. Hz, Sv) although the full name of the unit is not capitalized.

2.1.3 Radiation Quantities, Units and Measurements

Radiological quantities and units typically used in health physics are put forth by various national and international bodies such as the International Commission on Radiation Protection (ICRP) and national and international regulatory and standards agencies including the International Atomic Energy Agency (IAEA), International Standards Office (ISO) and International Electrotechnical Commission (IEC). Generally, the radiological quantities and units in international use are consistent with the SI, and as such can be quantified in terms of one of the three derived SI radiological units; the SI quantities and units themselves are based on and therefore fully consistent with the recent recommendations of the International Commission on Radiation Units and Measurements (ICRU 1993, 1998). Although the basic concepts have remained more or less the same for more than a half century, dose quantities have evolved and undergone considerable refinement over the past several decades. Table 2.3 summarizes the radiological quantities and units most commonly used in operational health physics practice. The table includes derived SI quantities and units as well as a number of traditional quantities and units used historically. Largely for historical reasons and a strong resistance to change, and in particular to the adoption of the SI in the United States, coupled with the unavoidable regulatory lag between the time a new or refined quantity is proposed and its adoption into national regulations have led to a somewhat confused and perhaps even chaotic situation in which both the old or traditional and new systems of units are used concurrently. The problem had been further exacerbated by American regulations which have not always been in consonance with the internationally accepted system of quantities and units, sometimes even to the extent of defining or using named quantities or terms differently or even of creating and defining new quantities and units inconsistent with the SI. Even today, more than a decade after the internationally agreed upon deadline for the adoption of the SI, American radiation safety regulations still use the conventional units, typically in conjunction or parallel with the new.

It should be noted that with few exceptions the dose quantities used in health physics are not directly measurable by instrument systems or dosimeters. What is in fact likely measured is the absorbed dose to the detector (or some surrogate thereof), which may be designed to respond in a manner like tissue and hence provide an output that can be measured and equated with the expected tissue response. Instrument readouts are the merely surrogates for the absorbed dose or dose equivalent incurred by an irradiated individual. This is true even for absorbed dose measurements with ionization chambers (Burlin 1968).

Table 2.2 Prefixes Used for Multiples and Submultiples of SI Units

Multiple	Prefix	Symbol	Submultiple	Prefix	Symbol
10^{24}	yotta	Y	10^{-24}	yocto	y
10^{21}	zetta	Z	10^{-21}	zepto	z
10^{18}	exa	E	10^{-18}	atto	a
10^{15}	peta	P	10^{-15}	femto	f
10^{12}	tera	T	10^{-12}	pico	p
10^{9}	giga	G	10^{-9}	nano	n
10^{6}	mega	M	10^{-6}	micro	μ
10^{3}	kilo	k	10^{-3}	milli	m
10^{2}	hecto	h	10^{-2}	centi	c
10^{1}	deka	da	10^{-1}	deci	d

Table 2.3 Common Quantities and Units Used in Health Physics

Quantity	Unit	Symbol	Brief Description	Comment
Activity	becquerel	Bq	reciprocal second (i.e. nuclear transformations per second)	SI quantity and unit
	curie	Ci	3.7 x 10^{10} disintegrations (nuclear transformations) per second	Historical quantity and unit
Exposure	coulomb per kilogram of air	C-kg^{-1}	Ionizing electromagnetic radiation (gamma and x-ray) in air	Not coherent with SI
	roentgen	R	2.58 x 10^{-4} C-kg^{-1}	Obsolete
Absorbed Dose	gray	Gy	1 J-kg-1	SI quantity and unit
	rad	rad	100 erg-g^{-1} = 0.01 J-kg^{-1} = 0.01 Gy	Traditional quantity and unit
	rep	rep parker	93 erg-g^{-1}	Obsolete

Quantity	Unit	Symbol	Brief Description	Comment
Kerma	gray	Gy	sum of initial kinetic energies of primaries from x-rays, gamma rays, and neutrons	SI quantity and unit
	rad	rad		Traditional quantity and unit
Dose equivalent	Sievert	Sv	Product of absorbed dose and radiation weighting factor	SI quantity and unit
	rem	rem		Traditional quantity and unit
Radiation weighting factor		w_R	For radiation safety application to account for different effects of various radiations	Dimensionless; tabulations available in handbooks and regulations
Quality factor		Q, QF	w_T predecessor	Obsolete
Relative Biological Effect		RBE	Compares dose of different radiations for same biological effect	Use reserved for radiobiology
Personal dose equivalent	sievert	Sv	Dose equivalent at a specified depth	Regulatory equivalents are deep dose equivalent and shallow dose equivalent
Effective dose equivalent	sievert	Sv	Product of absorbed dose, radiation weighting factor, and tissue weighting factor = $D w_R w_T$	Also known as effective dose.
Tissue weighting factor		w_T	For radiation protection application to account for different response of specific tissues	Dimensionless; tabulations available in handbooks and regulations

Quantity	Unit	Symbol	Brief Description	Comment
Committed effective dose equivalent	sievert	Sv	Time integrated organ or whole body effective dose equivalent from iradioactivity in the body	Abbreviated CEDE; time integration 50 yeas for adult unless otherwise specified
Total effective dose equivalent	sievert	Sv	Sum of external dose and CEDE in a specified one year period	Abbreviated TEDE; a regulatory construct
Dose commitment	sievert	Sv	Prospective calculated dose	Use limited to planning purposes
Collective absorbed dose	person-gray	person-Gy	Summed absorbed dose to a specified population	
Collective dose equivalent	person-sievert	person-Sv	Summed dose equivalent to a specified population	Traditional unit is person-rem or man-rem
Collective effective dose equivalent	person-sievert	person-Sv	Summed effective dose equivalent to a specified population	Traditional unit is person-rem or man-rem
Annual limit on intake	Becquerel	Bq	Specified amount of activity that will produce permissible dose if taken into body	Symbolized ALI
	Microcurie	μci		Traditional
Derived air concentration	becquerel per cubic meter	Bq-m-3	Specific limit for air concentration based ALI	Symbolized DAC; Traditional unit $\mu Ci\text{-}cm^{-3}$

Quantity	Unit	Symbol	Brief Description	Comment
Working level	Working level	WL	Release of 1.3 x 10^5 MeV of alpha energy from short lived radon progeny in 1 liter of air	Not directly relatable to dose; not coherent with SI
	Working level month	WLM	Product of WL and working months of exposure	Working month is defined as 170 hours measure of exposure to radon and its progeny

2.2 ACTIVITY AND RELATED QUANTITIES AND THEIR UNITS

2.2.1 Activity

The quantity *activity* refers to the amount of radioactive material in which a specified number of atoms are decaying per unit time. In the SI activity, is specified in units of *becquerel* (Bq), one of eight derived units for which a special name has been adopted. The Bq, named in honor of Henri Becquerel who discovered radioactivity, is defined as, one Bq is equal to one nuclear transformation per second, and its dimension is s^{-1}. The traditional pre-SI and still widely used unit of activity is the *curie* (Ci), which is defined as the amount of radioactive material in which there are exactly 3.7 x 10^{10} (37 billion) nuclear transformations per second. The reasons for this odd number are largely historic and are referenced to the activity of one gram of pure ^{226}Ra. Over the years, a number of other units have been proposed to specify activity. One, the rutherford (rd), defined as 10^6 (one million) disintegrations per second enjoyed a brief period of popularity and is occasionally encountered in the early literature.

In practice, radioactivity is often specified in units of *disintegrations per minute* (dpm) or *disintegrations per second* (dps). Assuming that a disintegration is synonymous with a nuclear transformation, these are exactly as stated and easily converted to the conventional units of activity, i.e. Bq in the SI and Ci in the old system. The terms dpm and dps should not be confused with nor are they interchangeable with *counts per minute* (cpm) or *counts per second* (cps). The latter refer to an instrument reading and not to an actual physical quantity; conversion to the standard unit for activity, the Bq, can be accomplished only if the efficiency of the counting system is known.

2.2.2 Fluence, Fluence Rate and Flux

Radiation fields are often characterized in terms of particle or energy fluence or fluence rate. *Fluence* is simply the number of particles (or photons in the case of electromagnetic radiation) or the total energy passing through or impinging upon a unit area and is generally expressed as particles (or photons) per square centimeter or $MeV\text{-}cm^{-2}$, as applicable; more rigorously, it is the time integral of particles (or energy) the enter a sphere of unit cross-sectional area. *Fluence rate*, or *flux* in the older terminology, is the time differential of fluence and is typically expressed in terms of particles per $cm^{-2}\text{-}S^{-1}$ or $MeV\text{-}cm^{-2}\text{-}S^{-1}$. Particle fluence and particle fluence rate can be converted to energy fluence by simply multiplying by the average particle energy. Since activity applies only to radionuclides, fluence and fluence rate are useful quantities for specification of ionizing radiation generated from other sources such as x-ray generators, particle accelerators, and nuclear reactors. Thus, thus neutron fields are typically specified in terms of particle fluence or fluence rate (i.e., $n\text{-}cm^{-2}$ or $n\text{-}cm^{-2}\text{-}s^{-1}$, respectively); x-ray fields may be characterized in terms of particle or energy fluence. Activity can also be converted to fluence if the geometry is known; for the simplest geometry, that of a point radionuclide source, the fluence rate ϕ follows the inverse square law and thus is simply

$$\phi = \frac{S}{4\pi r^2} \tag{2.1}$$

in which S is the source strength in Bq and r is the distance from the source in cm. Eqn. (1), of course, refers to the ideal point source or 4π geometry case, giving no consideration to air attenuation, self shielding, backscattering, or geometries other than the theoretical point source.

2.3 EXPOSURE

2.3.1 Exposure in the General and Special Sense

The quantity exposure has likely produced more confusion and misapplication than any other dosimetry construct, likely largely attributable to the fact that the word 'exposure' has two very different and distinct meanings and to the way the concept was historically used. As used in the general sense, exposure refers to the potential for, or actual delivery or accumulation of, a dose of ionizing radiation, and in this sense is synonymous with irradiation. It is also the legal definition of exposure as put forth in both federal and state radiation control regulations; the U.S. Nuclear Regulatory Commission in

Title 10, *Code of Federal Regulations* Part 20, defines exposure thusly: "Exposure means being exposed to ionizing radiation or to radioactive material" (USNRC 1999); the Department of Energy has no definition for exposure in their portion of Title 10 (USDOE 1999). Occupational exposure refers to exposure of a worker while at work.

In the quantitative technical sense, however, *exposure* is a measure of the intensity of an X- or gamma ray field and is a specific quantity based on to the production of ions in air by ionizing electromagnetic radiation (IER), defined in terms of the number of charges produced in air under specified conditions. To avoid confusion with the general usage and emphasize the distinction of the special sense of the term, the word *exposure* will be italicized throughout this chapter whenever it is used in the special sense. More globally, it is recommended that the term itself be limited to refer to *exposure* in the special sense, using the term irradiation for exposure in the general sense. In the SI, the units of exposure in the special sense are $C\text{-}kg^{-1}$ of air. In the traditional of quantities and units, *exposure* had a specific unit of its own, the roentgen (R), defined as the quantity of x- or gamma radiation that produced 1 statcoulomb of charge of either sign in 1 cm^3 of standard air (0.001293 g), which is equivalent to 2.58×10^{-4} $C\text{-}kg^{-1}$ of air. The roentgen was named in honor of Wilhelm Konrad Roentgen, the discoverer of x-rays and is still widely used, although officially is now an obsolete unit and is not used in the SI. In addition to being limited to electromagnetic radiation in air, other limitations were placed on *exposure* and on the use of the roentgen unit.. Largely because of a loss of charged particle equilibrium at high photon energies, and significant attenuation at low energies, *exposure* was limited to the energy range of 10 keV to 3 MeV.

2.3.2 Exposure-Dose Relationship

It cannot be too strongly stressed that *exposure* is not a dose quantity, but rather a specially defined quantity unto itself, and should under no circumstances be used as a surrogate for tissue dose nor for other than electromagnetic radiations. *Exposure* can be converted to dose in air if the average energy, W, required to produce one charge of either sign or one ion pair -- is known. Multiplying the charge C produced in air by a given level of irradiation by W will yield energy per unit mass, which are the units of the various dose quantities. Again, it is important to note that the dose thus calculated is the dose to air, and not the dose to tissue or any other form of matter. As a practical matter, *exposure* offers certain advantages, particularly in radiotherapy applications. The charge in a known volume (and hence mass) of air can be directly and accurately measured, and hence irradiations to patients can be precisely determined. *Exposure* can then easily be converted to tissue dose using published conversion factors available for that purpose, a practice now obsolete but formerly both common and accepted in medical and health physics.

Historically, the roentgen has occupied an important place in radiation dosimetry, and this historical usage has created much of the confusion that exists today with respect to *exposure*. At one time, some decades back, the roentgen was actually used as a measure of a quantity known as exposure dose; this quantity is now obsolete and wholly inconsistent with the SI or even the traditional of units and under no circumstances should *exposure* be used in lieu of dose. Another practice to be eschewed is to express dose measurements in terms of R, a practice actually permitted by both Federal and state radiation safety regulations a decade or so ago which equated the roentgen with the unit for absorbed x-ray dose and dose equivalent a practice inconsistent with the definition of the roentgen and the quantity that it quantifies. *Exposure* is still a highly useful concept and quantity, although it now has no special name, is now correctly expressed in terms of $C\text{-}kg^{-1}$.

2.4 ABSORBED DOSE AND KERMA

2.4.1 Absorbed Dose

Absorbed dose is the fundamental dosimetric quantity in radiation protection (ICRP 1990) and simply is the concentration of energy deposited or absorbed by a material as a result of irradiation; it is an all inclusive quantity that can be used for any ionizing radiation and any absorbing medium. The SI unit for absorbed dose is the joule per kilogram, which (Table 2.1.) has been given the special name gray (Gy) after the great British radiological physicist Lewis Harold Gray; 1 Gy is exactly equal to one joule of energy deposited in one kilogram of absorbing material ($1\ J\text{-}kg^{-1}$). In the previous or old system, the unit of absorbed dose is the rad, defined as $100\ erg\text{-}g^{-1}$. Since $100\ erg\text{-}g^{-1}$ is exactly equal to 0.01 J-kg, 1 Gy = 100 rad. As noted by the ICRP, for radiation safety purposes, the absorbed dose is typically taken to mean the average dose over a specified tissue or organ, and thus may be characterized by the symbol $D_{T,R}$ which refers to the average absorbed dose to tissue T by radiation R. This is also known as the tissue or organ average absorbed dose.

Another absorbed dose quantity that is sometimes used is the *absorbed dose index,* which was introduced by the ICRU in 1971 and designated by the symbol D_I. The absorbed dose index was defined by ICRU as the maximum absorbed dose in a 30 cm diameter tissue equivalent sphere (ICRU 1971). It was introduced in order to determine the maximum dose within the ody. The ICRU sphere is a model of the body that gives an estimate of the maximum dose in the trunk that is sufficiently accurate for radiation safety purposes.

Other units have been used in the past to quantify absorbed dose, most notably the *rep*, also known as the *parker* after the great British-American medical and health physicist Herbert M. Parker who introduced the unit in 1943 (Parker 1950). Over its relatively brief lifetime, the rep found broad application and also engendered some confusion. Initially values for energy deposition in soft tissue, ranging from 83 to 95 erg-g^{-1}, were assigned to the rep, with the value of 93 erg-g^{-1} finally agreed to. However, in 1953 the rep was replaced by the rad, which was defined as exactly 100 erg-g^{-1}.

2.4.2 Kerma

A quantity loosely related to absorbed dose is *kerma*, an acronym derived from the phrase "***k***inetic ***e***nergy ***re***leased in per unit ***ma***ss." Kerma is a dosimetric measure of the radiation field for uncharged ionizing radiations such as neutrons and photons and refers to the sum of the initial kinetic energy energies of all the primary charged ionizing particles released in a specific volume of matter by the interaction of an uncharged ionizing particle (ICRP 1969). Kerma cannot be measured and is thus a calculated quantity, being equal to the product of the mass energy transfer coefficient and the energy fluence for indirectly ionizing radiation such as photons or neutrons in a specific medium. Since this product yields energy per mass, kerma is typically expressed in the same units as absorbed dose, viz. Gy in the SI and rad in traditional terms. Kerma has found application in evaluating high energy photon and neutron fields, particularly around particle accelerators, and has been extensively used to characterize the irradiation received by the survivors of the Japanese atomic bombings and thus has found its way into the radioepidemiologic literature.

An important limitation of kerma is that it does not apply to all types of radiations, but is limited to photons and neutrons. It considers only the initial kinetic energies of the charged particles released in the interaction -- the so-called first collision dose -- and thus, although quantified in units of Gy or rad, kerma is not the same as absorbed dose nor is it necessarily numerically equal to absorbed dose, although it may approximate absorbed dose when charged particle equilibrium is present. Under nonequilbrium conditions, such as at or near the tissue-air interface, or for very high energy radiations, numerical values for kerma and absorbed dose may be quite different.

Recently, the ICRU defined a quantity named cema (ICRU 1998). Cema parallels kerma but applies to charged particles, and differs from kerma in that it is referenced to the energy lost in electronic collisions by the incoming charged particles, while kerma is referenced to the energy imparted to outgoing charged particles.

2.5 BASIC DOSE EQUIVALENT QUANTITIES

2.5.1 Evolution of the Concept of the Dose Equivalent

Since equal absorbed doses of different radiations do not necessarily produce equal risk or biological effect, the concept of *dose equivalent* was created. Dose equivalent is a special quantity that takes into account the efficacy of the particular radiation at producing biological damage, and as such is the absorbed dose weighted by those factors that affect the biological response to the radiation. Since the weighting factors are dimensionless, dose equivalent has the same physical units as absorbed dose, viz. J-kg^{-1} in the SI (Table 2.1) but is measured in terms of a special derived unit, the rem in the traditional system of units and the sievert SV, in the SI.

The concept of biological dosage was introduced in 1921 by Failla (Failla 1921) and put into practical usage as early as 1943 by Parker who devised the unit *rem* (for **r**adiation **e**quivalent **m**an) to quantify doses from any ionizing radiation that were biologically equivalent to one roentgen of "hard" or high voltage x-rays, all other factors such as dose rate, portion of the body irradiated, dose fractionation, etc. being equal (Kathren, Baalman and Bair 1986; Parker 1949, 1950). To accomplish this, Parker simply multiplied the absorbed dose in rep by a dimensionless factor known the Relative Biological Effectiveness (RBE) and this new quantity was quantified by a new unit which Parker called the rem, an acronym derived from **r**oentgen **e**quivalent **m**an. Thus the mathematical expression of what came to be known as 'RBE dose', or, more properly, biological dose, expressed mathematically, is

$$\mathbf{H = D}\ x\ \mathbf{RBE} \qquad (2.2)$$

in which *H* refers to the biological or RBE dose, is the absorbed dose, and *RBE* the Relative Biological Effectiveness. The concept of RBE dose was considered by the ICRP as early as 1950, and formally and officially adopted in 1955 (ICRP 1950, 1955).

In the 1959, the ICRP expressed misgivings over the dual use of the *RBE* in radiation biology and in radiation protection, and in 1960, jointly with the ICRU, the RBE was redefined and its use restricted to radiation biology. A new modifying factor, the *quality factor*, initially symbolized by *QF* and later *Q,* was defined for health physics purposes (ICRU 1962). Strictly speaking, *Q* was defined in terms of the Linear Energy Transfer (LET), or energy deposited along a unit of track length by the exposing radiation, but for practical health physics applications, discrete values were assigned to specific radiations based on their LET. High LET radiation, such as alpha particles, are more damaging, per unit dose, than low LET radiation, such as photons and beta particles.

Thus high LET radiation was assigned a high value of Q, while low LET radiation such as photons were arbitrarily assigned a Q of unity.

In addition to Q as a modifying factor, the 1962 report specified a *distribution factor* to account for nonuniform distribution of radionuclides incorporated into an organ of the body and introduced the concept of dose equivalent, which it abbreviated DE, and defined as the product of the absorbed dose and the modifying factors, and having the units of rem. Subsequently in its watershed 1977 recommendations, the ICRP adopted the SI and further recognized that other factors could modify the effect of radiation on an individual or population. The definition of dose equivalent *(H)* was further refined and formalized, and mathematically characterized as

$$H = DQN \tag{2.3}$$

in which D is the absorbed dose, Q is the quality factor, and N the product of all other factors that may modify or affect the biological response (ICRP 1977). N was assigned a value of 1, and Q was tied to collision stopping power in water for the radiation under consideration, although a table recommended values for various types of radiations was given. Since Q and N are dimensionless, the dose equivalent H has the same physical units as absorbed dose (i.e. $J\ kg^{-1}$), but is quantified in units of sievert (Table 2.1).

2.5.2. Deep, Shallow, and Personal Dose Equivalent

A quantity known as the *personal dose equivalent*, H_p(d) was formulated by the ICRU for application to operational personnel dosimetry of external fields (ICRU 1992). The parenthetical 'd' in the symbol for this quantity denotes the depth below a specified point on the surface of the body at which the dose equivalent is determined; for strongly penetrating radiations, a depth in tissue of 10 mm is employed and the quantity is then specified as $H_p(10)$. For a depth of 10 mm, the personal dose equivalent is identical to the *deep dose equivalent*, H_d which has been defined for regulatory purposes in the United States and is the dose equivalent at a depth of 1 cm in (soft) tissue from an external field (USDOE 1999, USNRC 1999). Similarly, the *shallow dose equivalent,* H_s, has been defined as the dose equivalent from external radiation at a depth of 0.007 cm in (soft) tissue (USDOE 1999), averaged over an area of 1 cm^2 (USNRC 1999). All of the above are dose equivalent quantities, and as such quantified in units of Sv or rem.

2.5.3 Effective Dose Equivalent

The risk based 1977 ICRP recommendations introduced the concept of a tissue weighting factor, designated as w_T, to represent the proportion of the stochastic risk resulting from irradiation of a specific tissue. In this manner, the risk from nonuniform irradiation could be equated to a uniform whole body dose, and nonuniform irradiation from radionuclides incorporated into the body could be combined with external radiation doses, giving rise to what was initially called the weighted equivalent dose and later came to be known as the *effective dose equivalent* (ICRP 1977). The effective dose equivalent, abbreviated as EDE and typically symbolized by the letter *H*, is a doubly weighted version of the average absorbed dose to the tissue of interest, weighted once for radiation quality by the quality factor and again by the tissue weighting factor to take into account the specific stochastic risk of the tissue of interest. Specific values for tissue weighting factors, based on the stochastic risk per unit dose to six specific tissues (gonads, breast, red bone marrow, lung, thyroid, bone surfaces) plus the remainder of the whole body, were put forth, with the sum total of all the various tissue weighting factors equal to unity. Thus, given the values of w_T, it was possible to equate the risk from irradiation of a single organ, say the thyroid, with whole body irradiation.

Mathematically, the whole body dose equivalent, H_{wb}, also termed the effective dose equivalent, was represented by which is simply the summation of the products of the

$$H_{wb} = \Sigma_T\, w_T H_T \qquad (2.4)$$

individual organ dose equivalents and the corresponding tissue weighting factor and is quantified in terms of sievert. The dose equivalent to any given tissue was simply the product of the tissue weighting factor and the average dose equivalent to the tissue of interest. Note that the same unit, the sievert, is used to specify the dose equivalent to any of a number of tissues, including the whole body, as well as to specify the effective whole body dose equivalent. Thus it is imperative to specify which tissues have in fact been irradiated and the dose equivalent to each if one is to calculate a whole body dose equivalent. And, it is important to note that once calculated, the single number given for the whole body dose equivalent does not given any indication of which tissues were in fact irradiated; it is impossible to know or to conclude from the single number specified for whole body EDE. For example, using the original weighting factors in ICRP Publication 26, a whole body dose equivalent of 1 mSv could be the result of a uniform external whole body irradiation with gamma rays, or could be the result of an external irradiation of 4 mSv to the gonads, or of a dose equivalent of 33 mSv from radioiodine in the thyroid, or of a combination of dose equivalents from a number of irradiated tissues. It cannot be too strongly stressed that without further information, it not possible to determine how the whole body dose equivalent was incurred. The 1977 ICRP recommendations also introduced or formalized a number of other dosimetric concepts and quantities, including collective dose equivalent, dose equivalent commitment, committed dose equivalent, quantities which are defined or described in the following sections.

The sweeping changes put forth by the 1977 ICRP recommendations were further expanded and clarified in the more comprehensive ICRP Publication 60, published in 1990 (ICRP 1991). The tables of tissue weighting factors and radiation weighting factors were expanded; weighting factors were given for 12 specified tissues plus the remainder of the body. A new modifying factor, the *radiation weighting factor,* w_R, was introduced and replaced Q. The term *effective dose* was put forth by the ICRP as a less cumbersome name for what was previously called the weighted dose equivalent or effective dose equivalent and given the symbol E in the 1990 recommendations, reserving the symbol H for dose equivalent. The mathematical definition of E given in ICRP Publication 60 is identical to H_{wb} as given in Eqn. 2.4.

In addition to the confusion that may result from the 1990 change in terminology and symbol (but not the quantity itself) recommended in ICRP Publication 60 is the regulatory definition in the United States. Both the Nuclear Regulatory Commission and the Department of Energy use the term effective dose equivalent and define this quantity as in Eqn. 2.4 above, but use the symbol H_E to designate this quantity(USDOE 1999; USNRC 1999). Thus, $H_E = H_{wb} = E$.

It bears mention that both dose equivalent and effective dose equivalent quantities, and their derivatives, cannot be directly measured by instrumentations systems, and are quantities intended for use in radiation safety practice and regulation. Although an instrument can be made and calibrated to read out in units of dose equivalent (i.e. Sv or rem), the instrument is not in fact measuring dose equivalent, even in the case of a broad external field, but rather a surrogate of absorbed dose (See Chapter 18) to which the appropriate tissue and radiation weighting factors have been applied.

2.6 DERIVATIVE OR SUBSIDIARY DOSE AND DOSE EQUIVALENT QUANTITIES

2.6.1. Committed Dose Quantities, Units and Symbols

The concept of committed dose was introduced in 1977 by ICRP Publication 26 in terms of the *committed dose equivalent*, and extended to effective dose in ICRP Publication 60 (ICRP 1977, 1991). The committed dose equivalent is simply the time integral of the dose equivalent rate from external radiation or from a single intake of radioactivity into the body. By extension, the committed effective dose equivalent and absorbed dose can be similarly defined. Although in theory the dose can be integrated over any time period subsequent to intake, the typical practice, codified in the American regulations, is to integrate over a 50 year period following intake. The ICRP stipulates that the time of integration be specified, and if not specified, is assumed to be 50 years for an adult and 70 years for children. The units for both the committed dose equivalent and the committed effective dose equivalent (CEDE) are the Sv in the modern system and rem in the traditional system. For committed absorbed dose, a quantity occasionally used, the units are Gy and rad.

Confusion may sometimes arise because of the different symbols used for the same collective dose quantities. The ICRP uses $H_T(\tau)$ for committed effective dose, and $E(\tau)$ for committed effective dose equivalent or CEDE (ICRP 1991). American regulations specify $H_{T,\,50}$ and $H_{E,50}$, respectively, for these quantities (USDOE 1999; USNRC 1999).

The ICRP has also defined a quantity known as *dose commitment* which is sometimes confused with committed dose. The dose commitment, symbolized in the ICRP system as $H_{c,\,T}$ or E_c is a calculational tool for determining the risk to a critical group or the entire world population from a specific event or practice, and is defined as the infinite time integral of the per caput dose rate from that event or practice. Dose commitment is primarily a planning device.

2.6.2 Collective Dose Quantities and Units

Collective dose is the summation of all the individual doses incurred by a specified population, or alternatively the mean dose multiplied by the number of persons exposed. The summed dose quantity may be absorbed dose, dose equivalent, effective dose, or one of the committed dose quantities. Collective dose should be used and applied with caution, and guidance is given in a recent NCRP publication (NCRP 1995). The preferred unit of collective absorbed dose is the person-Gy (e.g. dimensionally the product of the number of persons irradiated and the mean absorbed dose); likewise, the preferred unit for collective dose equivalent or collective EDE is the person-Sv. Alternatively, person-rad and person-rem are used in the traditional system.

2.6.3 Total Effective Dose Equivalent

The *total effective dose equivalent* (TEDE) is a regulatory construct defined by both the US Department of Energy and U.S. Nuclear Regulatory Commission but somewhat differently by the two agencies. The USDOE defines CEDE as the sum of the

EDE for external exposures and the CEDE for internal exposures (USDOE 1999); the USNRC defines CEDE as the sum of the deep dose equivalent (See Section 2.5.1) for external exposures and EDE for internal exposures. As neither definition specifies time, the period over which the accrued doses are incurred is implicitly assumed to be equal; thus the CEDE can be specified for any specific time frame -- e.g. a year, lifetime, or decade. There is a clear and potentially highly significant difference between the two definitions of TEDE; the USNRC definition would seem to exclude the dose to the skin, which could be an important component of the total external dose equivalent if the individual received a significant portion of his/her external dose equivalent from nonpenetrating radiations as might occur from external exposure to beta radiation of sufficient energy to deliver a large dose to the skin, but lacking sufficient energy to deliver a dose to the tissues at a depth of 1 cm. Although the 1990 ICRP recommendations do not define or even use the term TEDE as such, the permissibility and even desirability of combining both external and internal doses it is abundantly clear from the document itself that in the absence of qualifying statements stipulating specific tissue(s) involved, what the ICRP calls effective dose is in reality the same as TEDE.

2.6.4 Specification of Activity Concentration and Intakes of Activity

Concentrations of radioactivity in air or water are most simply and directly expressed in terms of activity per unit volume and, if appropriate units are chosen, such expression is then fully coherent with the SI and indeed qualifies as derived SI units. In practice, however, activity concentrations in air and activity in the body are often expressed in terms of fractions or multiples of the *Derived Air Concentration* (DAC) and *Annual Limit on Intake* (ALI), respectively. The ALI is quantified in units of activity, but is actually based on a dose limit, being that quantity of a radionuclide which, if taken into the body over the period of a year, would deliver to the individual a CEDE of 50 mSv or 500 mSv to any tissue or organ. The ALI is not a single fixed value for all radionuclides or even for a single radionuclide, but rather a function of many factors including the specific radionuclide, route of entry, and, in the case of inhaled particulate radioactivity, solubility class and particle size, and thus a single radionuclide may have several values of ALI. Normally, but by no means always the most restrictive ALI is chosen for the radionuclide in question.

Despite the lack of rigor, potential for ambiguity, and other criticisms that might be applied to specification of intakes of activity in terms of the ALI, the practice is well established, accepted, and highly convenient to operational health physics practice. It should be noted that in actual fact, specifying intake in terms of ALI is simply another way of specifying intake in terms of activity, albeit in a unit that is not coherent with the

SI. For regulatory purposes, one ALI is considered to be equivalent to 50 mSv (5000 mrem).

Logically devolving from the ALI is the DAC, which is simply the ALI for a specified nuclide divided by the breathing rate for a reference man in a working years (2400 m^3) and hence has the units of activity per volume. Like the ALI, air concentrations are often expressed in terms of fractions or multiples of DAC. However, in operational health physics practice use of the DAC-hour to specify inhalation exposure is more common. For any radionuclide or combination of radionuclides, the number of DAC-hours of exposure is simply the product of the average air concentration expressed in units of DAC and the number of hours of exposure. The DAC-hour is a very convenient unit for operational radiation safety; based on a work year of 2000 hours (50 weeks per year times 40 hours per week), the annual exposure limit is 2000 DAC-hours, or, putting it another way, an exposure to 2000 DAC-hours is equivalent to one ALI. Thus, an atmospheric exposure of 1 DAC-hr is considered to be equivalent to a dose of 0..25 mSv (25m rem).

Note that the DAC is the modern analogue of the obsolete quantity Maximum Permissible Concentration (MPC) and the ALI is related to the formerly used, and now obsolete, Maximum Permissible Body Burden (MPBB) described in the earlier literature.

2.6.5. The Working Level

In 1957, the U.S. Public Health Service, in an attempt to better characterize airborne radioactivity exposure to miners, introduced the concept of *working level* (WL) and defined it in terms of the concentration of alpha energy released by the progeny of radon in air, a quantity that could be related to the dose delivered to the lungs. The WL is defined as any combination of short lived radon progeny that will result in the emission of 1.3×10^5 MeV of alpha particle energy in one liter of air, which equates to 0.0208 J released in one cubic meter of air and to concentration of 3.7 $kBq\text{-}m^{-3}$ (100 pCi/L) of short lived ^{222}Rn progeny in equilibrium with ^{222}Rn. The WL is thus not coherent with the SI. Neither is not a unit or quantity of absorbed dose or activity concentration although it can be related to either of these quantities, although the conversion is not straightforward and additional information must be known. And, although usually thought of in conjunction with ^{222}Rn, consistent with the definition, it can be used with any isotope of radon, and indeed has been used to quantify ^{220}Rn, commonly known as thoron.

Exposure to radon and its progeny is commonly measured in terms of the *working level month*, WLM, a concept analogous to the DAC-hour. The number of WLM is product of the average WL and the number of months worked; a month is considered to be a working month and thus contains 170 (some authorities say 173) hours, based on a 40 hour work week. Continuing the analogy with DAC-hour, the WLM month is thus equal to 170 WL-hours. Total exposure is frequently expressed in *cumulative working level months* (CWLM).

2.7 A (VERY) FEW WORDS IN CONCLUSION

Historically, radiological quantities and units have undergone a more or less constant evolution which has beneficially led to more rigorous and exact definitions for quantities and units, as well as to confusion due to simultaneous use of different systems of quantities and units which may in fact not always be fully interchangeable. So too have the number of radiological quantities and units grown over the years, and changing concepts and regulatory requirements have sometimes added to the confusion Hopefully, the brief, semirigorous discussion provided above will help to clear the confusion and lead not only to better understanding of radiological quantities and units, but also to more consistent, accurate and unambiguous applications in the future.

2.8 REFERENCES

Bureau International des Poids et Mesures (BIPM). Le System International d'Unites (SI). Seventh Edition. Sevres: Bureau International des Poids et Mesures; 1998.

Burlin, T. E. Cavity Chamber Theory. In: Attix, F. H.; Roesch, W. C., eds. Radiation Dosimetry. Volume I. New York, NY: Academic Press; 1968: Chapter 8.

Failla, G. Dosage in Radium Therapy. Amer. J. Roentgenol. 8:674-678; 1921.

International Commission on Radiological Protection (ICRP). International Recommendations on Radiological Protection. Brit. J. Radiol. 24:46-53; 1950.

International Commission on Radiological Protection (ICRP). Recommendations of the International Commission on Radiological Protection. Brit. J. Radiol. Supplement 6; 1955.

International Commission on Radiological Protection (ICRP). Protection Against Ionizing Radiation from External Sources. ICRP Publication 15. Oxford: Pergamon Press; 1969.

International Commission on Radiological Protection (ICRP). Recommendations of the International Commission on Radiological Protection. Oxford: Pergamon Press; ICRP Publication 26; Ann. ICRP 1(3); 1977.

International Commission on Radiological Protection (ICRP). 1990 Recommendations of the International Commission on Radiological Protection. Oxford: Pergamon Press; ICRP Publication 60; Ann. ICRP 21(1-3); 1991.

International Commission on Radiological Units (ICRU). Radiation Quantities and Units. ICRU Report 10a. National Bureau of Standards Handbook 84. Washington, DC: U. S. Government Printing Office; 1962.

International Commission on Radiation Units and Measurements (ICRU). Radiological Units. ICRU Report 19. Washington, DC: International Commission on Radiation Units; 1971.

International Commission on Radiation Units and Measurements (ICRU). Measurement of Dose Equivalents from External Photon and Electron Radiations, ICRU Report 47, Bethesda, MD: International Commission on Radiation Units and Measurements; 1992.

International Commission on Radiation Units and Measurements (ICRU). Quantities and Units in Radiation Protection and Dosimetry, ICRU Report 51, Bethesda, MD: International Commission on Radiation Units and Measurements; 1993.

International Commission on radiation Units and measurements (ICRU). Fundamental Quantities and Units for ionizing radiation, ICRU Report 60, Bethesda, MD: International Commission on Radiation Units and Measurements; 1998.

Kathren, R. L.; Baalman, R. W; Bair, W. J. Herbert M. Parker, Publications and Other Contributions to Radiological and Health Physics. Columbus, OH: Battelle Press; 1986, 181.

National Bureau of Standards (NBS). The International System of Units (SI), translation approved by the International Bureau of Weights and Measures (BIPM) of its publication "Le Système International d"Unités". NBS Special Publication 330. Gaithersburg, MD: National Bureau of Standards; 1981.

National Council on Radiation Protection and Measurements (NCRP). SI Units in Radiation Protection and Measurements. NCRP Report No. 82. Bethesda, MD: National Council on Radiation Protection and Measurements; 1985.

National Council on Radiation Protection and Measurements (NCRP). Principles and Application of Collective Dose in Radiation Protection. NCRP Report No. 121. Bethesda, MD: National Council on Radiation Protection and Measurements; 1995.

Parker, H. M. Notes on Radiation Dosimetry. In: Kathren, R. L.; Baalman, R. W; Bair, W. J., eds. Herbert M. Parker: Publications and Other Contributions to Radiological and Health Physics. Columbus, OH: Battelle Press; 1986, 184-192.

Parker, H. M. Tentative Dose Units for Mixed Radiations. Radiol. 54:257-261; 1950.

Taylor, B. N. The International System of Units (SI). National Institute of Standards and. Technology Special Publ. 330, 1991 Edition Washington, DC: U.S. Government Printing Office; August 1991.

United States Department of Energy (USDOE). Occupational Radiation Protection, Title 10, Code of Federal Regulations Part 835. Washington, DC: U.S. Government Printing Office; January 1, 1999.
United States Nuclear Regulatory Commission (USNRC). Standards for Protection Against Radiation, Title 10, Code of Federal Regulations, Part 20. Washington, DC: U. S. Government Printing Office; January 1, 1999.

Chapter 3

GAS AND SCINTILLATION DETECTORS TYPICAL TO HEALTH PHYSICS INSTRUMENTS

Alan Justus

3.1 GAS DETECTORS TYPICAL TO HEALTH PHYSICS INSTRUMENTS

3.1.1 Introduction

Gas-filled detectors are among the oldest nuclear radiation detectors. They probably also comprise the most important group of health physics instrumentation. That group includes our ion chambers (pressurized low range, vented medium range, small high range), gas proportional detectors (multiplying ion chambers, α/β proportional counters, thermal neutron proportional counters), and GM detectors (very low- to very high-sensitivity energy-compensated probes, thin-windowed pancake and end-window probes).

All gas detectors consist of a gas-filled chamber with at a minimum a central electrode(s) (i.e., the anode or collection electrode) well insulated from the chamber walls (i.e., the cathode or polarizing electrode). To understand the relation among the three types of detectors (ion chambers, gas proportional detectors, and GM detectors), let us consider a detector of very typical geometry, i.e., that of two coaxial cylinders with a gas between them. Furthermore, let the inner cylinder be a fine (anode) wire held at a positive potential difference V relative to the outer (cathode) cylinder. Assume that the passage of a certain particle (e.g., α or β) produces a certain number of ion pairs. As the potential difference V increases, it is observed that the number of ion pairs collected also increases. The differences in the three types can be best explained through inspection of the resulting 'gas gain' curves of Fig. 3.1.

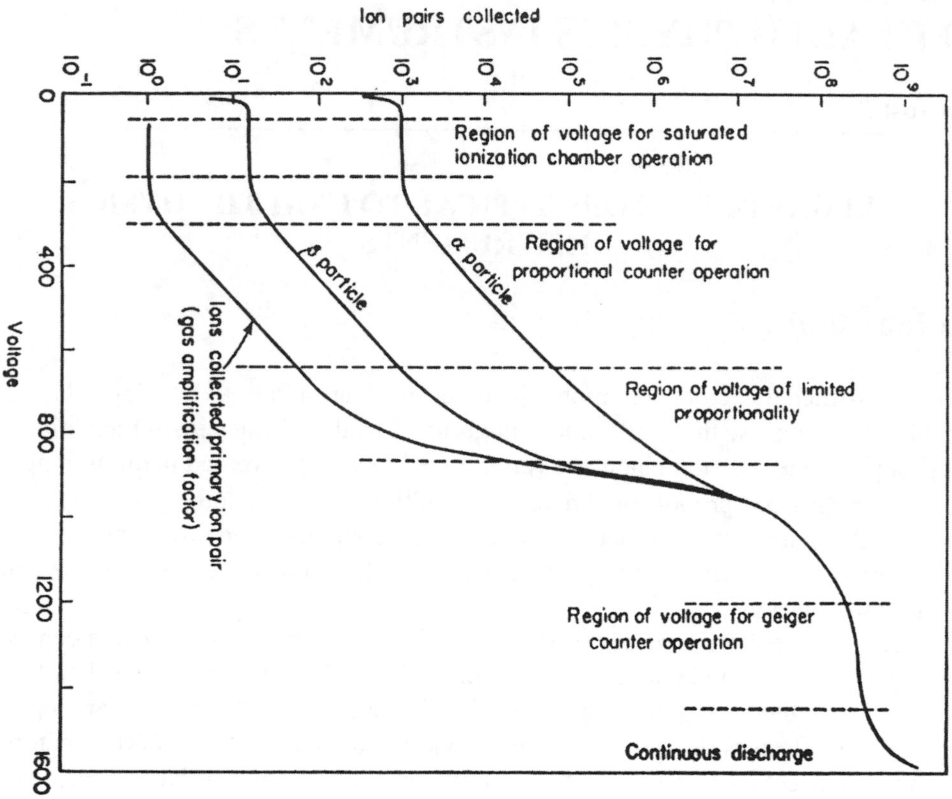

Fig. 3.1. Illustration of pulse-size versus applied voltage for two event types in some particular detector, gas, geometry, etc. combination. [From Handloser (1959), who in turn adapted the illustration from Montgomery et al. (1941). With permission.]

The first point to note in Fig. 3.1 is that pulse size is plotted and not count rate over some discrimination level or mean current. As the voltage is first applied, columnar recombination (so called because ionization is initially formed in a column along the path of the ionizing particle) and diffusion losses are overcome in the so-called recombination and ionization regions. The ionization region corresponds to the voltage range for ionization detector (or chamber) operation. If the negative charge carriers are essentially free electrons only (i.e., the gas is a noble gas such as argon), then as the voltage continues to be increased, the process of gas multiplication begins and we enter the proportional region.

The process of gas multiplication can be explained as follows: as the voltage is increased, the electrons drifting towards the anode can achieve high enough velocities (due to increased accelerating force acting for average distance equal to the mean free path), and therefore high energies, between collisions to ionize more gas, thus multiplying the number of ion pairs. The gas multiplication process therefore takes the form of a cascade, referred to as a Townsend avalanche. Please see Fig. 3.2 for an illustration of one typical avalanche near a 0.001" anode wire. It should also be noted that many excited atoms and molecules from the process decay to their ground state by the emission of visible or UV light, which could create more photoelectrons and therefore another avalanche, hence leading to loss of proportionality or spurious pulses. A polyatomic quench gas (such as methane) is added to absorb the visible and UV photons and dissipate the energy through radiationless transitions to the ground state. Referring back to Fig. 3.1, as long as the α and β curves are parallel to one another (on the log scale!), it can be said that a condition of true proportionality exists between the α and β events. This is the voltage range corresponding to proportional detector (or PC counter) operation. At higher voltages, it should be noted that the α and β curves are no longer parallel and a condition of limited proportionality exists, which is due to both anode end effects and space charge effects from an enlarging positive ion sheath.

Beyond the proportional regions lies the GM region. In GM detectors, the negative charge carriers are still essentially free electrons only (i.e., the gas is a noble gas). However, unlike PC detectors, the polyatomic quench gas (such as methane) is not added to absorb any of the visible and UV photons and, referring to Fig. 3.1, instead of approx. 10^1 and 10^3 Townsend avalanches per β or α pulse, a self sustained propagation of Townsend avalanches over many portions of the thin anode wire occurs until terminated by space charge effects due to the positive ion sheath essentially surrounding the entire anode (Montgomery 1940). This phenomena is referred to as the Geiger discharge. Both β and α pulses are the same size. If the voltage is increased, a larger positive ion sheath (and therefore larger pulse) will develop to terminate the discharge. Unfortunately, when the positive ion sheath arrives at the cathode, it is probable that electrons will be released from the cathode surface which create an additional discharge, ad infinitum. A halogen quenching gas (i.e., Br_2 or Cl_2) with a lower ionization potential than the noble gas is typically added to transfer the positive charge from the noble gas to the quench gas. When neutralized at the cathode, the halogens temporarily disassociate instead of liberating an electron, and multiple discharges are prevented.

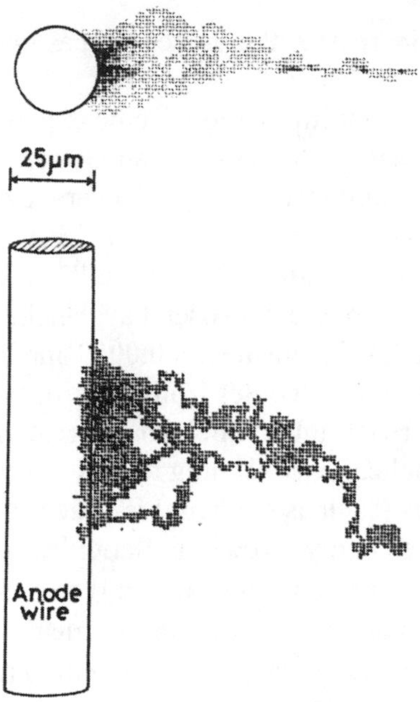

Fig. 3.2. Illustration of one Townsend avalanche as simulated by a Monte Carlo calculation. (From Knoll 1989. With permission.)

3.1.2 Ionization Chambers

The gases used within our health physics ion chambers are typically air, noble gas, or TE (tissue equivalent) gas. With air as the gas, the charge carriers are positive ions and essentially negative ions (due to rapid free electron attachment, the probability for which is near 10^{-4} per collision with oxygen and water vapor). With a noble gas or TE gas, the charge carriers are positive ions and essentially free electrons only (the probability in these cases being near 10^{-6} per collision). Under the influence of typical applied voltages, the net drift velocities for (positive/negative) ions and free electrons are on the order of 10^3 and 10^6 cm/sec, respectively, implying typical detector transit (collection) times on the order of milliseconds and microseconds, respectively. As mentioned earlier, the two major loss mechanisms are recombination and diffusion. When large numbers of simultaneous or overlapping events create ion pairs throughout the chamber volume, the recombination loss additionally involves volume recombination as well as the initial columnar recombination. The recombination loss rate is proportional to the positive and negative charge concentrations, and hence recombination loss is dominant when the charge carriers are positive and negative ions. (Furthermore, the constant of proportionality is also much smaller for free electrons.) The diffusion loss is large for free electrons only, by a factor of several hundred with respect to ion carriers, and represents an increased time for recombination and/or electron flow out of the active volume. What this all implies in regard to health physics ion chambers is that for pulsed fields or intense fields (assuming equal voltage and electrode separation), a 'free electron' chamber would experience less loss than a 'negative ion' chamber, and furthermore, a parallel plate geometry would experience less loss than a cylindrical geometry which in turn would experience less than a spherical geometry (Boag 1956, 1966).

The usefulness of ionization chambers in health physics is based on cavity theories due to Bragg-Gray, Spencer-Attix, Burlin, and Fano. These theories relate the dose in a cavity (gas) to the dose in the wall to the dose in a medium. The Bragg-Gray theory holds as long as the size of the cavity is small compared with the ranges of the secondary particles. Fano (1954) extended the Bragg-Gray theory to conditions where the range of particles is comparable to or even less than the mean chamber depth, as is often the case with neutron and perhaps soft x or β interactions, as long as the elemental composition of the walls and gas are identical, i.e., a homogeneous or matched chamber. Spencer-Attix introduced a modified mass stopping power ratio accounting for energetic δ rays (tertiary electrons capable of ionizing the gas). Burlin modified the Spencer-Attix theory allowing for cavities of any size (zero through infinity) through the introduction of yet another modified mass stopping power ratio. The general cavity theory due to Burlin can be written (Shani 1991) as:

$$D_{med} = \frac{(\mu_{en}/\rho)_{med}}{(\mu_{en}/\rho)_{wall}} \cdot \left[\frac{1}{d \cdot \frac{(S/\rho)_{gas}}{(S/\rho)_{wall}} + (1-d) \cdot \frac{(\mu_{en}/\rho)_{gas}}{(\mu_{en}/\rho)_{wall}}} \right] \cdot W \cdot J \qquad (3.1)$$

where d is a weighting factor that is equal to the average attenuation of the electron spectrum emerging from the walls as it crosses the cavity, W is the mean energy per ion pair of the gas, and J is the number of ion pairs per mass of gas. When the cavity size approaches zero, d = 1 and the expression in brackets reduces to the mass stopping power ratio of the wall to the gas. When the cavity approaches infinity, d=0 and the expression reduces to the mass energy absorption coefficient ratio of the wall to the gas. For large cavities there are of course other factors introduced to correct for photon fluence perturbation and geometric averaging over the large region occupied by the cavity. What this all implies in regard to (large!) health physics ion chambers is that for measurement of exposure, it's quite important for the wall to be air equivalent and the gas to be air so that both ratios become unity. Then $D_{wall} = D_{gas} = 0.873$ cGy (or rad) per R. If the medium is TE, then D_{TE} per R = 0.873 times the ratio of the mass energy absorption coefficient of TE to air. This is essentially 0.96 cSv (or rem) per R from about 3 MeV down to about 100 keV, where it gradually decreases to about 0.7 cSv per R at 10 keV. For β dose to TE medium (the wall), cavity theory (for a small cavity only) gives D_{TE} per R reading = 0.873 times the mass energy stopping power ratio of TE to air. This ratio ranges from about 1.1 for 2 MeV electrons to 1.15 for 15 keV electrons, hence cavity theory gives β D_{TE} per R reading as essentially unity. For measurement of TE dose where neutrons are present, it's extremely important for the wall to be TE equivalent and the gas to be TE gas, since the μ/ρ ratios for neutrons can deviate significantly from unity otherwise (ICRU 1989).

Our health physics ion chamber instruments are always used in the average current mode of operation. If used in a pulse mode, the pulse height would be position dependent due to changes in magnitude of induced charge with position of interaction between the electrodes. Only with very long time constants (positive ion collection included) or very fast time constants (but μV level pulses!) could the signal be position independent (Price 1964). Some of the general design concepts of ion chambers operated in the average current mode of operation are illustrated in Fig. 3.3. An ion chamber without a guard is shown in Fig. 3.3a; its effective (active) volume is not known and the potential for leakage current is significant. Guards are added in the chambers shown in Fig. 3.3b-d. The guard designs shown in Fig. 3.3b&c help define an active volume; those in Fig. 3.3c&d help eliminate leakage.

Some additional practical points that the health physicist should know relate to sensitivity and effective center. The sensitivity is obviously proportional to the wall area intercepted by the fluence, the wall being the source of the secondary electrons. These secondary electrons due to interactions in the wall become the main ionizing particles as they travel through the gas. Additionally, more ion pairs will be produced by secondary electrons in the gas with an increased depth of the gas. Hence the sensitivity is volume dependent (actually mass within that volume, so a density (i.e., P, T) correction is necessary for vented chambers). This sensitivity equates to 0.33 nC per R per cc, or only 93 fA per R/h per cc. Also, the nominal (or quoted) volume might not be the effective or active volume. Hence, leakage should be minimized and active volume maximized, stressing once again the importance of a 'good' guard design. Additionally, an appropriate balance between electronic equilibrium and xγ filtration (i.e., wall attenuation) must be achieved for the application at hand. Electronic equilibrium is established when the wall thickness is equal to or greater than the range of the maximum energy secondary electrons that are generated by the photon interactions with the wall. These electron ranges depend on the photon's energy, and thus electronic equilibrium wall thickness is energy dependent. For example, for 0.1 and 1 MeV photons, the required ion chamber wall thickness is 14 and 430 mg/cm^2 respectively. The practical requirement for an "energy independent" response is a wall thick enough for electronic equilibrium, but not thick enough to significantly attenuate the photon beam relative to other energies.

The effective center of a chamber is not necessarily the geometric center (see Fig. 3.4). It is dependent on the position of the 'source,' e.g., the Compton electrons, photoelectrons, and produced pairs. The angular distribution of Compton electrons is shown in Fig. 3.5 and that of photoelectrons, which could be of concern for lower xγ energies or higher Z wall material, is shown in Fig. 3.6. As can be surmised from the figures, the effective center moves not only more towards the front wall as the geometry becomes more planar and less spherical, but also as the incident photon energy increases. Hence, it can be exactly determined only empirically. Practically, this means that the accuracy of a measurement could be affected by the energy and non-uniformity of the radiation field.

Now let's look at a few specific examples of our health physics ionization chambers, including interleaved, pressurized, and specialized chambers. Some of our popular hand-held ion chamber instruments are based on an interleaved ionization chamber. This includes the Eberline RO-20, with a range of 50 R/h, and the Saint-Gobain (Bicron) RSO-5/50/500 ion chamber instruments. An illustration of the Eberline RO-20's ionization chamber is presented in Fig. 3.7. Note the conductive collection electrode plate (≈80 mg/cm^2) at the center of and surrounded by the conductive chamber (polarizing) electrode, each of air equivalent material. The chamber volume is 220 cm^3 and is vented to atmosphere through a desiccant. Note the position of the guard ring between the collection and chamber electrode electrical connections, similar to Fig. 3.3d. The front wall is an ≈7 mg/cm^2 conductive film with an ≈1 g/cm^2 'beta slide'. The Bicron chamber is almost identical, except it lacks the guard ring and relies on excellent insulators, similar to Fig. 3.3a. Both interleaved chambers can be thought of as a front plus a back chamber, the front chamber obviously having a better β-ray and a flatter x-ray response. The Eberline RO-7 utilizes a similar interleaved chamber of 7 cm^3 volume, vented to atmosphere, with a range of 200 R/h. The hand-held HPI 1030 TE ion chamber for penetrating neutron and γ absorbed dose measurement is also based on an interleaved design. However its ≈100 cm^3 chamber with 25 atm. of TE gas filling utilizes a total of 5 interleaved collector plates.

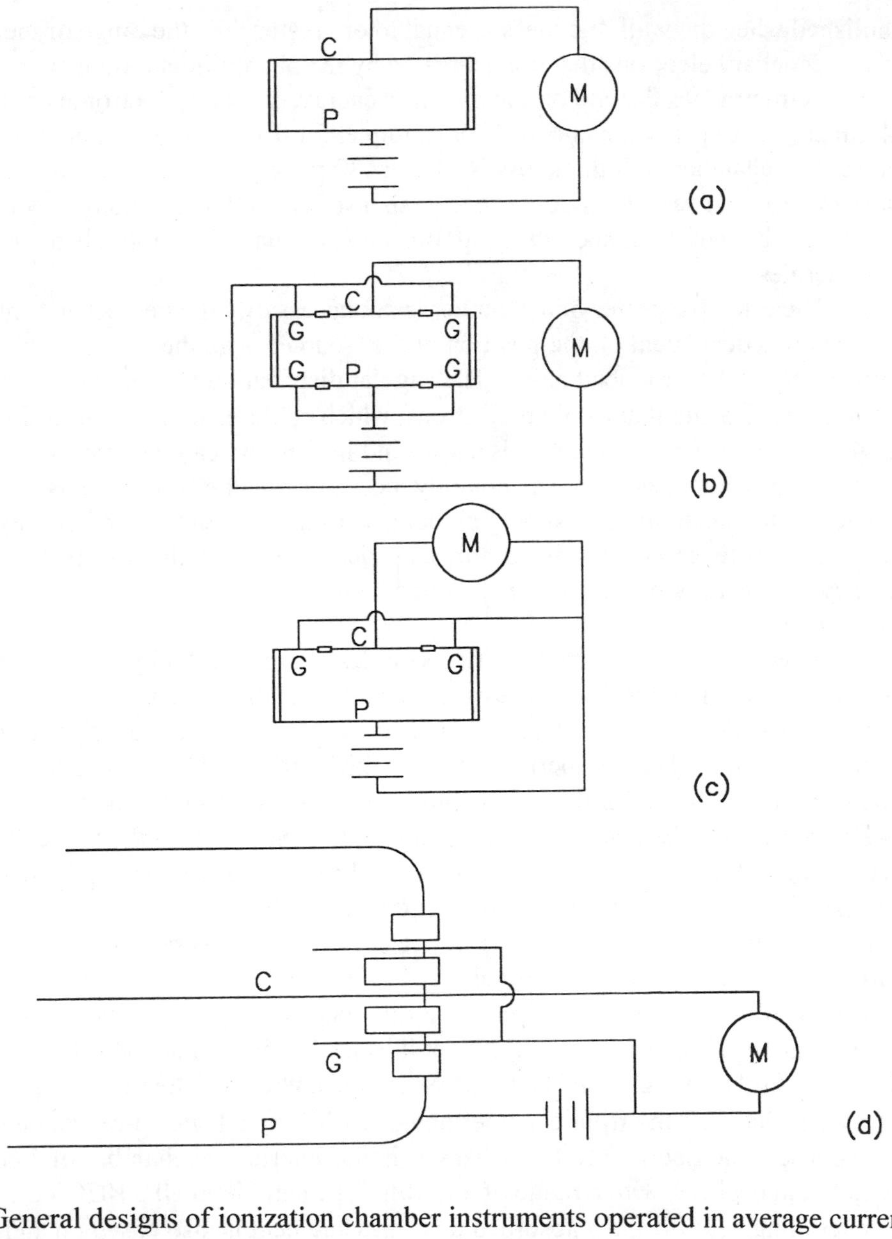

Fig. 3.3. General designs of ionization chamber instruments operated in average current mode. Note that C = collection electrode, G = guard electrode, and P = polarizing electrode.

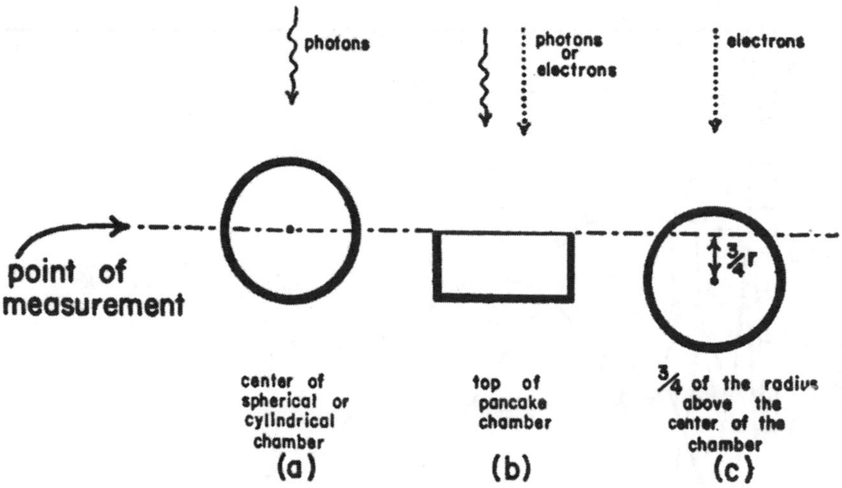

Fig. 3.4. Illustration of the positioning of chambers during calibration and measurement for parallel plate, cylindrical, and spherical chambers. [From Cacak (1985). With permission.]

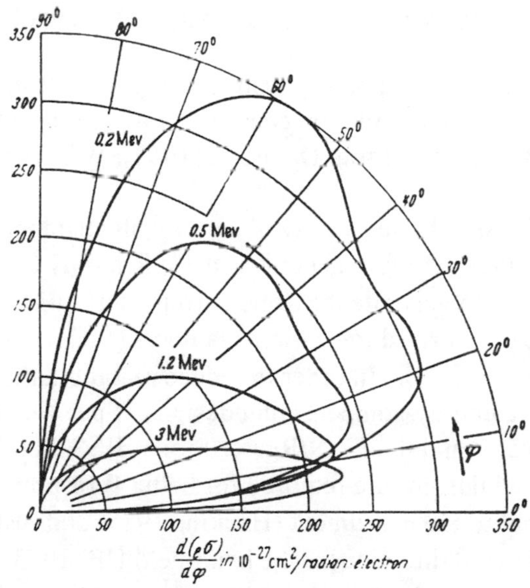

Fig. 3.5. The number of Compton recoil electrons per angle (in radians) as a function of angle (in degrees) for the primary photon energies shown in the curves. The photon is incident from the left. [From Evans (1958). With permission.]

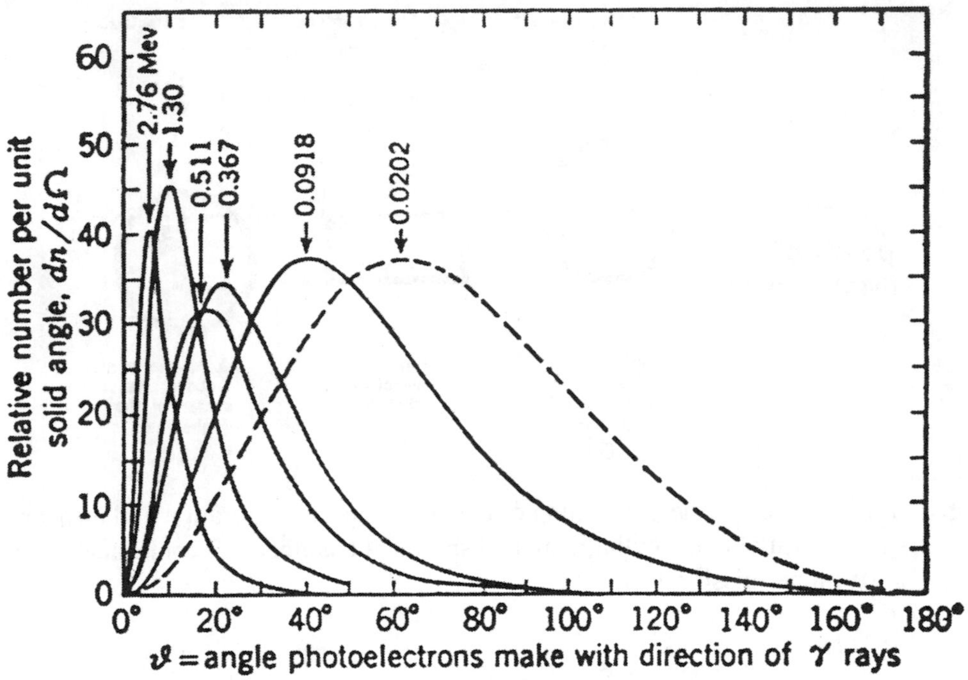

Fig. 3.6. The relative number of photoelectrons per angle as a function of angle (in degrees) for the primary photon energies shown in the curves. The photon is incident from the left. [From Davisson (1952). With permission.]

As noted previously, the sensitivity of an air wall chamber is only 93 fA per R/h per cc. A 220 cm^3 chamber therefore generates at 1 mR/h only 0.02 pA. It would require a 22 liter chamber to generate the same current at 10 μR/h, which is ridiculously large. Therefore, large pressurized ionization chambers (PICs) and sensitive electromers are utilized for μR-measurements. In order to reduce recombination, pressurized noble gas is used and therefore energy-compensating designs. Specific examples are the Inovision (Victoreen) 450P and the EML/Reuter-Stokes RSS-111/112.

Some specialized designs are required for better β response due to both air-scattering of β's external to the instrument (Hankins 1985) and instrument absorption, especially from the sideward directions. The hand-held HPI 1075 dual β/γ ion chamber is based on an interesting piggyback design with an exposed front βx chamber and a rear γ chamber. There are also specialized tritium β CAM chambers with interesting guard concepts and designs such as those from Overhoff Technology.

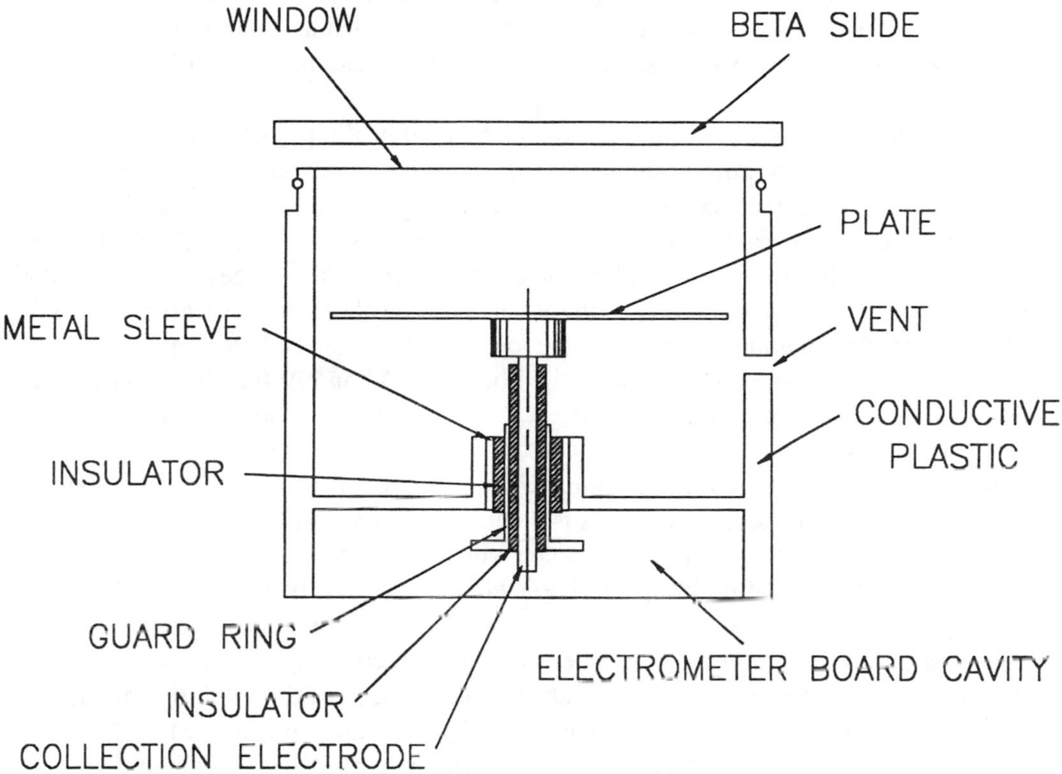

Fig. 3.7. Illustration of the Eberline RO-20's ionization chamber.

3.1.3 Gas Proportional Detectors

The gas used within our health physics proportional counters is (and must be as mentioned previously) a noble gas or TE gas, where the negative charge carriers are essentially free electrons only. In a proportional counter essentially all the charge can be thought of as having originated at one position, i.e., very close to the anode. The time dependence of the induced charge (and hence voltage) is determined utilizing Green's Reciprocation Theorem which relates the induced charge on an electrode to the location of charge carriers between the electrodes (Price 1964). Unlike pulse ionization counters, we find that the initial voltage is due overwhelmingly to the positive ions and not very position dependent (Wilkinson 1950). Even though the positive ion collection time is typically 1 to 2 msec, the time for the induced voltage to reach half-amplitude is typically 2 to 5 μsec, respectively, so proportional counters can be used for high count rates.

Now let's look at a few specific examples of our health physics gas proportional detectors, which includes multiplying ion chambers, α/β proportional counters, and thermal neutron proportional counters. First of all, every proportional detector is, or needs to be, a good (i.e., an efficient) ion chamber before it's a proportional detector. In fact, if an ion chamber with a thin 'collection electrode' (i.e., thin anode) is operated at a voltage beyond the ionization region such that the process of gas multiplication begins, then we enter the proportional region and the instrument is referred to as a multiplying ion chamber. This is the principle behind the Eberline DA1 and Aptec-NRC IP-100 'ion chamber' series of area monitors, which use a noble gas and therefore energy-compensating designs. This is also the mode of operation of the hand-held HPI 1010 TE 'ionization chamber' for penetrating neutron and γ absorbed dose measurements. One important concern with these instruments is the under response in pulsed fields that occurs, for a given operating voltage, at some given dose per pulse and is due to the cessation of gas multiplication by space charge effects due to a positive ion sheath essentially surrounding the entire anode.

Health physics gas proportional detectors for α/β counting usually utilize P-10 gas (90% Ar/10% methane) and a thin conductive window, typically 0.85 mg/cm^2, although 0.1 mg/cm^2 is also used. Multiple parallel anodes are typically centered between the window and the detector body, although one type utilizes an anode loop within a hemispherical chamber. Once again, every proportional counter needs to be a good ion chamber before it's a proportional detector (see Fig. 3.8). The β particles deposit just a sampling of energy based on each electron's specific energy loss dE/dx and the Δx traversed; the α particles can deposit from partial through full remaining energy based on each α's energy when entering the detector and the paths involved within a specific detector. Typical α/β spectra from the thin detectors utilized in our automatic planchet counting (APC) systems, such as those from Oxford and Canberra, are shown in Fig. 3.9 for a 0.1 mg/cm^2 window. Note the β dE/dx portion at the low pulse heights and the α dE/dx portion at the high pulse heights. Because of this difference in pulse heights, it is possible to obtain α or α+β plateaus, or even better, dual α/β channels simultaneously. Typical α/β spectra from the thicker large area detectors utilized with our portable α/β PC detectors (such as the Eberline HP-100GS and the Ludlum 43-68GS) and in our automatic H&S/IPCM systems (such as those from Berthold, Eberline, Saint-Gobain (NE), Ludlum, and Aptec-NRC) are shown in Fig. 3.10. Note the similarity with Fig. 3.9, the only difference being the appearance of an α remaining energy distribution due to a thicker window (i.e., 0.85 mg/cm^2) and a deeper chamber.

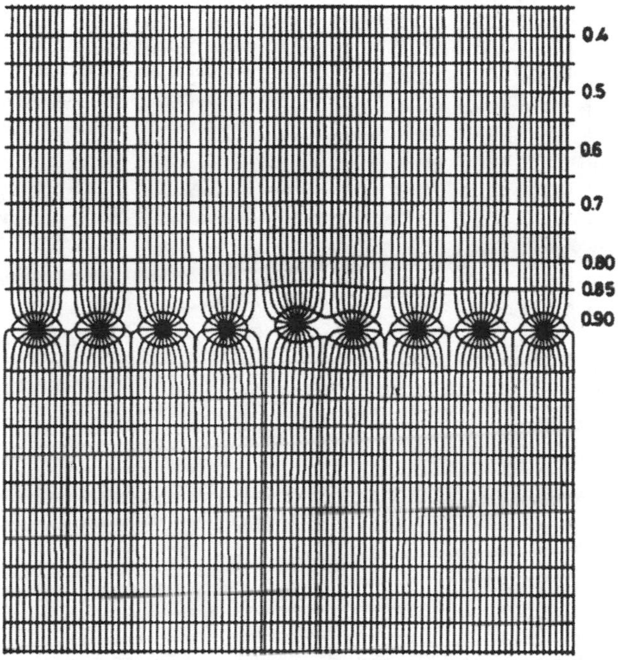

Fig. 3.8. Illustration of electric field lines created from a grid of anode wires placed equidistant between two cathode plates at the top and bottom of the figure. [From Knoll (1989). With permission.]

One of the components of most health physics neutron rem meters is a thermal neutron detector utilizing either gas proportional $^{10}BF_3$ or ^3He counters. The new WENDI wide-energy neutron rem meter uses a ^3He tube from Gamma Industries, whereas the Hankin's sphere (Eberline NRD), 10" sphere (Ludlum 42-30), and Anderson-Braun (Aptec-NRC and HPI) rem meters use various $^{10}BF_3$ tubes from Nancy Wood. A typical neutron plus gamma spectrum from a N. Wood $^{10}BF_3$ tube is shown in Fig. 3.11. Note the secondary electron dE/dx portion at the low pulse heights and the α plus Li energy deposition portion at the high pulse heights. Because of this difference in pulse heights, it is possible to discriminate effectively against gamma interactions. The neutron rem meter energy and temporal characteristics are beyond the present scope and could fill an additional chapter.

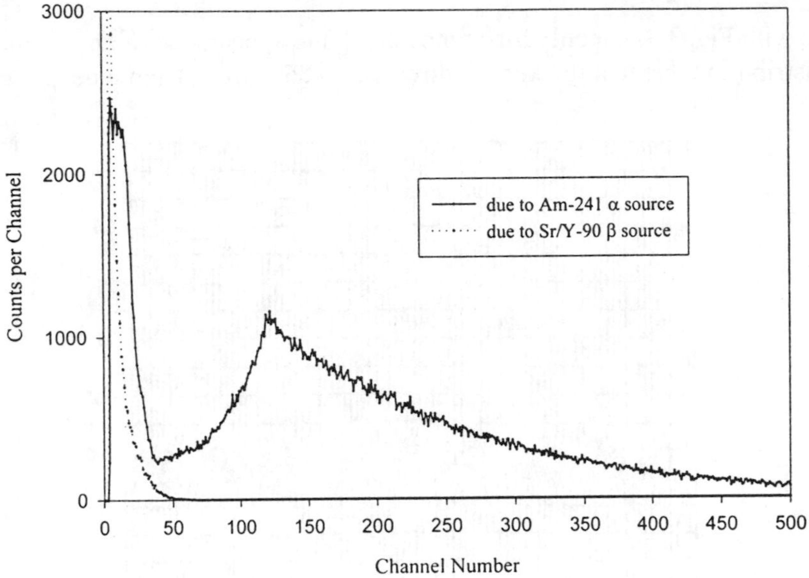

Fig. 3.9. Typical α and β spectra from an Oxford APC System.

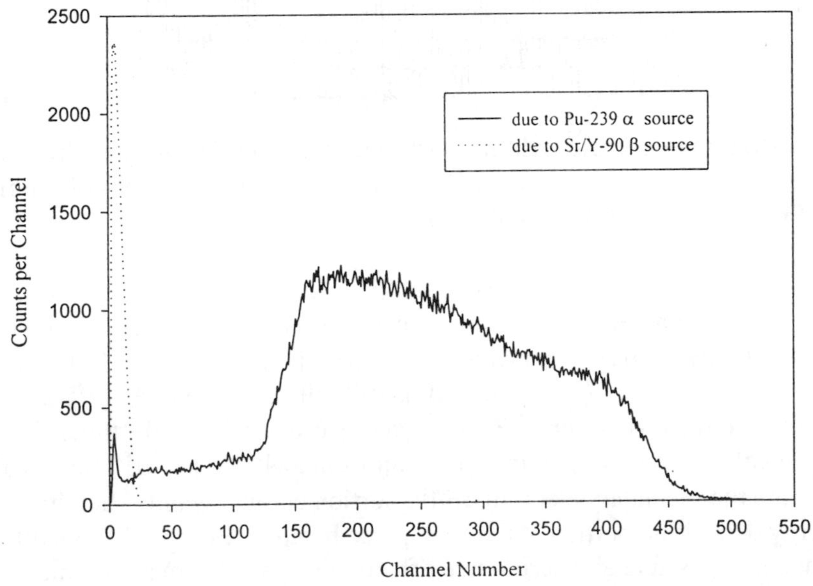

Fig. 3.10. Typical α and β spectra from a P-10 gas filled Ludlum 43-68GS 100 cm^2 Probe.

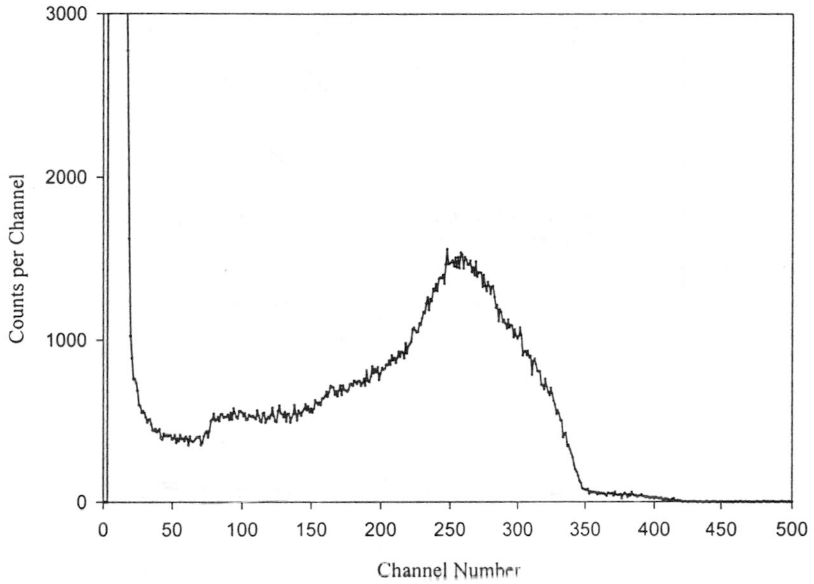

Fig. 3.11. A typical neutron plus gamma spectrum from a N. Wood G-10-2 $^{10}BF_3$ tube.

3.1.4 G-M Detectors

The gas used within our health physics GM counters is a noble gas, where the negative charge carriers are essentially free electrons only. After the Geiger discharge, all the charge can be thought of as having originated very close to the anode. The time dependence of the induced charge (and hence voltage), determined utilizing Green's Reciprocation Theorem, shows that the initial induced voltage is due to the positive ions (Price 1964). The positive ion collection times are typically 50 to 300 μsec, and even though the time for the induced voltage to reach half-amplitude is typically 0.5 to 3 μsec, respectively, the recovery from the Geiger discharge makes the counter dead for times comparable to the collection times and GM counters cannot be made to count very quickly. A GM's count rate over some discrimination level will increase with increasing applied voltage within the GM region. This rate increase (i.e., plateau) is not due to the increasing pulse size, but rather to increasing electron collection efficiency (remember also that every GM is an ion chamber before it's a GM) and to spurious discharges, see Fig. 3.12.

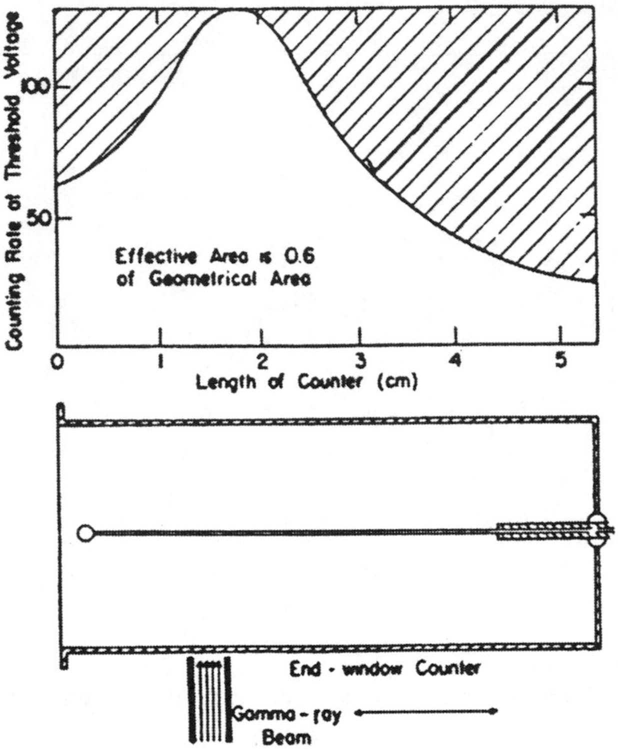

Fig. 3.12. Relative counting rate for a collimated γ source moved parallel to the axis of an end window GM. (From Sinclair 1956).

Now let's look at a few specific examples of our health physics GM detectors, which includes very low- to very high-sensitivity energy-compensated probes, thin-windowed pancake and end-window probes, and specialized probes. Since the gas is Ar and the wall stainless steel, exposure measurements are energy dependent.)see Emery 1966). An energy compensating design is therefore required to make the counter approximately energy independent. Energy-compensated probes, such as the TGM N378S; Eberline HP-270, and small LND or TGM tubes, are used within our xγ portables, area monitors, EPDs, and teledetectors. When a thin mica window is incorporated into the GM's design, the result is an αβx sensitive pancake or end window probe. One specialized use of GM detectors is the Ag-foil-wrapped GM within the HPI 2080 rem meter for pulsed neutron fields.

3.2 SCINTILLATION DETECTORS TYPICAL TO HEALTH PHYSICS INSTRUMENTS

3.2.1 Introduction

Scintillation (sparkling) detectors also comprise an important group of health physics instrumentation. This group includes our NaI(Tl) γ detectors, α/β plastic- or liquid-scintillation-based counters, and fast or thermal neutron scintillation detectors. Some are among the first nuclear radiation detectors, whereas others are of a more recent vintage. They are classified as either inorganic scintillators (where light emission derives from the deexcitation of energy levels of the whole crystal lattice) or organic scintillators (where light emission derives from the deexcitation of excited molecular energy levels).

The inorganic scintillators used in health physics instruments are typically NaI(Tl), ZnS(Ag), or ^{6}LiI(Eu) scintillators of various thickness and extent. These scintillators are crystals of inorganic salts with certain added impurities called activators. The normal energy band structure of the crystal is modified at the activator sites in such a way as to create energy states (i.e., activator ground state and excited states) within the normally forbidden gap separating the valence and conduction bands. Some of the energy (i.e., typically about 1/3) deposited to the scintillator crystal by charged particles can create electron-hole pairs (comparable to ionization in gases). These may diffuse through the crystal, the hole to the activator ground state, thus ionizing the activator site, and the electron to such an ionized activator site, neutralizing it but leaving it in an excited state. This excitation energy can be reemitted in the form of photons of visible or ultraviolet light The emitted pulse is found to decay exponentially with time constants typically on the order of hundreds of nanoseconds. An important consequence of this luminescence through activator sites is the fact that the light is not strongly self-absorbed, since it typically represents only about one-half of the forbidden gap energy.

The organic scintillators used in health physics instruments are various types, thickness, and/or extent of liquid or plastic scintillators. These scintillators are mostly organic molecules with conjugated multiple bonds which give rise to a so-called π-electron structure (Justus 2000)*. Some of the energy deposited to the scintillator material by charged particles can excite molecular levels of the constituent organic molecules. This excitation can undergo energy transfer from molecule to molecule until a fraction of it is emitted in the form of photons of visible or ultraviolet light. The emitted photons of light are not strongly self-absorbed, and are found to decay

* Justus, B.L. Personal Communications, 2000.

exponentially with time constants typically on the order of nanoseconds. This prompt fluorescence represents most of the light. A longer-lived fluorescence component (i.e., order(s) of magnitude increase in lifetime) is also present in some scintillators, corresponding to delayed fluorescence. Although it does not represent a significant amount of the light output, its amount relative to the fast component increases with increasing LET, and forms the basis of the PSD (pulse shape discrimination) techniques utilized for fast neutron spectrometry. (Please see, for example, Birks (1964) for more detail regarding the exact mechanisms at work in organic scintillators.)

Because the emission is via the deexcitation of individual molecules, organic scintillators can still function as scintillators when dissolved in organic liquid or plastic (e.g., polyvinyltoluene) solutions. The energy deposited by charged particles is dissipated in the liquid or plastic solvent and subsequently transferred to the dissolved molecules of the organic scintillator. Wavelength-shifting molecules are often added in order to accept the energy from the scintillator molecules and reemit photons of lower frequency, which more nearly match the PMT photocathode spectral response (Justus 2000)[†].

The light pulse from both inorganic and organic scintillators must be collected at the PMT, ideally with a high light collection efficiency. This light must then be converted to photoelectrons, ideally by a high sensitivity photocathode spectrally matched to the wavelength of the scintillation light. The photoelectrons are then multiplied by the dynodes and the charge pulse collected by the anode.

3.2.2 NaI(Tl) Detectors

One of the most common inorganic scintillators used in health physics instruments is NaI(Tl). It is a hygroscopic detector and must be sealed, usually in an aluminum container with a transparent window. As discussed above, the average energy for electron-hole production is about 3 times the 5.9 eV gap energy or about 18 eV. The migration, transfer to, and quantum efficiency of activator luminescence sites is near 70%, with 3 eV average energy photons emitted (about 50% of the 5.9 eV gap). This corresponds to about 50 eV per light photon emitted. With a light collection efficiency of say 95% and a PMT photocathode quantum efficiency of 25% (bialkali), an average energy of 213 eV would be required for every photocathode electron produced. Therefore, in comparison to HPGe (with an average energy of 2.96 eV per electron-hole pair produced and a Fano factor near 0.1), NaI(Tl) is a low resolution xγ detector with its average energy per photocathode electron of about 210 eV and its Fano factor near unity.

[†] Justus, B.L. Personal Communications, 2000.

(The Fano factor is a measure of the detector's energy resolution. The lower the value of the Fano factor, the better is the energy resolution.) It is a very sensitive xγ detector, however, due to its high Z (53 for iodine) and reasonably high density (ρ=3.67). The linear attenuation μ (cm^{-1}) and energy absorption μ_{en} (cm^{-1}) coefficients for NaI(Tl) are plotted in Fig. 3.13. The photoelectric and Compton effects are near equal contributors to the attenuation coefficient at 0.3 MeV. Therefore, at higher energies, one or more Compton interactions are typically required to 'shift' a gamma's energy downward to where the photoelectric effect can 'snap' it up, contributing a so-called photopeak to the spectrum at the expense of what would have been Compton continuum. [Note that this process would effectively raise the μ_{en} curve up to the μ curve.] For an instrument operating in the 'gross' mode, however, any portion of a spectrum above the discrimination level will contribute a count. Hence, by varying the thickness and/or discrimination level, a multitude of sensitivity characteristics can be achieved. When the discrimination level is 10 keV, the term 'total' is used, as in Fig. 3.14, which is a family of curves of the intrinsic total efficiency for various thicknesses of NaI(Tl). Filtration (window attenuation) can modify these curves in a comparable manner to discrimination level.

Now let's look at a few specific examples of NaI(Tl) detectors utilized in our health physics instruments. These include very thin to very thick probes, with thin or thick windows and of various extent. Specifically, there are thin (i.e., 2 mm) mini-FIDLER's such as the Ludlum 44-17 and Eberline PG-2, intermediate 0.5 inch-thick probes such as the Eberline SPA-9 and Ludlum 44-62, and thick (i.e., 2-inch) probes such as the Ludlum 44-10 and Eberline SPA-3. Note that the thin detectors, obviously intended for low energy detection, have 0.001-0.005" thin Al entrance windows.

3.2.3 ZnS(Ag)/Thin Plastic Detectors

ZnS(Ag) is an inorganic scintillator, available only as a polycrystalline powder and usable only as a thin detector due to scintillator light self-absorption. ZnS(Ag) is used as an α scintillator in health physics instruments as a thin 5-10 mg/cm^2 layer in our portable α survey instruments and inert α (i.e., Rn) CAMs. A useful new application is the dual α/β scintillator, utilizing a thin 4-5 mg/cm^2 ZnS(Ag) α scintillator on a thin (0.4-1 mm) β dE/dx plastic scintillator. The α particles deposit essentially their full energy in the ZnS(Ag) α scintillator; the β particles deposit just a sampling of their energy in the thin plastic β scintillator based on each electron's specific energy loss dE/dx and the Δx traversed. Typical α/β spectra from dual α/β scintillators used with our portable α/β probes (such as the Saint-Gobain (NE) DP6 series, Eberline HP-380AB, and Ludlum 43-

89) are shown in Fig. 3.15. Note the β dE/dx portion at the low pulse heights and the α full energy portion at the high pulse heights. Because of this difference in pulse heights, it is possible to obtain α or α+β plateaus, or even better, dual α/β channels simultaneously. One important characteristic of these dual α/β scintillators is the under response to low energy β's due to their attenuation by the near 6 mg/cm^2 of window plus ZnS(Ag) material.

3.2.4 Liquid Scintillation/Thick Plastic Detectors

Several health physics instruments utilize thick organic scintillator detectors for either β, neutron, or γ counting. For the counting of soft β's, internal source liquid scintillation counting is used. This involves dissolving the sample to be counted directly in a liquid scintillator and its counting in a liquid scintillation counter (LSC) such as those available from Canberra-Packard and Beckman. Neutron spectrometry systems use PSD (pulse shape discrimination) techniques and a special liquid scintillator such as the Saint-Gobain (Bicron-NE) BC501 series (includes NE-213) for high-resolution/ high-energy neutron spectrometry. Plastic scintillation detectors are sometimes used as γ detectors, such as the 2-inch thick Eberline SPA-6 detector. However, plastic is not that great a γ detector, due to its very low Z and low density (ρ=1.032) as well as its low scintillation efficiency (i.e, about 200 eV per light photon emitted). Please refer once again to the linear attenuation μ (cm^{-1}) and energy absorption μ_{en} (cm^{-1}) coefficients for NaI(Tl) and polyvinyltoluene-based plastic scintillators in Fig. 3.13 above. For the plastic scintillator, the photoelectric and Compton effects are near equal contributors to the attenuation coefficient at 0.02 MeV. Notice how far removed the plastic scintillator total absorption curve is from the corresponding curve of NaI(Tl), from about three orders of magnitude at the low energies to less than one at the higher energies. Additionally, since the γ interactions are predominantly Compton down to 20 keV, it is the Compton electrons that are typically responsible for the energy absorption, i.e., the spectrum is essentially a low resolution Compton continuum only. Compton electron energy distributions are shown in Fig. 3.16. For a low energy γ, such as 60 keV, the distribution would be similar to the somewhat symmetric distribution seen at the left, except with an upper end point energy of only 11.4 keV. Hence, a finite counter lower discrimination level can have drastic effects on counting efficiencies for low energy γ's. Therefore, plastic scintillators should be used with caution as a γ detector; they are only good at higher γ energies for coverage of large areas such as with whole-body and truck monitoring instruments.

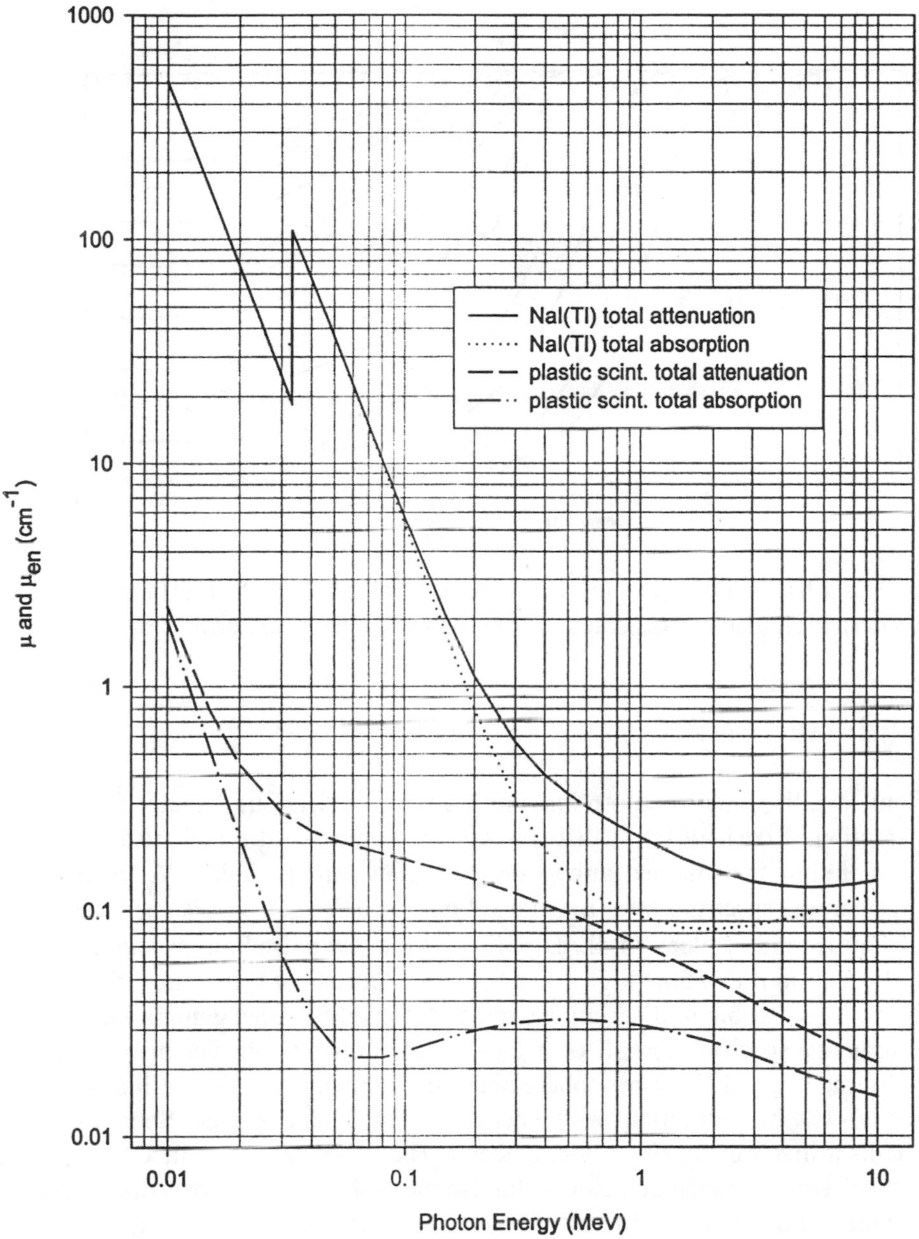

Fig. 3.13. Plot of linear attenuation μ (cm^{-1}) and energy absorption μ_{en} (cm^{-1}) coefficients for NaI(Tl) and polyvinyltoluene-based plastic scintillator.

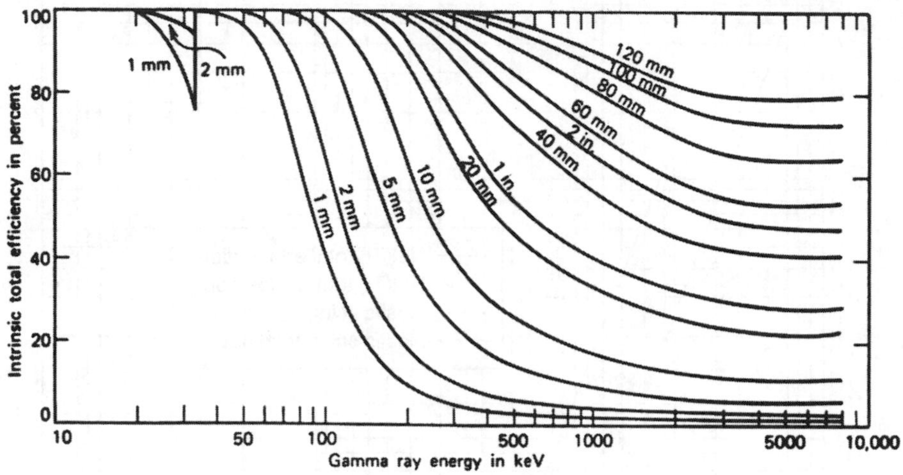

Fig. 3.14. The intrinsic total efficiency of various thicknesses of NaI(Tl) for γ rays perpendicular to its surface. (From Knoll 1989. With permission.)

3.2.5 ^{6}LiI(Eu) Detectors

Some health physics neutron instruments utilize a ^{6}LiI(Eu) detector for thermal neutron detection. Like NaI(Tl), ^{6}LiI(Eu) is a hygroscopic detector and must be sealed, usually in an aluminum container with an optically transparent window A typical neutron plus gamma spectrum from a 4 mm x 4 mm ^{6}LiI(Eu) detector is shown in Fig. 3.17. Note the secondary electron dE/dx portion at the low pulse heights and the α plus ^{3}T energy deposition peak at the high pulse heights. Because of this large difference in pulse heights, it is possible to discriminate quite effectively against gamma interactions. Now let's look at a specific example or two of use in our health physics instruments. This includes the original Hankin's sphere neutron rem meter, which is still available as the Ludlum 42-4. Low-resolution/ wide-energy neutron Bonner Sphere Spectrometry (BSS) systems utilize various sensitivity detectors (i.e., sizes from 4 mm x 4 mm to 0.5" x 0.5") as the thermal neutron detector within Bonner spheres of assorted diameters. A BSS set comprised of a 4 mm x 4mm ^{6}LiI(Eu) detector with spheres up to 12 inch diameter is available as the Ludlum 42-5. As mentioned previously, the neutron rem meter and BSS energy and temporal characteristics are deemed far beyond the present scope.

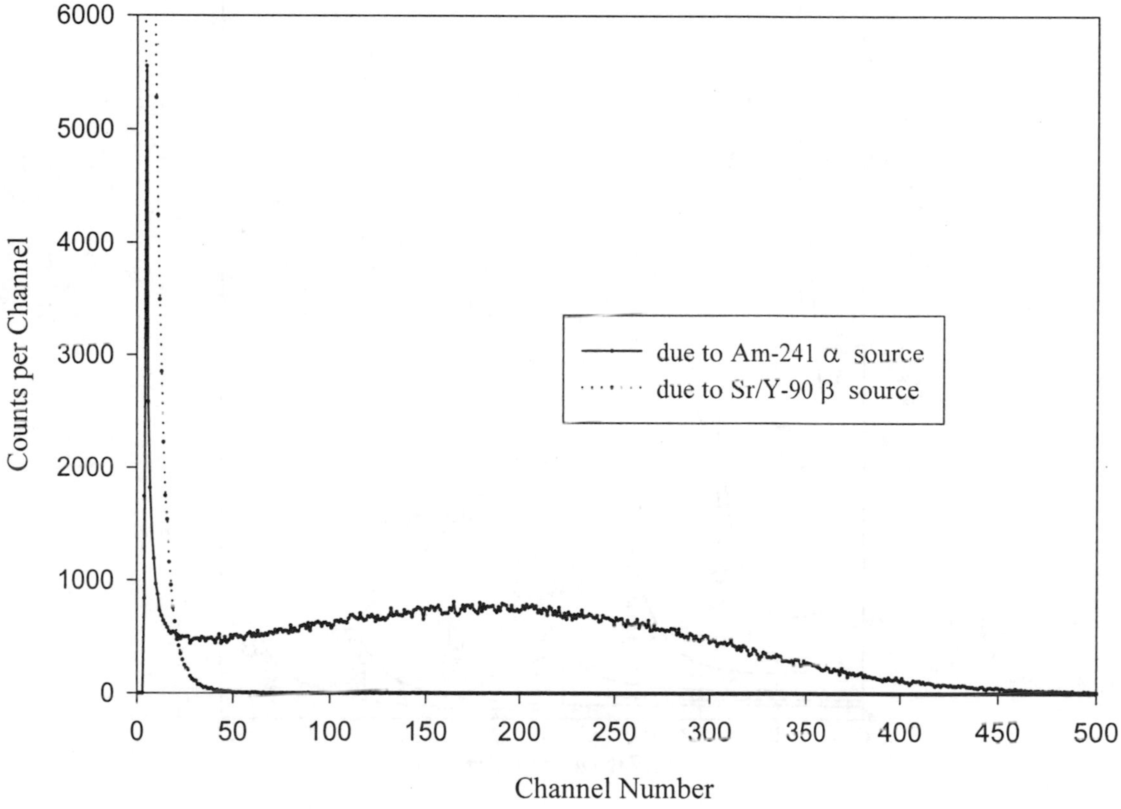

Fig. 3.15. Typical α and β Spectra from an NE DP6 Dual α/β Scintillator Probe.

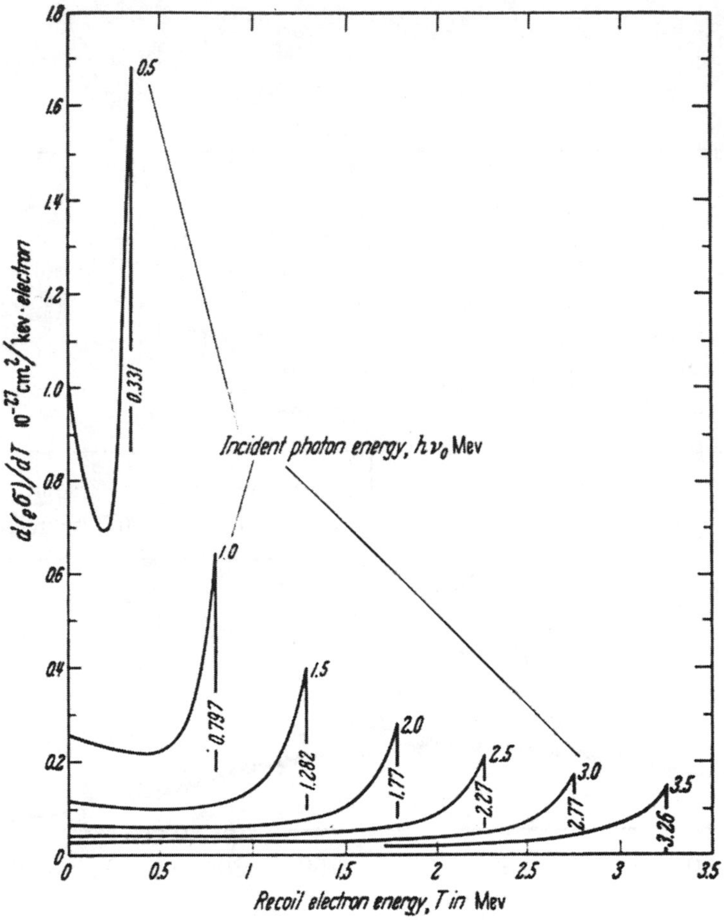

Fig. 3.16. The energy of Compton recoil electrons for various primary photon energies. (From Evans 1958. With permission.)

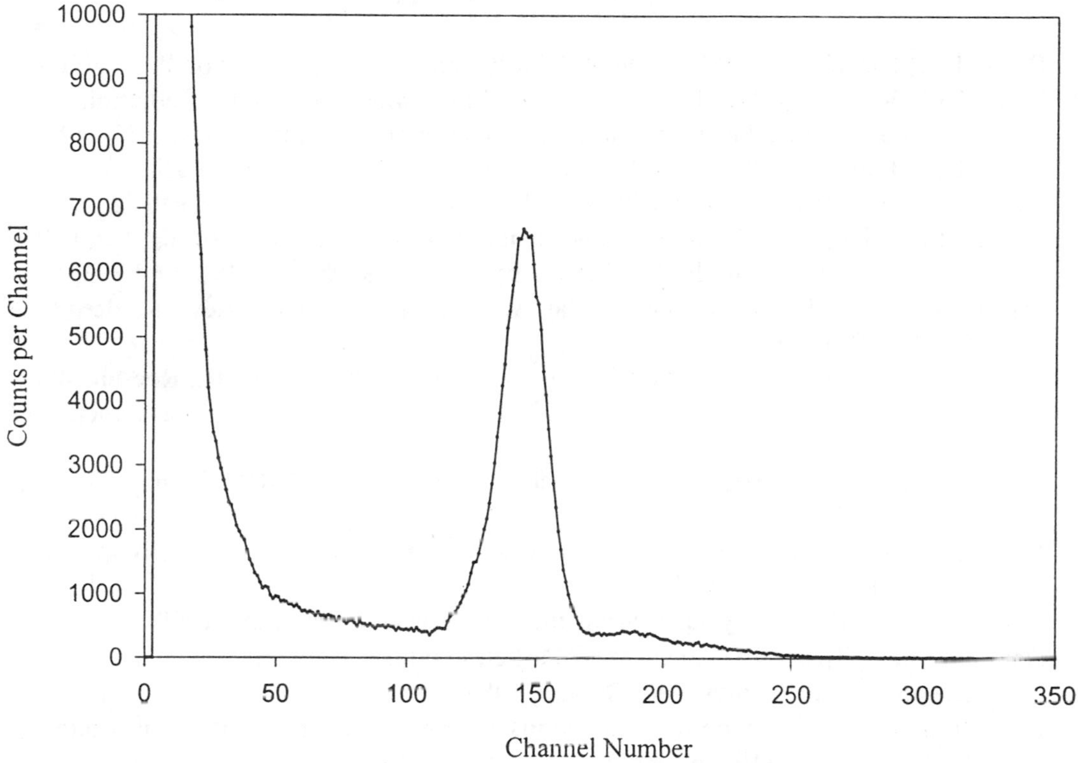

Fig. 3.17 A typical neutron plus gamma spectrum from a 4 mm x 4 mm ^{6}LiI(Eu) detector.

3.3 ACKNOWLEDGMENTS

The author is indebted to Mr. Dave Reilly for his AutoCad drawings of the general ion chamber electrical configurations and the rendering of the Eberline RO-20's ionization chamber. The author is also indebted to Mr. Scott Borkowski and Mr. Jerry Letizia for their patient scanning and editing of the figures accompanying the text. Thanks are also extended to Mr. McLouis Robinet, Mr. Ron Cooper (ORNL/SNS), and Dr. Brian Justus (NRL) for their review and comments.

3.4 REFERENCES

Birks, J.B. The Theory and Practice of Scintillation Counting. Pergamon Press. 1964.

Boag, J.W. Ionization Chambers. In: Hine, G. J.; Brownell, G. L., eds. Radiation Dosimetry. Second Printing. New York :Academic Press, Inc.; 1956: 153-212.

Boag, J.W. Ionization Chambers. In: Attix, F.H.; Roesch, W. C., eds. Radiation Dosimetry. Second Edition. New York :Academic Press, Inc.; 1966: 1-72.

Cacak, R.K.; Hendee, W.R. Ionization Chamber Dosimetry. In: Mahesh, K.; Vij, D.R., eds. Techniques of Radiation Dosimetry. John Wiley & Sons; 1985: 67-113.

Davisson, C.M.; R.D. Evans. Gamma-Ray Absorption Coefficients. Revs. Modern Phys. 24:79-107, 1952.

Emery, E.W. Geiger-Mueller and Proportional Counters. In: Attix, F.H.; Roesch, W.C., eds. Radiation Dosimetry. Second Edition. New York :Academic Press, Inc.; 1966: 73-122.

Evans, R.D. Compton Effect. In: Handbuch der Physik, Vol. XXXIV. Springer-Verlag; 1958: 218-298.

Fano, U. Note on the Bragg-Gray Cavity Principle for Measuring Energy Dissipation. Radiat. Res. 1:237-240, 1954.

Handloser, J.S. Health Physics Instrumentation. Pergamon Press, Inc. 1959.

Hankins, D.E. Effect of Air-Scattered β Particles on Instrument and Dosimeter Response. Health Phys., 49: 435-441, 1985.

International Commission on Radiation Units and Measurements. Tissue Substitutes in Radiation Dosimetry and Measurement. Report 44, 1989.

Knoll, G.F. Radiation Detection and Measurement. Second Edition. John Wiley & Sons; 1989.

Montgomery, C.G.; D.D. Montgomery. The discharge Mechanism of Geiger-Mueller Counters. Phys. Rev. 1030-1040, 1940.

Montgomery, C.G.; D.D. Montgomery. Geiger-Mueller Counters. J. Franklin Inst., 231:447-467, 1941.

Price, W. J. Nuclear Radiation Detection. Second Edition. McGraw-Hill, Inc.; 1964.

Shani, G. Radiation Dosimetry Instrumentation and Methods. CRC Press, 1991.

Sinclair, W.K. Geiger-Mueller Counters and Proportional Counters. In: Hine, G.J.; Brownell, G.L., eds. Radiation Dosimetry. Second Printing. New York :Academic Press, Inc.; 1956: 213-243.

Wilkinson, D.H. Ionization Chambers and Counters. Cambridge University Press. 1950.

Chapter 4

SEMICONDUCTOR DETECTORS

Charles Kent

4.1 FUNDAMENTALS OF SEMICONDUCTOR DETECTORS*

4.1.1 Semiconductors

A semiconductor is a solid or liquid material able to conduct electricity at room temperature more readily than an insulator but less easily than a metal. Electrical conductivity, which is the ability to conduct electrical current under the application of a voltage, has one of the widest ranges of values of any physical property of matter. Metals such as copper, silver, and aluminum are excellent conductors, but such insulators as diamond and glass are very poor. At low temperatures, pure semiconductors behave like insulators. Under higher temperatures or light or with the addition of impurities, however, the conductivity of semiconductors can be increased dramatically, reaching levels that may approach those of metals.

4.1.2 Crystals and Crystal Structure

Semiconductors consist of atoms placed in an ordered form called a crystal. Crystals are identified based on their lattice structure. For instance, the crystal structure of silicon is like that of diamond and is referred to as the diamond lattice, shown in Fig 4.1. Each atom in the diamond lattice has a covalent bond with four adjacent atoms forming together a tetrahedron.

Compound semiconductors such as GaAs and InP have a lattice structure similar to that of diamond. However the lattice contains two different types of atoms. Each atom still has four covalent bonds, but they are bonds with atoms of the other type.

---

* Van Zeghbroeck, 1998, Unpublished Technical Notes, Bart.VanZeghbroeck@Colorado.edu/Bart

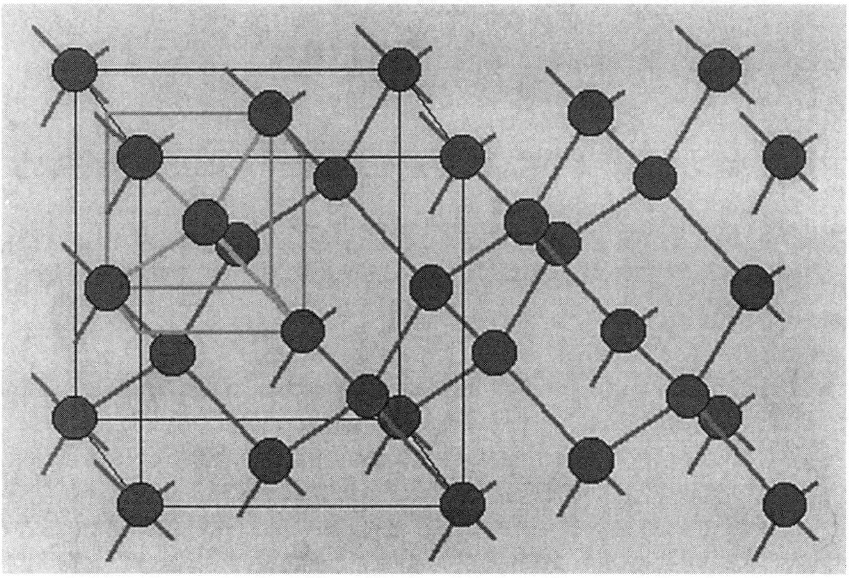

Fig. 4.1 - The diamond lattice of silicon and germanium.

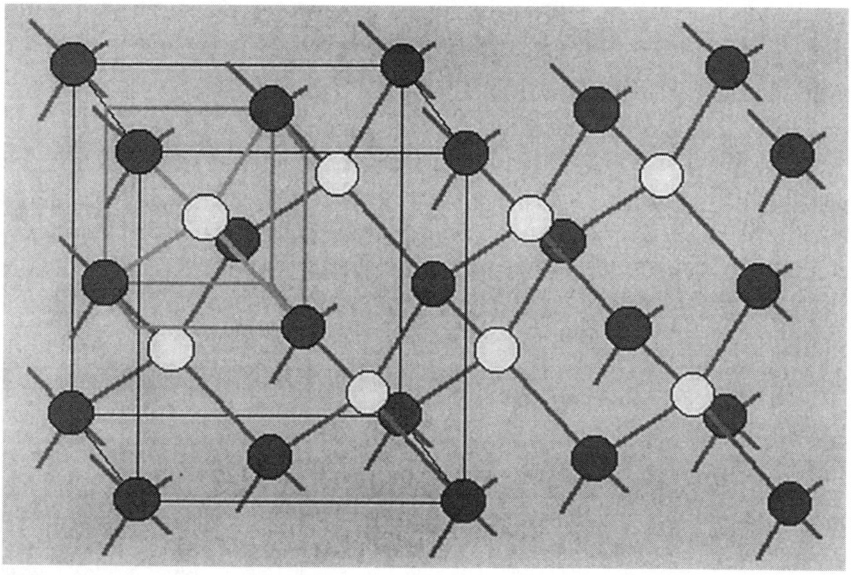

Fig. 4.2 - The lattice structure of GaAs and InP.

4.1.3 Energy Bands

Energy bands occur in solids where the discrete energy levels of the individual atoms merge into bands that contain a large number of closely spaced energy levels. In an isolated atom, the allowed energy levels are very sharp and defined as shown from the discrete emission and absorption spectral energy lines of atoms. When a collection of atoms are brought together in a crystal, Pauli's exclusion principle requires that no two electrons can have exactly the same set of quantum numbers. The atomic energy levels therefore change and the sharply defined energy levels broaden into bands of energy levels within the material. It is the detailed band structure of a given material that determines its conducting, insulating or semiconducting behavior.

Once the band structure of a given material is known, it is necessary to determine which energy levels are actually occupied and whether specific bands are empty, partially filled or completely filled.

Empty bands do not contain electrons and therefore are not expected to contribute to the electrical conductivity of the material. Partially filled bands contain electrons as well as unoccupied energy levels that have a slightly higher energy. These unoccupied energy levels enable carriers to gain energy when moving in an applied electric field. Electrons in a partially filled band therefore contribute to the electrical conductivity of the material.

Completely filled bands contain plenty of electrons but do not contribute to the conductivity of the material. This is due to the fact that the electrons cannot gain energy since all energy levels are already filled.

In order to identify the filled and empty bands, it is necessary to determine how many electrons can be fit into one band and how many electrons are available. Since one band is due to one or more atomic energy levels, it can be concluded that the minimum number of states in a band equals twice the number of atoms in the material. The reason for the factor of two is that even a single energy level can contain two electrons with opposite spin.

To further simplify the analysis, it is assumed that only the valence electrons (the electrons in the outer shell) are of interest, while the core electrons are assumed to be tightly bound to the atom and are not allowed to wander around in the material.

Four different possible scenarios are shown in the figure below:

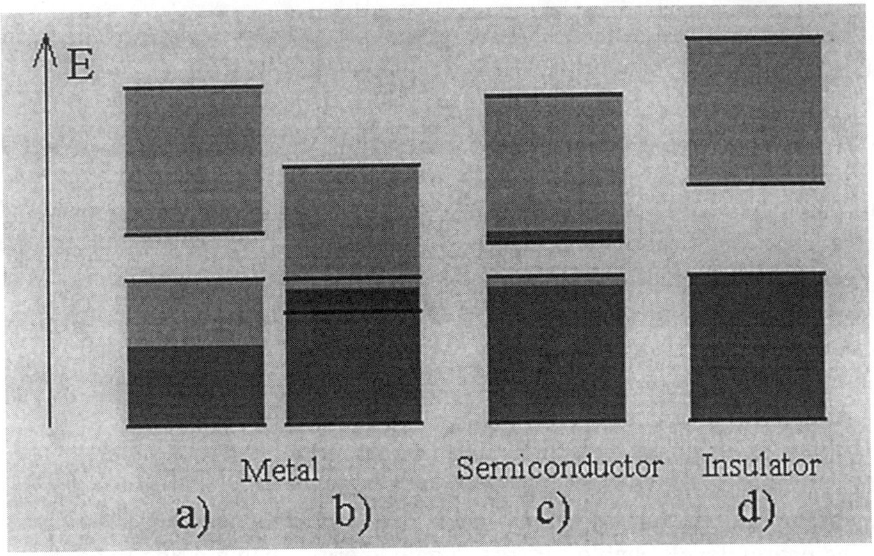

Fig. 4.3 - Possible energy band diagrams containing one filled or partially filled band and one empty or partially empty band. Shown are **a)** a half filled band, **b)** two overlapping bands, **c)** an almost full band separated by a small band gap from an almost empty band and **d)** a full band separated by a large band gap from an empty band.

A half-filled band is shown in a). This situation occurs in materials consisting of atoms that contain only one valence electron each. A lot of highly conducting metals satisfy this condition. Materials consisting of atoms that contain two valence electrons can still be highly conducting, provided that the resulting filled band overlaps with an empty band. This scenario is shown in b). No conduction is expected for scenario d) where a completely filled band is separated from the next higher empty band by a larger energy gap. Such materials behave as insulators. Finally scenario c) depicts the situation in a semiconductor. The completely filled band is now close to the next higher empty band so that electrons can make it into the next higher band yielding an almost full band below an almost empty band. The almost full band is called the valence band, since it is occupied by valence electrons; the almost empty band will be called the conduction band, as electrons are free to move in this band and contribute to the conduction of the material.

4.1.4 Energy Bands of Semiconductors

As semiconductors are of primary interest in this text, a simplified energy band diagram for semiconductors will be introduced for semiconductors and some key parameters defined. The diagram is shown in the figure below:

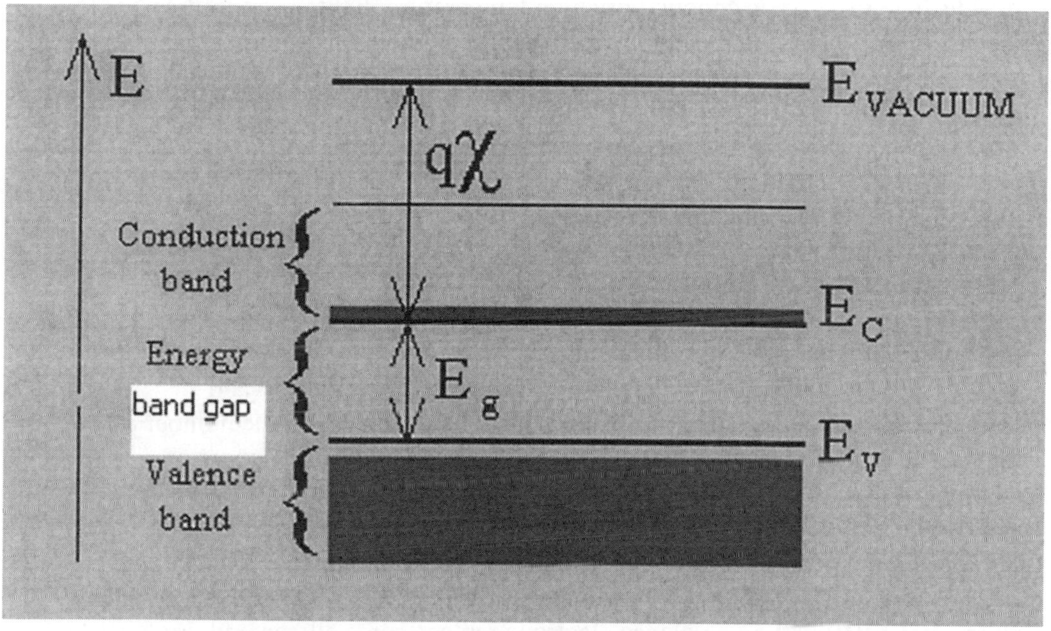

Fig. 4.4 - A simplified energy band diagram used to describe semiconductors. Shown are the valence and conduction band as indicated by the valence band edge $\mathbf{E_v}$ and the conduction band edge $\mathbf{E_c}$. The vacuum level E_{VACUUM} and the electron affinity χ are also indicated on the figure.

The diagram identifies the almost-empty conduction band simply by a line that indicates the bottom of the conduction band and is labeled E_c. Similarly, the top of the valence band is indicated with a line labeled E_v. The energy band gap is between the two lines and separated by the band gap energy E_g. The distance between the conduction band edge E_c and the energy of a free electron (called the vacuum level labeled E_{vacuum}) is quantified by the electron affinity χ multiplied by the electronic charge q.

4.2 CURRENT CARRIERS IN SEMICONDUCTORS

4.2.1 Electrons and Holes in Semiconductors

The common semiconductors include chemical elements and compounds such as silicon, germanium, selenium, gallium arsenide, zinc selenide, and lead telluride. The increase in conductivity with temperature, light, or impurities arises from an increase in the number of conduction electrons, which are the carriers of the electrical current. In a pure, or intrinsic, semiconductor such as silicon, the valence electrons, or outer electrons, of an atom are paired and shared between atoms to make a covalent bond that holds the crystal together. These valence electrons are not free to carry electrical current. To produce conduction electrons, temperature or light is used to excite the valence electrons out of their bonds, leaving them free to conduct current. Vacancies, or "holes," are left behind that also contribute to the flow of electricity. (These holes are said to be carriers of positive electricity.) The energy required to excite the electron and hole is called the energy gap.

Another method to produce free carriers of electricity is to add impurities to, or to "dope," the semiconductor. The difference in the number of valence electrons between the doping material, or dopant (either donors or acceptors of electrons), and host gives rise to negative (n-type) or positive (p-type) carriers of electricity. This concept is illustrated in the accompanying diagram of a doped silicon (Si) crystal. Each silicon atom has four valence electrons; two are required to form a covalent bond. In n-type silicon, atoms such as phosphorus (P) with five valence electrons replace some silicon and provide extra negative electrons. In p-type silicon, atoms with three valence electrons such as aluminum (Al) lead to a deficiency of electrons, or to holes, which act as though they are positive electrons. The extra electrons or holes are the carriers of electrical charge in the conduction of electricity.

As pointed out before, semiconductors distinguish themselves from metals and insulators by the fact that they contain an "almost-empty" conduction band and an "almost-full" valence band. This also means that we will have to deal with the transport of carriers in both bands.

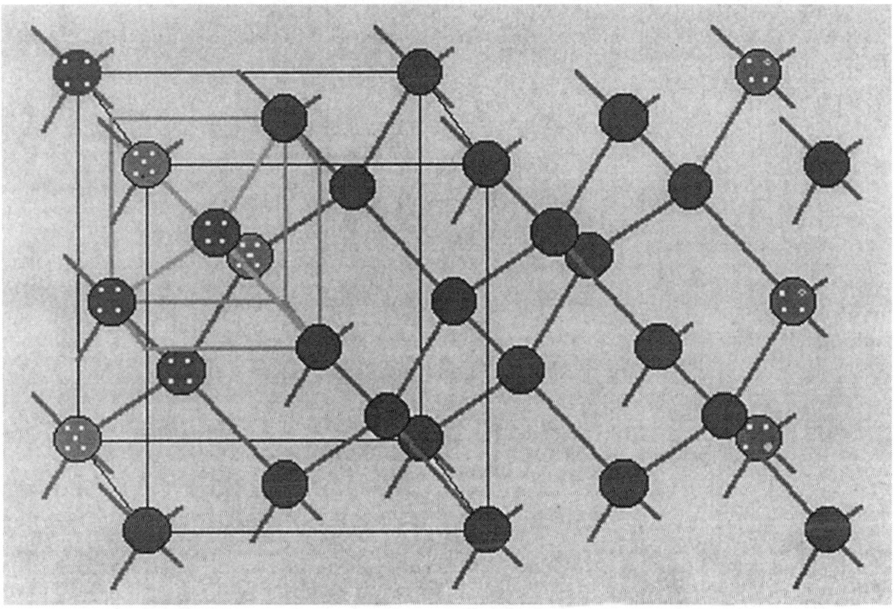

Fig. 4.5 - The lattice structure of "Doped" Si with both "p" and "n" type regions.

The concept of holes in a semiconductor is introduced to facilitate the discussion of the transport in the "almost-full" valence band. It is important to understand that one could deal with only electrons (since these are the only real particles available in a semiconductor) if one is willing to keep track of all the electrons in the "almost-full" valence band. The concepts of holes is introduced based on the notion that it is a whole lot easier to keep track of the missing particles in an "almost-full" band, rather than keeping track of the actual electrons in that band.

Holes are missing electrons. They behave as particles with the same properties as the electrons would have occupying the same states except that they carry a positive charge. This definition is illustrated further with the figure below depicting the simplified energy band diagram in the presence of an electric field.

A uniform electric field is assumed which causes a constant gradient of the conduction and valence band edges as well as a constant gradient of the vacuum level. The gradient of the vacuum level is associated with the potential energy of the electrons outside the semiconductor. However the gradient of the vacuum level represents the electric field within the semiconductor.

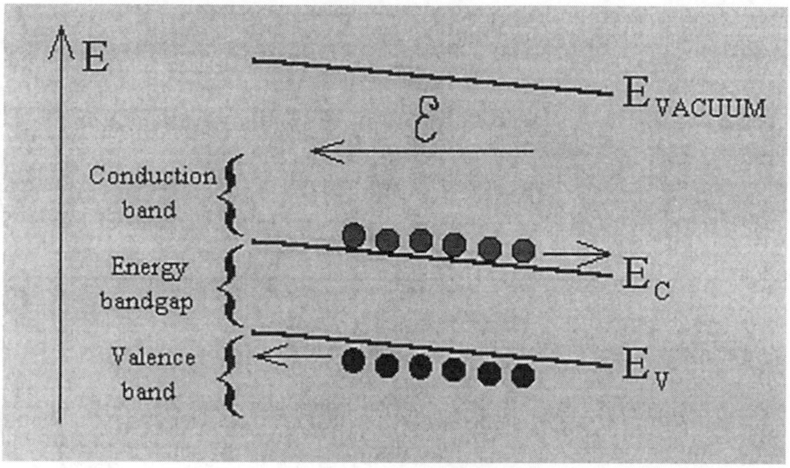

Fig. 4.6 - Energy band diagram in the presence of a uniform electric field. Shown are electrons (conduction band circles), which move against the field, and holes (valence band circles), which move in the direction of the applied field.

The electrons in the conduction band are negatively charged particles that therefore move in a direction that opposes the direction of the field. Electrons therefore move down hill in the conduction band.

The current in the valence band is due to positively charged particles associated with the empty states in the valence band. These particles are called *holes*. Keep in mind that there is no real particle associated with a hole, but rather that the combined behavior of all the electrons which occupy states in the valence band is the same as that of positively charge particles associated with the unoccupied states.

As illustrated by the above figure, the holes move in the direction of the field (since they are positively charged particles). They move upward in the energy band diagram similar to air bubbles in a tube filled with water which is closed on each end.

4.3 SEMICONDUCTOR JUNCTIONS

4.3.1 Creation Of p, n Junctions

When p-type and n-type semiconductor regions are adjacent to each other, they form a semiconductor diode, and the region of contact is called a p-n junction. (A diode is a two-terminal device that has a high resistance to electric current in one direction but a low resistance in the other direction.) The conductance properties of the p-n junction depend on the direction of the voltage, which can, in turn, be used to control the electrical nature of the device. Series of such junctions are used to make transistors and other semiconductor devices such as solar cells, p-n junction lasers, rectifiers, radiation detectors, and many others.

A p-n junction consists of two semiconductor regions with opposite doping type as shown in the figure below. The region on the left is p-type with an acceptor density N_A, while the region on the right is n-type with a donor density N_D. The electron (hole) density in the n-type (p-type) region is approximately equal to the donor (acceptor) density.

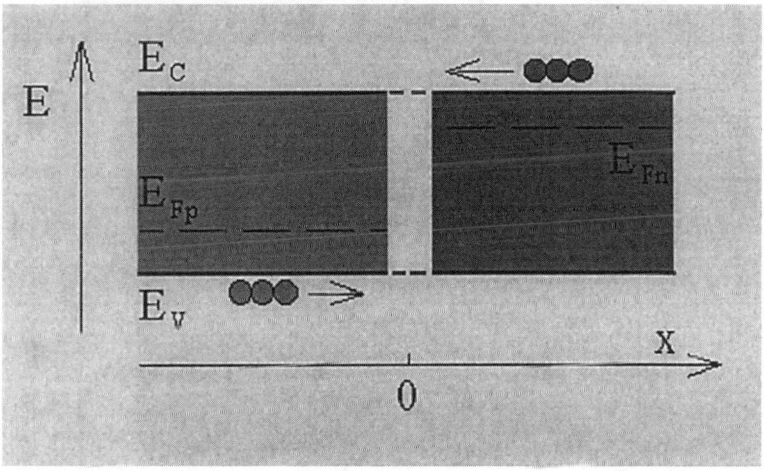

Fig. 4.7 - Cross-section of a p-n junction.

It will be assumed, unless stated otherwise, that the doped regions are uniformly doped and that the transition between the two regions is abrupt. This structure will be referred to as being an **abrupt** p-n junction.

Frequcntly, one side of the p-n junction is distinctly higher doped than the other. In such case, only the low-doped region needs to be considered, since it primarily determines the device characteristics. This structure will be referred to as a **one-sided** abrupt p-n junction.

The junction is biased with a voltage V_a. This will be call the junction **forward-biased** junction if a positive voltage is applied to the p-doped region and **reversed-biased** if a negative voltage is applied to the p-doped region. The contact to the p-type region is also called the **anode**, while the contact to the n-type region is called the **cathode**, in reference to the anions or positive carriers and cations or negative carriers in each of these regions.

The principle of operation can be explained by imagining that one can simply bring both regions together, aligning both the conduction and valence band energies of each region as shown in the figure below.

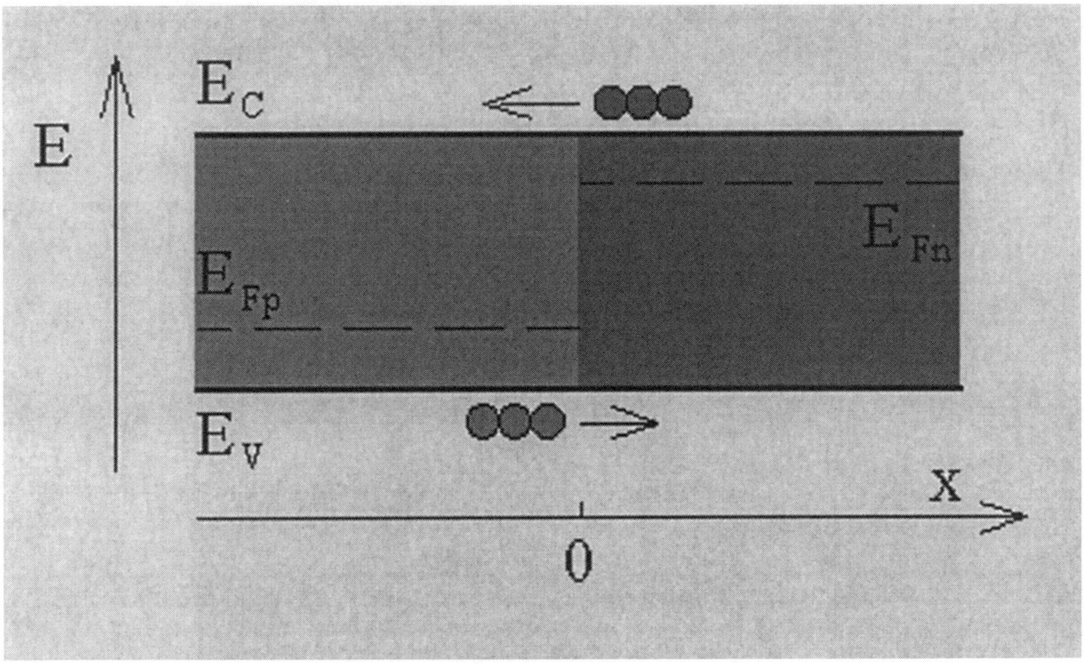

Fig. 4.8 - Energy band diagram of a p-n junction just after merging the n-type and p-type regions.

The electrons and holes close to the junction diffuse across the junction into the p-type and n-type region respectively. This process leaves the ionized donors and acceptors behind, creating a region around the junction that is depleted of mobile carriers. The depletion region is indicated as the region around "**O**" in Fig. 4.4. The charge due to the ionized donors and acceptors causes an electric field that in turn causes a drift of carriers in the opposite direction. The diffusion of carriers continues until the drift current balances the diffusion current, thereby reaching equilibrium.

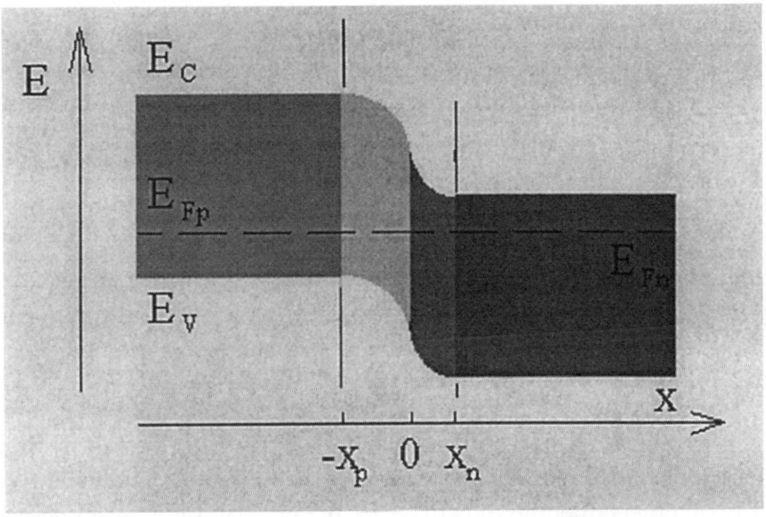

Fig. 4.9 - Energy band diagram of a p-n junction in thermal equilibrium showing depletion region.

While in thermal equilibrium with no bias voltage difference existing between the n-type and p-type material, there is an internal potential that is caused by the fermi, electrochemical potential, energy difference between the n-type and p-type semiconductors. Through the application of a reverse bias voltage the depletion region can be extended significantly creating the basic conditions necessary for radiation detection.

4.3.2 Characteristics of Semiconductors Detectors†‡

Semiconductor detectors are fabricated from either elemental or compound single crystal materials having a band gap in the range of approximately 1 to 5 eV. The group IV elements Silicon and Germanium are by far the most widely used semiconductors, although some compound semiconductor materials are finding use in special applications as developmental work on them continues.

† Canberra Industries, 2000, Technical Product Literature.

‡ Princeton Gamma Technologies, 2000, Technical Product Literature.

Table 4.1 shows some of the key characteristics of various semiconductors as detector materials:

Table 4.1: Element vs. Band Gap

Material	Z	Band Gap (eV)	Average Energy per e-h pair (eV)
Si	14	1.12	3.61
Ge	32	0.74	2.98
CdTe	50	1.47	4.43
GaAs	32	1.43	5.2

Semiconductor detectors have a P-I-N diode structure in which the intrinsic (I) region is created by depletion of charge carriers when a reverse bias is applied across the diode. When photons or charged particles interact within the depletion region, charge carriers (holes and electrons) are freed and are swept to their respective collecting electrode by the electric field. The resultant charge is integrated by a charge sensitive preamplifier and converted to a voltage pulse with amplitude proportional to the original photon/particle energy.

Since the depletion depth is inversely proportional to net electrical impurity concentration, and since counting efficiency is also dependent on the purity of the material, large volumes of very pure material are needed to ensure high counting efficiency for high energy photons.

Prior to the mid-1970s, the required purity levels of Si and Ge could be achieved only by counter-doping P-type crystals with the N-type impurity, lithium, in a process known as lithium-ion drifting. Although this process is still widely used in the production of Si (Li) X-ray detectors, it is no longer required for germanium detectors since sufficiently pure crystals have been available since 1976.

The energy band gap of semiconductors tends to decrease as the temperature is increased. This behavior can be better understood if one considers that the inter-atomic spacing increases when the amplitude of the atomic vibrations increases due to the increased thermal energy. This effect is quantified by the linear expansion coefficient of a material. An increased inter-atomic spacing decreases the potential seen by the electrons in the material, which in turn reduces the size of the energy band gap

A plot of the resulting band gap versus temperature is shown in the figure below for germanium, silicon and gallium arsenide.

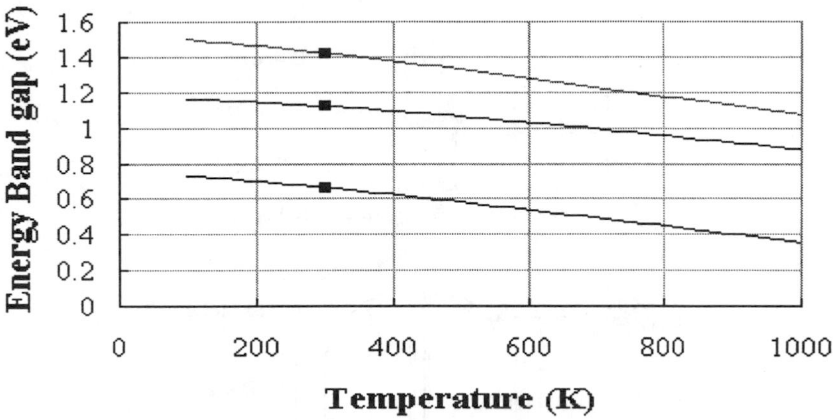

Fig. 4.10 - Temperature dependence of the energy band gap of germanium (bottom curve), silicon (middle curve) and GaAs (top curve).

The band gap figures in Table 4.1 signify the temperature sensitivity of the materials and the practical ways in which these materials can be used as detectors. As a practical matter both Ge and Si photon detectors must be cooled in order to reduce the thermal charge carrier generation (noise) to an acceptable level. This requirement is quite aside from the lithium precipitation problem that made the old Ge(Li), and to some degree Si(Li) detectors, perishable at room temperature.

The most common medium for detector cooling is liquid nitrogen. However, recent advances in electrical cooling systems have made electrically refrigerated cryostats a viable alternative for many detector applications.

In liquid nitrogen (LN_2) cooled detectors, the detector element (and in some cases preamplifier components), are housed in a clean vacuum chamber that is attached to or inserted in a LN_2 Dewar. The detector is in thermal contact with the liquid nitrogen that cools it to around 77 °K or –200 °C. At these temperatures, reverse leakage currents are in the range of 10^{-9} to 10^{-12} amperes.

In electrically refrigerated detectors, both closed-cycle Freon and helium refrigeration systems have been developed to eliminate the need for liquid nitrogen. Besides the obvious advantage of being able to operate where liquid nitrogen is unavailable or supply is uncertain, refrigerated detectors are ideal for applications requiring long-term unattended operation, or applications such as undersea operation, where it is impractical to vent LN_2 gas from a conventional cryostat to its surroundings. A cross-sectional view of a typical liquid nitrogen cryostat is shown in Fig. 4.11.

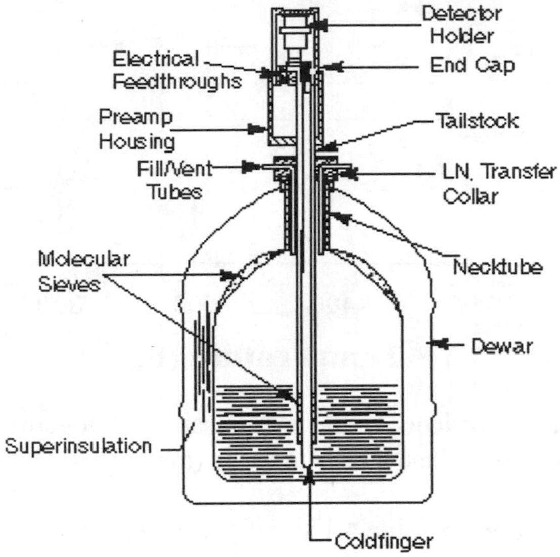

Fig. 4.11 - Model 7500SL Vertical Dipstick Cryostat.

4.4 PARTICLE DETECTION

4.4.1 Charged Particle Detection

Silicon Charged Particle detectors have a P-I-N structure in which a depletion region is formed by applying reverse bias, with the resultant electric field collecting the electron-hole pairs produced by an incident charged particle. The resistivity of the silicon must be high enough to allow a large enough depletion region at moderate bias voltages. A traditional example of this type of detector is the Silicon Surface Barrier (SSB) detector. In this detector, the n-type silicon has a gold surface-barrier contact as the positive contact, and deposited aluminum is used at the back of the detector as the ohmic contact.

A modern version of the charged particle detector is called PIPS, an acronym for Passivated Implanted Planar Silicon. This detector employs implanted rather than surface barrier contacts and is therefore more rugged and reliable than the SSB detector it replaces.

At the junction there is a repulsion of majority carriers (electrons in the n-type and holes in p-type) so that a depleted region exists. An applied reverse bias widens this depleted region that is the sensitive detector volume and can be extended to the limit of breakdown voltage. Detectors are generally available with depletion depths of 100 to 700 μm.

Detectors are specified in terms of surface area and alpha or beta particle resolution as well as depletion depth. The resolution depends largely upon detector size, being best for small area detectors. Alpha resolution of 12 to 35 keV and beta resolutions of 6 to 30 keV are typical. Areas of 25 to 5000 mm^2 are available as standard, with larger detectors available in various geometries for custom applications. Additionally, PIPS detectors are available fully depleted so that a dE/dx energy loss measurement can be made by stacking detectors on axis. Detectors for this application are supplied in a transmission mount, (i.e. with the bias connector on the side of the detector).

A chart of the energies, MeV, of various particles that will be strapped within several depletion depths is shown in Table 4.2.

Table 4.2 Particle Energies, MeV, and PIPS Depletion Depth

Depletion Depth (Range) in μm	Electron	Proton	Alpha
100	0.15 MeV	7 MeV	15 MeV
300	0.31 MeV	15 MeV	55 MeV
500	0.45 MeV	21 MeV	85 MeV
700	0.52 MeV	27 MeV	105 MeV
1000	0.73 MeV	33 MeV	130 MeV

Note that even the thinnest detector is adequate for alpha particles from radioactive sources, but that only very low energy electrons are fully absorbed. However, for a detector viewing a source of electron lines, such as conversion electron lines, sharp peaks will be observed since some electron path lengths will lie fully in depleted region. Fig. 4.12 shows ranges of particles commonly occurring in nuclear reactions.

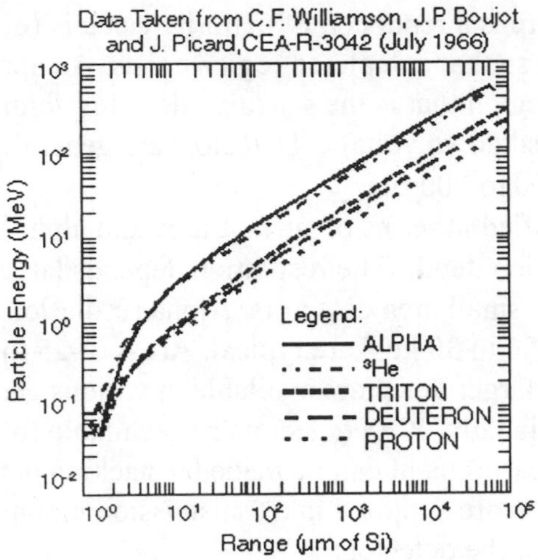

Fig. 4.12 - Range-Energy Curves in Silicon

Since charge collected from the particle ionization is so small that it is impractical to use the resultant pulses without intermediate amplification, a charge-sensitive preamplifier is used to initially prepare the signal.

Fig. 4.13 illustrates the electronics used in single-input alpha spectroscopy application. Note that the sample and detector are located inside a vacuum chamber so that the energy loss in air is not involved.

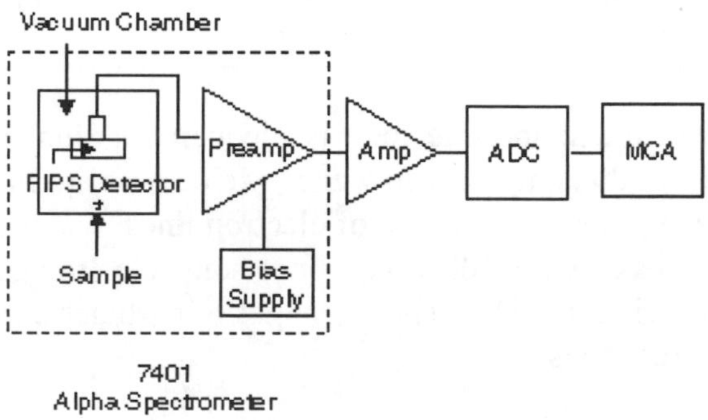

Fig. 4.13 - Single-Input Alpha Spectroscopy Application.

A sample alpha spectral plot is shown below.

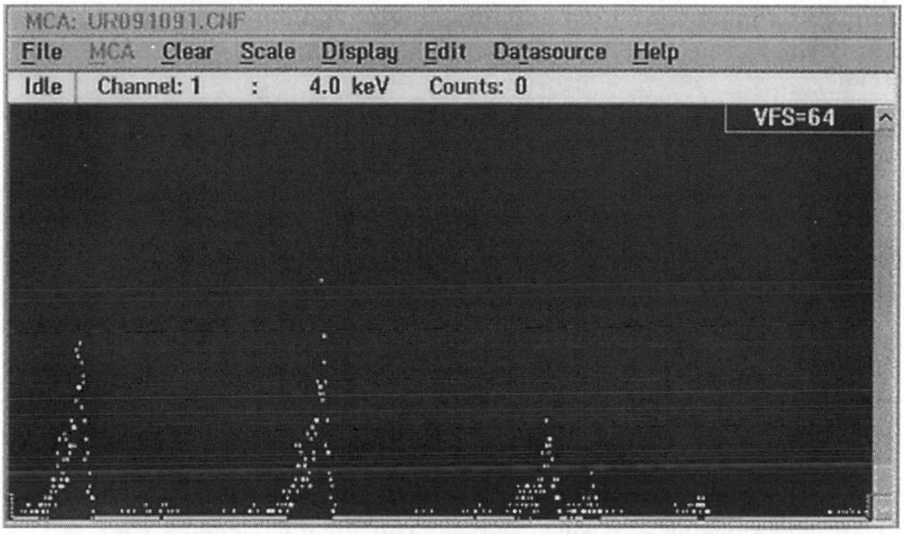

Fig. 4.14 - Alpha Spectral Plot.

4.5 PHOTON DETECTION

4.5.1 Photon Semiconductor Detector Performance

Semiconductor detectors provide greatly improved energy resolution over other types of radiation detectors for many reasons. Fundamentally, the resolution advantage can be attributed to the small amount of energy required to produce a charge carrier and the consequent large "output signal" relative to other detector types for the same incident photon energy. At 3 eV/e-h pair (see Table 4.1) the number of charge carriers produced in Ge is about one and two orders of magnitude higher than in gas and scintillation detectors respectively. The charge multiplication that takes place in proportional counters and in the electron multipliers associated with scintillation detectors, resulting in large output signals, does nothing to improve the fundamental statistics of charge production.

The resultant energy resolution in keV (FWHM) vs. energy for various detector types is illustrated in Table 4.3.

Table 4.3 Energy Resolution (keV FWHM) vs. Detector Type

Energy (keV)	**5.9**	**1.22**	**1.332**
Proportional Counter	1.2	----	----
X-ray NaI(Tl)	3.0	12.0	----
3 x 3 NaI(Tl)	----	12.0	60
Si(Li)	0.16	----	----
Planar Ge	0.18	0.5	----
Coaxial Ge	----	0.8	1.8

At low energies, detector efficiency is a function of cross-sectional area and window thickness, while at high energies total active detector volume more or less determines counting efficiency. Detectors having thin contacts, e.g., Si(Li), Low-Energy Ge and Reverse Electrode Ge detectors, are usually equipped with a Be cryostat window to take full advantage of their intrinsic energy response.

Coaxial Ge detectors are specified in terms of their relative full-energy peak efficiency compared to that of a 3 in. x 3 in. NaI(Tl) Scintillation detector at a detector to source distance of 25 cm. Detectors of greater than 100% relative efficiency have been fabricated from germanium crystals ranging up to about 75 mm in diameter. About two kg of germanium is required for such a detector.

The first semiconductor photon detectors had a simple planar structure similar to their predecessor, the SSB detector. Soon the grooved planar Si(Li) detector evolved from attempts to reduce leakage currents and thus improve resolution.

The coaxial Ge(Li) detector was developed in order to increase overall detector volume, and thus detection efficiency, while keeping depletion (drift) depths reasonable and minimizing capacitance. Other variations on these structures have come and some have gone away, but there are several currently in use. These are illustrated in Fig. 4.14 with their salient features and approximate energy ranges.

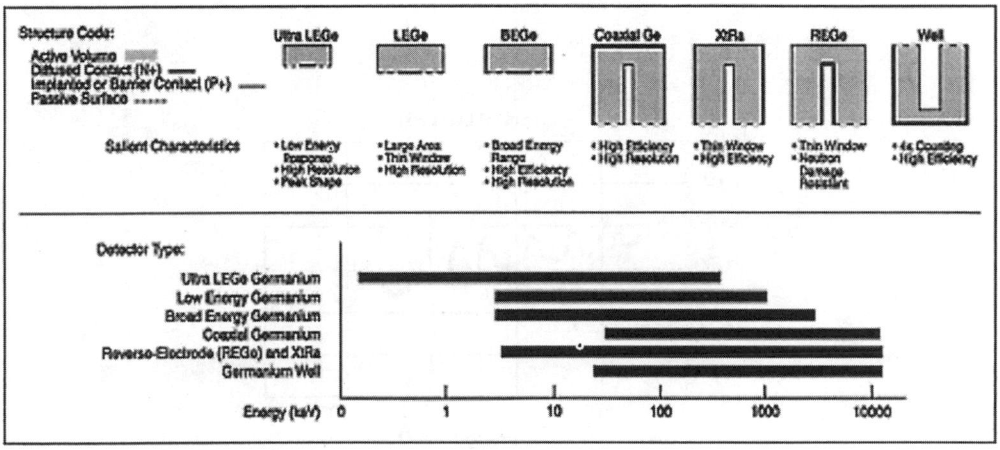

Fig. 4.15 - Coaxial Ge Detector Configurations.

For comparison purposes, sample spectral plots are sown below for semiconductor and scintillation detectors of the same source.

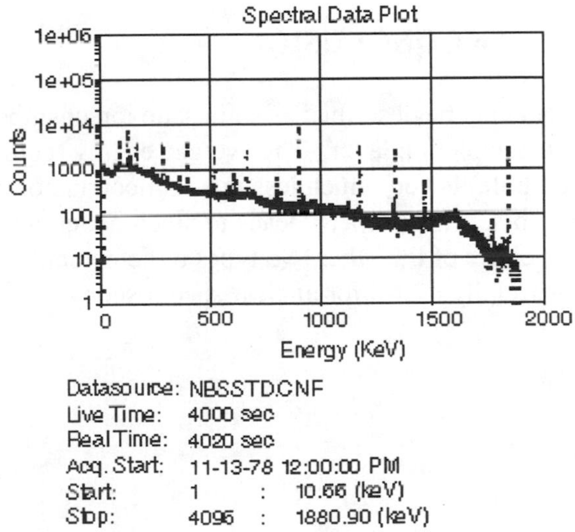

Fig. 4.16 - Ge Semiconductor Spectral Plot.

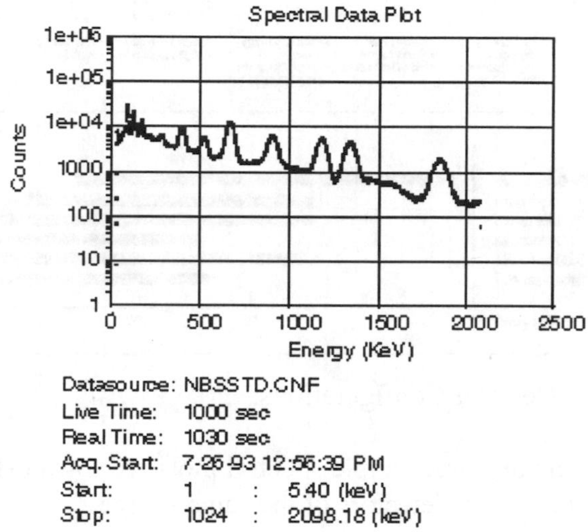

Fig. 4.17 - NaI(Tl) Scintillation Spectral Plot.

4.6 CONCLUSION

Semiconductor detectors may be described as solid state ion chambers, in which the sensitive volume is called the depletion layer. The average energy required to produce an "ion-pair" (electron-hole) is very much less in a semiconductor detector than in scintillation or gaseous detectors. This property leads to much better energy resolution than is possible with either of the other two types of detectors. Thus, semiconductor detectors are especially useful for nuclear spectroscopy.

Chapter 5

PERSONAL MONITORING

R. Craig Yoder

5.1 INTRODUCTION

The acute health effects that appeared soon after the discovery of x rays created a need for methods to measure the relative amount of radiation received by those exposed. The current approaches to personal monitoring arose from years of attempting to answer the following questions:

- What is to be measured?
- What strategies and methods will generate a meaningful measurement?

The remainder of this chapter addresses these questions in the context of the current radiation protection culture in the United States for controlling doses received from external radiation sources.

Personal monitoring to assess occupational exposures to radiation represents an unusual scoring system. Devices worn by individuals provide information that a series of assumptions and computations transform into numerical values accepted by radiation protection professionals as a metric related to risk. The current precepts regarding radiation risks preclude direct measurement. The personal dosimetry system establishes a process where an observable effect of the radiation energy absorbed in an inanimate object becomes a surrogate for the dose equivalent, and consequently, the risk to the exposed individual.

The importance of the personal monitoring information varies among individuals and groups according to their perspectives and motivations. For the regulator or enforcement official, the score indicates an organization's compliance with the rules set forth by society's governing institutions. For the radiation safety professional, the score provides information useful for making decisions about radiation control needs, effective control methods and resource allocations. For the exposed individuals, the score may occupy a place among the other measures of their health status that affect their perceptions of well being. A strong tradition of personal monitoring practices has arisen from these different applications of the scoring system.

5.2 QUANTITIES SPECIFIC FOR PERSONAL MONITORING

5.2.1 Personal Dose Equivalent

The International Commission on Radiation Units and Measurements (ICRU 1992; ICRU 1993) developed the *personal dose equivalent* as an operational quantity that could be implemented pragmatically as an estimator of the *effective dose equivalent,* H_E, the immeasurable risk-related quantity used to establish limits on exposure to radiation. The *personal dose equivalent* is expressed with the same units as the *effective dose equivalent*, namely Sieverts (Rem) but differs in that several forms of the *personal dose equivalent* exist to satisfy different operational emphases. Symbolized as $H_P(d)$, the *personal dose equivalent* simplistically represents the dose equivalent in soft tissue at specified depths, d, below a point on the body. Radiation protection professionals focus on depths of 0.07 mm, 3 mm and 10 mm corresponding to various doses as shown in the following table.

Table 5.1 Commonly used names for the personal dose equivalent.

Personal Dose Equivalent	Common Regulatory Name	Colloquial Name
$H_P(0.07)$	Shallow dose equivalent	Skin dose
$H_P(3)$	Dose to the lens of the eye	Lens of the eye dose
$H_P(10)$	Deep dose equivalent	Whole body dose

The *personal dose equivalent* at 10 mm depth, $H_P(10)$, measured on the torso estimates the risk based quantity, *effective dose equivalent,* which concerns with the effects resulting from chronic exposure as compared to $H_P(0.07)$ and $H_P(3)$ which deal with effects associated with both acute and chronic exposure. The dose at a depth in an individual cannot be directly measured. Instead, dosimeters are calibrated to assess the dose at various depths in a phantom composed of a tissue substitute, thus giving an element of legitimacy to the protocol used for personal monitoring.

The relationship between the $H_P(10)$ value assessed by a dosimeter and the *effective dose equivalent* depends on the geometrical shape of the phantom used to represent the human torso, the type and quality of radiation, and the geometrical configuration of the incident radiation field (ICRU 1998). While a detailed examination of these factors exceeds the scope of this chapter, the issues should be recognized as they ultimately influence the practical steps by which personal monitoring devices are developed, calibrated and evaluated. Various computations using sophisticated techniques to model radiation fields and mathematically approximate the human body show the relationship between $H_P(10)$ and the *effective dose equivalent* under very special conditions. However, the physical differences among people make the actual relationship indeterminate and highly speculative. Nonetheless, radiation safety professionals have agreed on an intelligible system in which the results of personal radiation monitors offer a relative means for ranking individuals according to their exposure to radiation.

Physicists link the response of the personal dosimeter to the dose equivalent occurring at given depths in a phantom. International standards recommend a thin-walled, slab phantom filled with water and having dimensions of 30 cm x 30 cm x 15 cm as a suitable substitute for the human torso (ISO 1996). The water-filled phantom presents a few inconveniences; so instead, many use an identically sized, solid polymethylmethacrylate (PMMA) phantom. Other phantom shapes and compositions exist for other parts of the body such as the wrist and finger. In summary, the practical meaning of the *personal dose equivalent* becomes the dose equivalent at a specified depth in a phantom approximating soft tissue.

5.2.2 Effective Dose Equivalent

Several methods of using personal monitoring results to estimate the *effective dose equivalent* lend confusion to a complicated system of radiation protection quantities when the operational conditions deviate significantly from the rudimentary conditions under which $H_P(10)$ is thought to adequately estimate the *effective dose equivalent*. An implicit assumption of irradiation uniformity exists when an individual is monitored with a single dosimeter. Uneven irradiation of the body alters the basic relationship between the *effective dose equivalent* and $H_P(10)$ used for the basic calibration assumptions.

Scientists and various authoritative bodies have developed various means to assess the *effective dose equivalent* when the internal organs receive materially different doses resulting from an uneven exposure of the body. General regulations require a dosimeter be located on the body where the maximum annual *personal dose equivalent* is expected to occur. If the maximum deep and shallow doses appear at different locations, each location would be monitored. Many radiation protection professionals consider this approach to yield an unreasonably conservative estimate of the *effective dose equivalent,* especially when the exposed area is confined to a small portion of the whole body.

The National Commission on Radiological Protection and Measurements has recommended several strategies for using the results of one or two dosimeters to yield a reasonable estimate of the *effective dose equivalent,* particularly for special situations in diagnostic radiology where the use of a leaded apron protects the body such that only the head and perhaps neck become significantly exposed (NCRP 1995). The following table summarizes the recommendations.

Table 5.2 Recommendations to assess the effective dose equivalent from a measurement of the personal dose equivalent.

Conditions	Number of dosimeters	Formula to estimate
Photons > 40 keV incident from the: front to back, side (lateral), equally from all directions (isotropic)	1	$H_E = H_P(10)$
Photons > 40 keV from an unknown or uncertain angle of incidence	2, one in the front; one in the back	$H_E = 0.55H_P(10)_{front} + 0.5H_P(10)_{back}$
Protective apron used in diagnostic and interventional medical fluoroscopy	1 at neck outside of the apron	$H_E = 0.18H_P(10)$
Protective apron used in diagnostic and interventional medical fluoroscopy	2, one under apron at chest or waist; one outside of apron at neck	$H_E = 1.5H_P(10)_{waist} + 0.04H_P(10)_{neck}$

The use of a leaded apron significantly changes the radiation field enveloping the dosimeter and adds another factor influencing the selection and assessment of personal dosimeters.

ANSI N13.41 provides a more general treatment for combining the results of multiple dosimeters into a single estimate of the *effective dose equivalent* (Health Physics Society 1997). The standard extends the tenet that the dose assessed from a dosimeter represents the dose to the local or underlying tissues. The standard divides the torso into various compartments, each having a composite tissue weighting factor based on the organs present in the compartment. Dosimeter results are modified by the composite weighting factors of the compartments for which the various dosimeters are assigned.

Table 5.3 Compartment factors for personal monitoring with multiple dosimeters.

Area of the Body	Compartment Factor
Head and neck	0.10
Thorax, above the diaphragm	0.38
Abdomen, including pelvis	0.50
Upper right/left arm	0.005 each
Right/left thigh	0.005 each

5.2.3 Summary

The definition and practical use of the *personal dose equivalent* influences the design, calibration and interpretation of the data elicited from personal dosimeters. As the operational radiation field deviates from the metrological conditions used to calibrate and evaluate dosimeters, the uncertainty increases in how well the apparent dosimeter result relates to the limiting quantity. Adding the impact of other influencing factors discussed later, the results from personal dosimeters represent more a score than an absolute measurement. Experience since the discovery of x-rays has yielded a system where the consistency of the score in repeated situations has instilled cconfidence in our ability to assess radiation exposures received by individuals so that radiation may be safely used.

5.3 THE PHYSICAL ELEMENTS OF A DOSIMETRY SYSTEM

5.3.1 The Radiation Sensors

The radiation sensitive elements receive much focus because their radiological properties influence many of the other components of the dosimetry system. The radiation sensitive elements or sensors possess properties whereby the energy absorbed from radiation interactions can be transformed into observable and quantifiable attributes. Surprisingly, most radiation sensors used for personal monitoring exhibit optical attributes such as optical density, luminescence of phosphors or visible tracks in solids. Traditionally, passive monitors receive preference because they require no external power to detect or preserve the effects of the radiation interactions until the measurable attribute can be quantified in a laboratory.

Dosimeters to monitor *personal dose equivalent* contain multiple radiation sensitive elements in order to obtain information needed to assess the dose at different depths and detect different radiation types. This can be accomplished in many ways such as:

- using structures in the holder that effectively place the sensors at different depths,
- using two or more sensors that have different interaction properties, or
- by using similar sensors of different thickness which together reveal differences in the volumetric distribution of absorbed dose.

The elemental composition of the sensors and their physical form impart the greatest influence on the observed radiological properties. Those composed of high atomic number elements exhibit large changes in response as a function of photon energy. Thick sensors exhibit large changes in dose response as a function of beta particle energy because attenuation losses in the sensor result in an uneven distribution of absorbed dose throughout the sensor volume.

Detectors that provide a relatively large two dimensional profile can indicate qualitative features about the radiation field that may aid investigations about the exposure situation. The added dimension may reveal uneven irradiation of the dosimeter as might occur if the dosimeter was exposed close to or in contact with a source. Uneven irradiation can result from objects shielding the dosimeter thereby altering the quality of the information.

Many health physicists seek information about the types and energies of the radiation contributing to the dose. Judicious use of multiple sensors can create a dosimeter that offers more than a simple measurement of the *personal dose equivalent*. Additionally, the use of several sensors provides some redundancy should a single sensor fail to yield appropriate data. Physical properties of the sensors will dictate packaging requirements to protect them from damage, to identify them from other units and to configure the sensors for efficient handling.

5.3.2 The Dosimeter Housing or Holder

The dosimeter housing or holder contains filters that alter the radiation field to compensate for the non-tissue energy response exhibited by many types of sensors. Combinations of low and high atomic number filters located in the holder attenuate and alter the radiation spectrum impinging on these detectors. Examining the differential attenuation offered by the filters, physicists may discern radiation quality and energy, and implement calculations that convert the detector data into tissue doses.

Table 5.4 Common filters and their application in personal dosimeters.

Filter material	Measurement application	Radiological Effect
Aluminum	Beta/photon dosimetry	Attenuates beta particles and very low energy photons.
Copper	Beta/photon dosimetry	Attenuates beta particles and moderate energy photons.
Lead; Tin	Beta/photon dosimetry	Attenuates beta particles and higher energy photons. Will create added charged particles from pair production interactions with very high energy photons.
Plastic	Beta/photon dosimetry	Attenuates beta particles and provides added depth for photons.
Polyethylene	Neutron dosimetry	Provides additional recoil protons to increase sensitivity and add depth.
Cadmium	Neutron dosimetry	Influences thermal neutron on albedo type dosimeters; Provides a capture gamma ray that can be used to assess thermal neutron dose using gamma sensitive sensors.

In addition to altering the radiation field reaching the detector, filters place the detectors at different effective depths so that the deep, shallow and lens of the eye doses can be estimated. Dosimeters are worn on the body surface and the radiation physicist needs information about the radiation field and the propagation of dose at various depths in tissue. The attenuation characteristics of the holder are important when the incident radiation strikes the body from the sides or at large angles. Dosimeter calibrations and performance tests require the dosimeter to indicate the dose at a given point 10 mm deep in the center of the slab phantom. As the angle of incidence increases from normal incidence, the true depth of the reference point changes as a function of the cosine of the angle.

The holder protects the sensors and provides room for information about the anticipated user, the radiation qualities that can be assessed by the sensors, and period of use. Additionally, holders include a means of attachment to the user. Various attachment devices can be employed depending on the types of clothing the user may wear, the stresses received by the dosimeter to dislodge it from the user and the need to integrate with other identification badges the user may wear.

The relationship between the holder and radiation sensors is so close that interchanging different radiation sensor technologies with a given holder design is likely to introduce great error. That is, holders for film are likely to be inappropriate for thermoluminescent crystals.

5.3.3 The Processing System

The processing system acts on the radiation sensors to reveal the attribute related to absorbed dose. Examples include the developing process for film, the heating process for thermoluminescent dosimeters, the etching process for plastic track detectors and the optical stimulation process for optically stimulated luminescent dosimeters. Each process depends on parameters that must be precisely controlled to achieve a consistent radiation related response. Processes involving wet chemistry such as film processing and plastic etching usually place strict limits on chemical concentration, temperature and processing time. The relationship between optical density and dose can be altered by changing the developing time by as little as 30 seconds or changing the temperature by more than $2^{O}C$. Over etching plastic track detectors can obliterate radiation induced tracks, disrupt the surface clarity to add non-radiation induced tracks or alter the shape of the tracks so that they escape detection by the counting system.

The sensitivity, dose response and stability of thermoluminescence dosimeters (TLD) depend on a consistently replicated time-temperature profile. Even the heating method can influence the observed radiation response by modifying the luminescent intensity. Optically stimulated luminescence systems depend on light sources that provide stable power and consistent spectral emission.

Luminescence based systems usually permit a variety of adjustments to the process parameters to elicit radiation response properties that favor the intended use of the dosimeter. The processing systems for TLD frequently prepare the dosimeter for reuse. The heating process provides a means of annealing the crystals and may be performed as a part of the process to determine dose or as a subsequent treatment. Optically stimulated luminescence systems may alter the optical power or color spectrum to change the amount of luminescence created to better match the detection characteristics of the analytical system.

The chain of custody issues pertinent to processing center on establishing the relationship between the processing parameters that existed and the sensors that were subject to those parameters. When processing can be performed on groups of sensors - for example, pieces of film developed as a large collection in a common tank - irradiated calibration sensors or controls may accompany the sensors exposed to unknown doses so that processing variables can be properly accounted.

5.3.4 The Analytical System

The analytical system quantifies the attribute related to the radiation dose absorbed by the sensor. Film systems employ densitometers to determine the amount of darkening due to the radiation exposure. Luminescence systems use photomultiplier tubes and other light quantification methods. Particle tracks in plastics may be counted manually or with automated vision or image analysis systems.

The analytical system for film and plastic track dosimeters can be separated from the processing system because the radiation related attribute is preserved. That is, the attribute can be evaluated at any time after processing. In contrast, luminescence based systems require the processing and analytical systems to be merged into a common instrument because of the transient nature of the luminescence phenomenon.

The same general stability requirements apply to the analytical system as apply to the processing system. Precision results from a constant geometrical relation between the sensor and analysis system as well as the quality of electrical and optical systems. Calibration of the analytical system involves evaluating sensors irradiated to know doses under the same processing parameters and analytical instrumentation configurations as to be used for sensors with unknown doses.

The analytical system should provide a means of linking the quantitative data to a uniquely identified dosimeter because this system is the only one required to examine dosimeters singly. In addition, the analytical systems benefit from provisions to record the quantitative data through electronic methods. Electronically stored data can be accurately input to dose assessment algorithms and archived for later examination should questions or anomalies appear.

5.3.5 The Dose Assessment System

The dose assessment system features the mathematical steps that convert the quantified attribute data into the *personal dose equivalent* values. Most assessment systems begin by transforming the quantified attribute measurement for each sensor in the dosimeter into an intermediate, normalized dose expression. For example, a film assessment system would use a characteristic curve or equation that relates density to air kerma for a specific energy spectrum such as that emitted by ^{137}Cs. This step is completed for each area of the film under the filters present in the holder. This initial converted dose data are input to logical and mathematical formulas that determine the *personal dose equivalent* at the various depths of interest. Frequently, the formulas include logical statements that indicate relationships among the sensors that fail to conform with those expected from properly used dosimeters.

The robustness of the assessment formulas depend on the variety of radiological conditions used to test and calibrate the dosimeter. Debate often focuses on the relative merits of complicated systems that can accurately assess a few very specific conditions compared to simpler formulas that may be less accurate for a given condition but will be more accurate over a larger variety of conditions. The assessment system becomes a matter of judgment based on the anticipated demands and expectations placed on the dosimetry system.

5.4 DOSIMETRY TECHNOLOGIES FOR PERSONAL MONITORING

5.4.1 Personal Monitoring Film

Films for personal monitoring consist of a thin base coated with an emulsion of silver halide grains suspended in a gelatin matrix. Other chemicals are added to achieve certain physical traits and control the chemical processes that occur during the formation of latent images following irradiation and their development during processing. The latent image arises from the aggregation of silver atoms in the grain created by the ionization events associated with radiation interactions. The developing process converts the silver in the grains with latent images to metallic silver which interferes with the transmission of light through the film base. More latent images form with increasing dose which leads to more developed crystals. As the number of developed crystals increases, the film appears darker and transmits less and less light. The optical density of the film is proportional to the number of developed grains which is proportional, over a finite dose range, to the absorbed dose (Becker 1966).

The relationship between dose and optical density is often depicted in a semi-logarithmic curve called the characteristic curve or H & D curve after Hurter and Driffeld. Fig. 5.1 shows the characteristic curve for two emulsions of different sensitivity. The plateau indicates saturation of the film and delineates the upper dose that can be measured with given processing and analysis parameters. Monitoring films must yield information over a very large range of doses which exceeds the capability of a single emulsion. Two emulsions of differing sensitivity enable an extended measurement range. The emulsions may be coated on separate film bases or coated on opposite sides of a common base. The Kodak Type 2 film is an example of the latter. For this film system, the optical density measured at low doses is actually the combined density of the two emulsions. As the faster emulsion reaches saturation, it must be removed from the film base so that the density from the slower emulsion can be independently evaluated. The cross-over region where one transitions from the faster to slower emulsion depends on the densitometer capabilities, the radiation energy spectrum, the influence of filters and processing conditions. The crossover between emulsions can cause the precision of the system to shift at intermediate doses because the certainty in the density measurement changes. The curves in Fig.5.1 were acquired with very high performance densitometers. Less capable densitometers would usually become unreliable at densities exceeding 5.0 so that the plateau would appear at a lesser dose.

Film systems possess extreme response variations as a function of photon energy and spectral composition. Fig. 5.2 shows the response per unit of air kerma for films exposed free in air with no holder or filtration other than the film packaging paper. The rapidly decreasing response between 60 and 100 keV greatly influences the design and performance of the film dosimetry system. High atomic number filters are commonly used to mitigate the energy response; however, the hardening of x ray beams by such filters and the heterogeniety of x ray spectra results in performance that can vary with small changes in the photon spectra reaching the film.

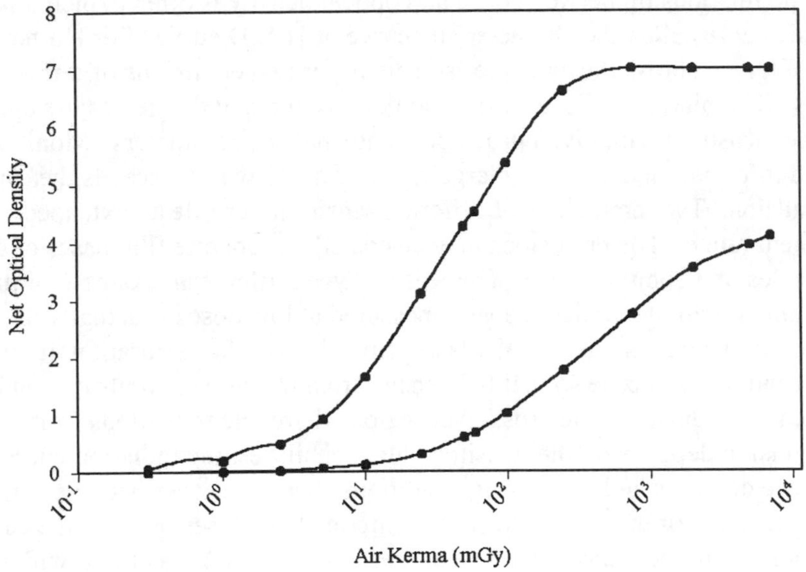

Fig. 5.1 - The characteristic curve for Kodak Type 2 film.

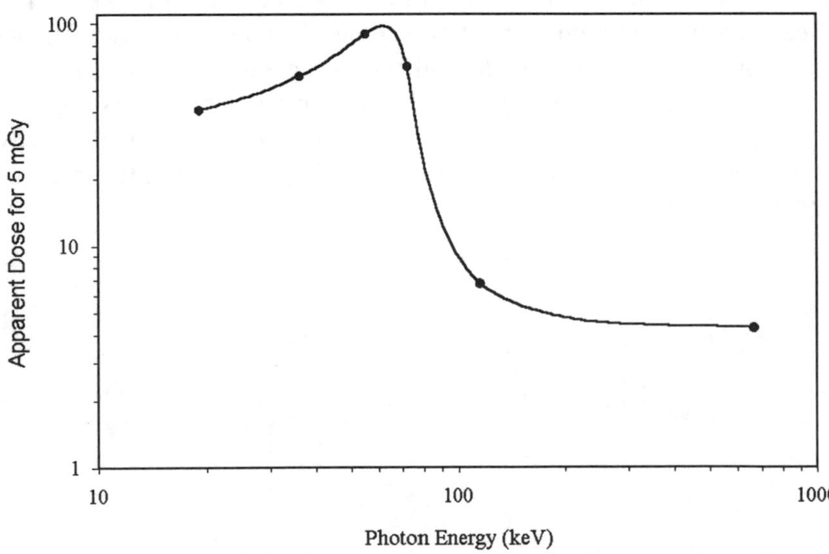

Fig 5.2. - The apparent response of Kodak Type 2 film exposed free-in-air.

The angular response characteristics of film also represent a challenge for the dosimetry system. The apparent thickness of filters increases as the cosine of the incident angle altering the relative proportion of low energy photons reaching the film. At very large angles, the photons will enter more from the sides of the dosimeter and could avoid proper filtration if there is a relatively large gap between the filter and film, or if the filter presents too small of a cross section. The location of the filter image on the film will move much like shadows move during the day. The analytical system must account for such changes in the location of the area to be evaluated.

Scientists at the Physikalisch-Technische Bundestalt developed an innovative design to improve the angular response of film dosimeters using a concept called the gliding shadow (Ambrosi et al., 1994). Two circular filters of different diameter are stacked, one on top of the other, to create a concentric density pattern on irradiated film. As the incident angle changes the circular images become elliptical and their degree of overlap changes. The resulting density information allows better interpretation of the *personal dose equivalent*.

The imaging properties of film remain one of its strongest features. The appearance of filter images, blotches of density from radioactive contamination of the dosimeter and other artifacts that may be imaged onto the film offer important information about the conditions that existed during the irradiation of the dosimeter. The most common use of the film image is to assess whether the exposure occurred under dynamic or static conditions, corresponding to whether the dosimeter was attached to the user during exposure or left alone in a stationary position fixing the geometrical relationship between the dosimeter and source.

The conversion of absorbed dose in silver halide grains to optical density involves a series of chemical and physical processes that are susceptible to various environmental factors. Pressure can induce both positive and negative optical density to the film depending on how the pressure is exerted and for what period of time. Concentrated force can depress the emulsion creating a thinner coating with fewer total grains. With fewer grains, the optical density per unit dose is reduced. Other types of pressure such as that experienced in winding and packaging can fracture the grains changing the total grain surface area which impact latent image formation resulting from x ray irradiation, usually by increasing the optical density per unit dose.

Heat is another factor that will alter the physical chemistry of latent image formation. Film dosimeters left in car windows during the summer frequently exhibit fogging or increased background optical density. The susceptibility to heat depends on the grain size. The smaller grains of the more insensitive emulsions tend to be more affected by heat than the larger grains of the more sensitive emulsions.

Time is another variable that must be controlled. Once coated, film begins to experience a growth of base fog, largely due to background radiation. Calibration films and those used to subtract the base fog value should be of similar age as the film distributed to the users.

Brief mention should be given to nuclear track emulsions such as NTA film for neutron dosimetry. Protons recoiling from elastic neutron interactions create a path of latent images in a series of grains. The path appears as a line of dark spots in the developed film. Track length corresponds to the energy of the recoil proton and consequently the energy of the incident neutron. The number of tracks per unit area reflects the fluence which can be converted to dose equivalent. NTA film exhibits a relatively high energy detection threshold of nearly 1 MeV to create recoil protons able to yield tracks with sufficient length to be discernible. Track counting is accomplished with microscopes set up for very high magnification which causes a narrow depth of focus. As a result, skilled technicians must constantly refocus to perceive tracks in three dimensions because they can be oriented in almost any direction in the emulsion thickness.

Track fading over time has been a major adverse feature that has been controlled in part with special packaging. Such fading is accelerated with heat so that even packaging cannot assure a reliable measurement. Gamma ray doses that usually accompany neutron exposures will uniformly darken the film and may obscure the neutron tracks, particularly if fading conditions were present. Presently, NTA film is not used in the U.S. to any meaningful degree.

5.4.2 Thermoluminescence Dosimeters

Thermoluminescence dosimeters (TLD) have become more popular over the past 25 years because of the development of compact, automatic and easily operated processing and analysis instruments. Improvements in the uniformity of TLD crystal production along with increased sensitivity also aided the growth in popularity. Unlike film where few manufacturers offer similar products, TLD providers offer more varieties of materials, shapes and processing approaches allowing dosimetry organizations to assemble systems that address specific operational demands.

Most TLD materials exceed film's robustness and durability allowing for longer monitoring periods without risking a degradation in measurement performance. Fading and time dependent shifts in sensitivity can be minimized with proper handling to permit monitoring periods of several months or more.

TLD materials are electrical insulators in that a wide band gap exists between the valence and conduction bands in the crystal. The band gap is the energy potential required to move a valence electron into the conduction band where it possesses a moderate amount of mobility. The ionization created by radiation interactions in the crystal impart energy to some of the valence electrons raising them to the conduction band. Once in the conduction band, the electrons migrate through the crystal until losing energy so that they must return to the valence band or become trapped in defects created by dopants introduced during the crystal manufacturing process. These traps lie at energy levels between the valence and conduction bands. The electrons can be freed from the traps if sufficient heat is applied to the crystal to cause the electrons to rise to the conduction band where they can migrate in search of a lower energy state. Some of the freed electrons will combine with an anion vacancy in the crystal to emit light. The frequency of the light is related to the energy differential between the conduction band and the luminescence center.

At normal temperatures, the kinetic energy imparted from the thermal heat transfer may be sufficient to free the electrons from shallow traps which lie at energy levels close to those of the conduction band. When this occurs, the evidence of the radiation exposure is lost from these traps. Shallow traps are one source of the fading trait often examined for TLD's. Fortunately, deeper dosimetric traps require much higher temperatures to release the trapped electrons, usually in excess of 150°C. Once released, the electrons can combine with a luminescence center.

In short, the number of trapped electrons can be related to the radiation absorbed dose. As the amount of luminescence is a function of the number of trapped electrons, the luminescence itself can be quantitatively related to absorbed dose in the crystal.

Common TLD materials include LiF, $LiBO_4$, CaF, $CaSO_4$, and Al_2O_3. The lower atomic number materials exhibit better tissue equivalence but none are perfect. Fig. 5.3 shows the energy response of bare LiF crystals exposed free-in-air. With a peak over-response of about 1.5, good dosimetry can be accomplished without the need for thick, high atomic number absorbers. Usually, thin amounts of copper or tin are sufficient for LiF based systems. Lithium borate requires even less attention to energy dependence but its fading characteristics introduce practical issues that need attention. The higher atomic number materials exhibit greater energy response dependencies but this also translates to greater low energy photon sensitivity and a better ability to decipher the energies contributing to the exposure. Abundant literature exists that describes the features of various materials and their utility (Oberhofer and Scharmann, 1979).

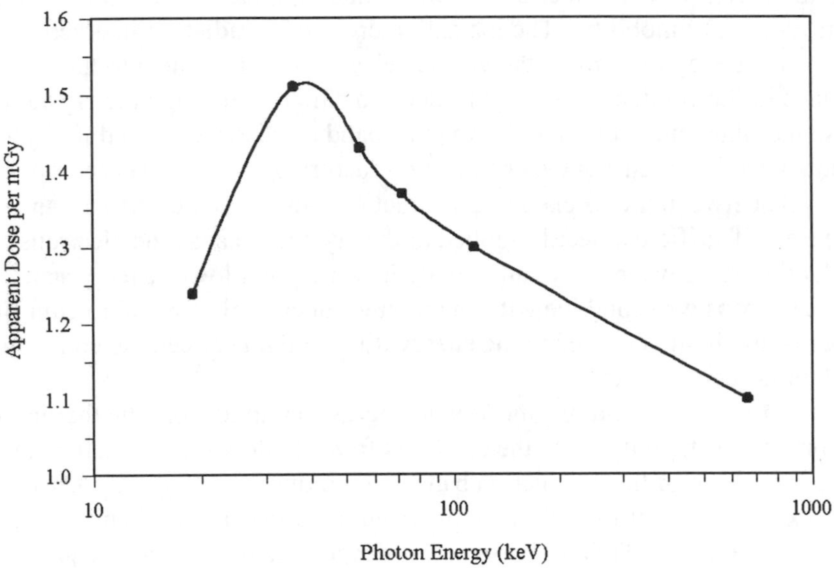

Fig. 5.3 – The energy response of lithium fluoride crystals free-in-air.

The elemental composition of the dopants used to create the electron traps influences the overall sensitivity. Additionally, different isotopes of the fundamental elements can alter the TLD response to radiation by taking advantage of special nuclear interactions. For example, the use of ^{6}Li in lithium based crystals will impart high sensitivity to thermal and very low energy neutrons because of the $^6Li(n,\alpha)^3H$ reaction.

The rate of light emission emitted depends on the heating rate and the residence times a crystal is kept at various temperatures. The duration at the maximum temperature is set to assure release of all charges from the dosimetric traps. Plotting the luminescence rate as a function of temperature or time creates a distinctive glow curve. The glow curve serves as a quality indicator because luminescence from oils, transparent packaging and foreign contaminants or incandescence from hot objects present in the heating zone present different curve shapes.

Hot planchets, hot gas streams, infrared lamps and lasers have been used to effect TLD heating. Each presents unique factors that must be controlled to assure consistent results. The faster methods enhance productivity and can provide a better signal to noise ratio; however, fast heating rates may affect material sensitivity and impart thermal stresses that can fracture crystalline TLD's. The heating rate will impact the appearance of the glow curve and may hide nuances in the curves of TLD's having several trap energy depths. Fast heating rates must be compatible with the luminescence measurement technique. Photomultiplier tubes operated in a photon counting configuration can be saturated when the luminescence becomes great.

TLD dosimetry has achieved an excellent reputation for accuracy and precision in the laboratory. The extended monitoring periods allowed by TLD's create opportunity for operational efficiency in the conduct of a personal monitoring program. Common complaints about TLD's include:

- the inability to re-evaluate an irradiated crystal because the heating process frees much of the trapped charges;
- the absence of visual data to answer the qualitative questions that arise following an unexpected dose;
- the requirement to calibrate TLD sensors individually to account for different absolute sensitivities that sensors exhibit for a given absorbed dose.

5.4.3 Optically Stimulated Luminescence

Optically stimulated luminescence (OSL) is an adaptation of techniques arising from archeological dating, retrospective radiation dosimetry and computerized tomography. The technology relies on many of the same electronic processes that occur in TLD's to trap charge released during radiation interactions. The energy from visible light releases the trapped charges allowing them to combine with luminescence centers to emit light in proportion to the absorbed dose. The only OSL system adapted for personal monitoring uses specially prepared aluminum oxide crystals. Aluminum oxide has been used for many years as a very sensitive TLD material so that its radiation dosimetry properties have been well researched. The electronic properties of aluminum oxide fortuitously favor optical stimulation with relatively short wavelengths of light. Once its OSL properties were discovered and crystal manufacturing methods perfected, application of the processing technique to personal monitoring became an obvious means to combine the advantages of film and TLD while avoiding many of the processing complexities.

OSL processing occurs at room temperature and the analysis methods are nearly identical to TLD. Therefore, OSL dosimetry affords the measurement excellence of TLD without the complexities of using a highly regulated heating system. The elimination of the heating requirement permits dosimeter designs based on powders of OSL crystals deposited on clear, plastic films. The deposition of a thin layer of aluminum oxide powder generates a large number of detectors having the same sensitivity to radiation . Additionally, the film of OSL powder creates a two dimensional radiation detector similar to radiographic film and imaging of filters or contamination can be achieved to provide qualitative information about certain irradiation conditions.

A unique trait of OSL is the ability to reprocess and analyze an irradiated dosimeter to confirm an initial analysis. The stimulation light releases only a small amount of the trapped charge so that much of the energy deposited by the radiation remains available for subsequent stimulation. In fact, one stimulation method uses a series of light pulses whose duration and frequency are timed to be compatible with the luminescence lifetime for aluminum oxide (Akselrod and McKeever, 1999). In essence a single dosimetric evaluation consists of a series of quick analyses, each taking less that one millisecond. Pulsed lasers operating at a pulse repetition frequency of 4000 Hz and delivering about one milliwatt of power can generate enough luminescence to measure a dose of 1 mSv in less that one-third of a second.

The amount of luminescence per unit dose generated in a given amount of aluminum oxide depends on the optical energy delivered to the crystals; the more energy, the more luminescence. This feature permits the processing system to create the amount of luminescence necessary to achieve a given lower limit of detection and to maintain a constant precision across the full dynamic range of dose measurements. The ability to perform repeated processing and analysis depends on the optical energy delivered. As more luminescence is created, the amount of trapped charge remaining for subsequent analysis is reduced. The ratio between two successive dose determinations at 100 milliwatts of pulsed green light delivered over 250 milliseconds is about 0.85; however less than 1% of the trapped charge is released when 1 milliwatt of green light is used.

The luminescence yield per unit dose also depends on the frequency of light used for stimulation. The luminescence yield maximizes with blue light having a wavelength of about 440 nanometers. Unfortunately, the luminescence wavelength is also blue making the separation of the stimulation and luminescence light difficult. While less efficient at generating luminescence, stimulation with green light allows optical filtration methods to easily separate the blue luminescence light from the stimulation light. Further separation can be obtained by using gated photon counting methods so that light measurements only occur when the stimulation light source is off.

The aluminum oxide film is extremely durable. Aluminum oxide is one of the hardest naturally occurring materials and is unaffected by water, pressure and high temperatures. Exposure to the heat found on the dashboard of a car in summer induces no fading or change in luminescence yield. The effect of blue light on the retention of trapped charge requires the aluminum oxide detector to be packaged in a light-tight wrapper. Exposure to red light has no effect. The light sensitivity can be used advantageously to eliminate the buildup of background radiation dose during the storage of the detectors prior to use. Detectors stored in blue light are kept in an optically annealed state until the moment when the dosimeters are packaged into the light-tight wrapper.

Foreign substances that interfere with the transmission of the luminescence light to the photo-detection system affect the accuracy of both TLD and OSL systems. However, unlike TLD systems where foreign substances burning in the heating chamber can contribute unwanted light and cause inaccuracies, OSL systems are relatively immune to this problem because very few materials emit blue light when exposed to green light. This is an upward transition, since green light is less energetic than blue light.

A current OSL dosimeter for personal monitoring includes a special filter designed to distinguish static from dynamic irradiation conditions (Akselrod et al., 2000). As indicated earlier, a unique feature of radiographic film is its imaging properties and their application to qualitative evaluation. The OSL dosimeter advances this feature by using a copper screen containing regularly spaced holes. The copper attenuates low energy photons to create a two dimensional dose distribution. This distribution is revealed with a CCD camera that responds to the luminescence. Fast Fourier transform protocols operate on the data to mathematically delineate static from dynamic irradiations. This approach removes much of the subjectivity involved with visually examining radiographic film.

5.4.4 Solid State Nuclear Track Detectors

Atomic particles with a high linear energy transfer (LET) coefficient create ionization densities able to disrupt the chemical bonds found in certain families of plastics and cellulose compounds that sometimes are called solid state nuclear track detectors (SSNTD). The sites of the disrupted bonds etch faster than the undamaged areas when the plastic is immersed in a caustic environment. The etched damage sites appear as small pits or tracks when viewed under a microscope. The number of tracks per unit area relates to the particle fluence and thus the dose equivalent. Beta particles and the secondary electrons from photon interactions cannot create tracks because their LET is insufficient. As a result, SSNTD's become very useful in measuring only the high LET component in a field containing mixtures of high and low LET radiation.

SSNTD's designed to assess neutron dose equivalent rely on the elastic scattering of neutrons with the protons or hydrogen nuclei found in most plastics. The high LET nature of the recoil protons disrupts the cross-linked monomers and polymers making up the plastic. The recoil particles require a threshold energy to elicit a damage site large enough to be visible after processing which essentially establishes the minimum neutron energy that can be detected. Irradiators such as polyethylene that are rich in hydrogen may be pressed against the surface of the SSNTD to enhance the sensitivity by increasing the fluence of protons. The etched pits are permanent so that SSNTD's can be retained for reanalysis.

A popular SSNTD material is polyallyl diglycol carbonate (PADC) commonly referred to by the commercial name of the starting monomer, CR-39®, a plastic used commonly for corrective lenses. Polycarbonate, commercially obtained as Lexan®, is another SSNTD plastic but its use has declined because CR-39® has a lower energy threshold and consistently better surface quality that offers a better lower limit of detection. With proper casting techniques, sheets of CR-39® can be obtained with smooth, clear surfaces that offer an excellent background for viewing the etched pits due to recoil protons.

The detection properties of CR-39® depend on the following factors:

- the purity of the starting monomer,
- the chemicals used to initiate the cross-linking of the monomer,
- the temperature profile maintained in the casting as the cross-linking reactions progress,
- the cleanliness of the casting forms,
- the thickness of the casting,
- the etching protocols used to reveal the damage sites, and
- the methods used to count tracks.

Differences among the manufacturers of CR-39® sheets and among the etching methods used by various laboratories create different measurement abilities of various CR-39® based neutron dosimetry systems. The first three factors above influence the etching properties and energy required to disrupt the cross-linked monomer molecules. Surface quality affects the minimum detectable dose equivalent as imperfections will contribute to the overall background density of pits that must be subtracted to determine the dose from occupational activities.

Etching protocols range from purely chemical processing to electrochemical processing with different etch times, temperatures and chemical concentrations. These parameters affect the conversion of the damage sites into visible pits. Chemical processing reveals pits of different sizes that can be related indirectly to the energy of the neutron field. The pits range from a few microns to 25 microns depending on the specific plastic casting and chemical conditions. Larger pits arise from alpha particles and heavy charged nuclei. The pits appear as conical depressions with the cone axis at various angles from the plastic surface indicating the path of the recoil particles. Etching conditions that yield smaller pits generally offer a greater measurement range because higher track densities can exist before overlapping occurs to interfere with accurate counting. Unfortunately, smaller pits are difficult to count with automated vision systems as the increased magnification introduces a variety of quality and productivity issues so that manual counting using microscopes is still frequently used. Automated track counting systems will often use a pit size discrimination setting so that only pits above a certain size will be counted. This approach reduces the range of energies detected because chemically etched pits have sizes that relate to the neutron energy.

Electrochemical etching introduces more complexity in the processing steps but creates very large tracks that possess intricately branching track lines to produce an image much like looking down onto the top of a bare tree. This treeing arises from an alternating electric field that is applied across the thickness of the plastic detector during the chemical etch process. Electrochemical etching produces relatively uniform, large tracks that are easily counted with automated vision systems. The large tracks limit the measurement range and some systems use multiple detectors etched slightly differently to achieve an expanded range.

Both chemical and electrochemical systems favor measurements of high energy neutrons encountered in well logging, neutron radiography and other industrial environments as well as those environments around high energy accelerators. Detection efficiencies deteriorate for neutron energies less than a few tens of kiloelectron volts. Special irradiators have been used to measure thermal and epithermal neutrons by creating charged particles though n,p or n,α reactions. Alpha particles create large pits that can be distinguished from the proton recoil pits so that appropriate energy response adjustments can be made. The prime shortcoming of SSNTD's is their extreme angular dependence. The response of CR-39® falls by nearly 90% at very large incident angles.

To summarize, SSNTD's and especially CR-39® offer methods to assess neutron dose equivalents without the potential for photon interference in mixed fields, a problem facing NTA and TLD albedo dosimeters. The tracks in the carbonate family of plastics do not fade and can be elicited through etching methods years after exposure. As a plastic material, mechanical durability and environmental robustness standout as practical traits. The detectors require no special packaging but the surfaces need to be protected from scratches that would make the identification of pits difficult.

5.5 EVALUATING PERSONAL MONITORING SYSTEMS

5.5.1 Measurement Performance

Many health physicists and those responsible for regulating occupational doses often consider measurement performance as simple evaluations of accuracy and precision under controlled and reproducible conditions created in the laboratory. This metrological approach uses precisely defined test environments to enable an objective evaluation of a particular dosimetry system. Different philosophies of testing have arisen, with the U.S. favoring a system of repetitive proficiency tests which examine a subset of possible operational conditions. While the test results are expressed in terms of bias and relative standard deviation, the data do not establish the overall measurement uncertainty associated with routine use. Other countries implement pattern or type test programs that fully characterize a system's performance under a wide set of circumstances in an attempt to assess an overall uncertainty or determine conditions where the system would be inadequate. Following the type test, periodic follow up tests examine a very abbreviated set of radiation fields and doses only as a means to trend calibration consistency.

In the U.S., the National Voluntary Laboratory Accreditation Program administers a proficiency testing program based on ANSI N13.11, a standard that examines dosimeter performance in eight measurement categories. Dosimeters are irradiated to a limited number of conditions over a series of three monthly measurement cycles. Variables in the test include dose level, radiation quality, energy and angle of incidence for higher energy photons. The Department of Energy operates a similar testing program but omits any angular irradiation conditions. Both programs define a performance quotient as the sum of the mean bias of the dosimeter results and the relative standard deviation, an approach similar to a tolerance statistic. The two programs define adequate performance differently but in essence, systems that can achieve a mean bias of less than 0.2 and a relative standard deviation less than 0.2 will be found acceptable for a given category. This statistical approach has no explicit bias limit for a single measurement; that is, only the group statistics are considered.

The proficiency tests examine performance in terms of the *personal dose equivalent* in a tissue equivalent slab with dimensions 30 cm x 30 cm x 15 cm. Irradiation procedures place the dosimeters against the central surface of the phantom, no closer than 10 cm from the edge. The detectors are usually displaced from the surface by the thickness of the holder and any clips used to attach the dosimeter to the user. The backscatter contribution from x rays makes this distance an important influence in the calibration so that measurements made at different dosimeter to phantom spacing may impart systematic biases.

5.5.2 Influences That May Affect Routine Performance

Many influences exist that could affect the consistency of routine personal dosimetry measurements. It is difficult to evaluate the true accuracy of a personal dosimeter when worn on a person because the use environment differs greatly from the calibration or test conditions imposed by the quantity definitions. That is, the degree to which a dosimeter measurement reflects the dose at 1 cm depth in a person is unknown. In addition to the geometrical differences distinguishing a person from a phantom, users of dosimetry systems should be aware the following issues.

5.5.2.1 Time Dependencies. The performance of a dosimetry system may be influenced by several time dependent changes. Some of these derive from changes in the radiation sensitivity and permanency of the radiation induced attribute. Others relate to durability in various work environments. Table 5.5 lists various time dependent influences.

5.5.2.2 Environmental Dependencies. Environmental conditions such as heat, humidity, light and mechanical stresses can alter the detector attributes or durability of the dosimeter and its related packaging. Many industrial settings encounter temperature extremes that are difficult to simulate in the laboratory. For example, oil field service workers such as well loggers and industrial radiographers may work in arctic conditions with temperatures reaching -40° to -50° F. Alternately, workers commonly leave dosimeters on the dashboards of cars and trucks where temperatures can exceed 150° F. Dosimeters can be dropped into pools of water or solvents, can travel through industrial washing machines, be dropped from great heights and subject to all kinds of physical pressures in the mail delivery systems.

5.6.2.3 Spacial and User Dependencies. The user imparts the greatest influence on the performance of the dosimeter. Dosimeters can be worn backwards, attached to a belt loop on the side of the body, placed in a pocket and dangled from a necklace around the neck. Some wear dosimeters on top of or under leaded aprons. All of these conditions differ from the basic calibration. Some dosimeters are designed to be relatively insensitive to these variables but others cannot. Photons backscattered from the body can contribute 30% of the total fluence irradiating a dosimeter. Albedo TLD dosimeters for neutron dosimetry depend heavily on neutrons reflected from the body. The reflected field changes as a function of location of the body and distance from the body. Inconsistent placement and use procedures introduce significant inaccuracies and uncertainties. In contrast, many CR-39® dosimeters are unaffected by backscatter so placement and distance from the body are not as crucial; however, CR-39® dosimeters exhibit extreme angular dependence so their geometrical relation to the neutron source must be controlled.

Table 5.5 Time dependent influences.

Influence	Description	Dosimeter types influenced
Fading	Loss of the radiation induced attribute.	Luminescence systems. Notably low atomic number materials.
Sensitivity drift	Change in the response per unit absorbed dose as a function of time after calibration.	Luminescence systems. Notably TLD. Film may experience spectral sensitivity changes over longer periods.
Background	Accumulation of background dose overtime which changes detection limits.	Film and luminescence systems.
Instrument drift	Ability of processing and analysis instrumentation to preserve their states as existed during calibration.	All systems.

5.6 CONCLUSION

Personal monitoring devices to assess occupational radiation dose include film, thermoluminescent and optically stimulated dosimeters, and plastic solid state nuclear track detectors. The response of these devices is proportional to personal absorbed dose, thus allowing them to serve as surrogates for the absorbed dose to the individual wearers. Meeting the accepted (ANSI 13.11) proficiency standards allows the dosimeters to be used with confidence to measure the wearer's absorbed dose for the purpose of radiation safety evaluations.

5.7 REFERENCES

Akselrod, M.S.; McKeever, S.W.S. A radiation dosimetry method using pulsed optically stimulated luminescence. Rad. Prot. Dosim. 81:167-176; 1999.

Akselrod, M.S., Agensnap Larsen, N., McKeever, S.W.S. A procedure for the distinction between static and dynamic radiation exposures of personal radiation badges using pulsed optically stimulated luminescence. Rad. Meas. 32:215-225; 2000.

Ambrosi, P.; Böhm, J.; Hilgers, G.; Jordan, M.; Ritzenhoff, K. The gliding shadow method and its application in the design of a new film badge for the measurement of the personal dose equivalent $H_P(10)$. PTB-Mitt. 104:334-338; 1994.

Becker, K. Photographic film dosimetry. London; Focal Press Ltd.; 1966.

Health Physics Society. American National Standard. Criteria for performing multiple dosimetry. Maclean, VA: Health Physics Society; HPS N13.41; 1997.

International Commission on Radiation Units and Measurements. Measurement of dose equivalents from external photon and electron radiations. Bethesda, MD; ICRU; ICRU Report 47; 1992.

International Commission on Radiation Units and Measurements. Quantities and units in radiation protection dosimetry. Bethesda, MD; ICRU; ICRU Report 51; 1993.

International Commission on Radiation Units and Measurements. Conversion coefficients for use in radiological protection against external radiation. Bethesda, MD; ICRU; ICRU Report 57; 1998.

International Standards Organization. X and gamma reference radiation for calibrating dosementers and dose rate meters and for determining their response as a function of photon energy - part 3: area and personal dosimeters. Geneva, Switzerland; ISO; ISO DIS 4037-3; 1996.

National Council on Radiation Protection and Measurements. Use of personal monitors to estimate effective dose equivalent and effective dose to workers for external exposure to low-let radiation. Bethesda, MD; NCRP; NCRP Report No. 122; 1995.

Oberhofer, M.; Scharmann, A. eds. Applied thermoluminescence dosimetry. Bristol; Adam Hilger; 1981.

Chapter 6

ELECTRONIC DOSIMETERS

Michael W. Lantz

6.1 ABSTRACT

Advanced-feature electronic personnel dosimeters (EPDs) have been incorporated into nuclear power plants as replacements for the outdated, self-indicating dosimetry technology. With the addition of these sophisticated features, electronic dosimeters became an integral part of most radiological protection programs as incremental dosimetry, and it seemed to be a natural progression to evaluate further uses for these dosimeters. About five years ago, several nuclear power stations proposed that EPDs could also replace the passive dosimeters in use as record dosimetry at their facilities. This article, originally authored in 1996 and now updated to the current status in the industry, presents technical information related to this proposal. The research presented here demonstrates that the best and most appropriate personnel dosimetry can be achieved utilizing passive TLD monitoring as the primary dosimeters in conjunction with electronic dosimeters regularly processed as incremental dosimetry.

6.2 INTRODUCTION

When this article was originally published in Radiation Protection Management five years ago (September 1996), the title was, "Should Electronic Dosimeters Be Used as Primary Dosimetry?" It was presented at a time when a number of nuclear power facilities had expressed their intention to replace the site passive personnel radiation monitoring, principally the TLD programs, with electronic personnel dosimetry, or EPDs. The facilities gave various reasons for this potential change:

- The costs of operating an EPD program may be less than a TLD program.
- The costs of performing NVLAP testing and assessments could be eliminated. The NRC even now (2001) does not require NVLAP accreditation for EPDs regardless of whether these dosimeters would be used as primary dosimeters at facilities regulated by 10 CFR Part 20.
- The need for in-depth Quality Assurance (QA) in a TLD program can be burdensome.
- The general difficulty of troubleshooting TLD systems had weighed heavily on some programs.

- They believed that there was no need for beta dosimetry.
- In their opinions, the EPDs worked adequately with no processing and only simple calibrations.
- Most EPD to TLD comparisons matched quite well in the high-energy photon environments of nuclear power stations. (*Note: this high-energy photon environment was the only in-field testing being presented anywhere at that time*).
- The current standard for comparison of incremental EPD results to TLD measurements was written in 1982 and was in need of update. Across the U.S., only about 5% of the nuclear power plant personnel dosimetry results are reviewed with a criterion that is not very stringent (+/- 25% if the TLD or EPD result is greater than 100 mrem).

The substitution of EPDs for TLDs for primary dosimetry was logical to many people within the radiation protection (RP) programs. Only a few people within the RP staffs knew much of the in-depth QA that had been developed for TLD programs. In fact, it would be rare for anyone outside the site dosimetry section to know much about TLD calibration, let alone the complicated dosimetry algorithms in use. Electronic dosimeters can be described more as single-detector instruments than dosimeters, and they fit easily into a facility's general instrument calibration program. The average RP technician would feel much more comfortable calibrating a portable instrument, such as an EPD, than any TLD reader. So, when faced with TLD programs that require separate staffs, routine dosimeter changeouts, and rigorous NVLAP testing and assessments, and with very few people within the program understanding the underlying benefits of their passive monitoring programs, it was no wonder that this change was proposed by RP management groups.

The primary point of this paper five years ago was to stem the tide of this movement. It was sincerely believed that RP staffs were moving too quickly from passive monitoring programs that had served them well for many years: dosimetry programs that were well-established and well-documented with over fifty years of peer-reviewed information bases. In suggesting this change, RP staffs relied heavily upon vendor-supplied EPD information, in addition to their own limited initial experience, to make this decision.

While not implying that EPD manufacturers were untruthful about their products, in my opinion it was not likely that all of the EPD problems were being loudly and distinctly announced, let alone published. The EPD manufacturers were in a race for market share, and any product defect or problem could have been perceived as a loss or a step backwards. In my opinion, it was left to the users to determine whether or not these dosimeters were actually ready for primary dosimetry. The article was therefore given its title, "Should Electronic Dosimeters Be Used as Primary Dosimetry?" In reality, it was asking, "Were EPDs ready for primary dosimetry?"

And this determination was a significant one. If radiation protection staffs were to remove the primary TLD dosimetry from workers and go solely with EPDs, and if previously undocumented problems were then discovered, plant radiation workers would have the right to question all of their dosimetry measurements during that time. The goal of this 1996 article was to provide the industry with as much knowledge as had been acquired relative to EPD performance, so that more informed decisions could be made about the dosimetry systems.

In the past five years, what has happened?

Several U.S. sites went forward with experiments or trial periods of the TLD-EPD substitution. All have subsequently switched back. It was always my hope that this article was part of the information base that helped bring about the correct decisions at the sites that considered the switch. The article and associated presentations may have assisted American Nuclear Insurers (ANI) in development of their firm stance that personnel dosimetry always be performed with two dosimeters. It is clear that their decision finally put the brakes on it. Since preferably the two personnel dosimeters should have separate, independent failure modes, ANI's policy made any thought of eliminating TLDs very doubtful.

ANI's fundamental principle was that adequate dosimetry must be provided to every worker at all times, not just when their primary dosimeter functions correctly. There must be a backup dosimeter and a comparison process even when both dosimeters are thought to have functioned correctly. With this philosophy, many people then recognized that they already had the best and most appropriate dosimetry programs - EPDs processed after every incremental job and TLDs processed regularly. The EPDs could be treated as instruments providing an enormous base of radiological safety features, with the in-depth QA and rigorous algorithms of the TLDs providing a solid foundation of personnel dosimetry for all radiological environments (β, γ, x, and n).

The remainder of this report is an update of the more relevant issues from the 1996 article as they pertain to the use of EPDs, then and now. Information relative to the substantial progress made by the manufacturers in producing better EPDs and, in turn, superior safety devices is also provided.

6.3 UPDATE OF 1996 EPD INFORMATION

The nuclear power industry has been using electronic alarming dosimeters for dose control in high radiation areas for many years. Several years ago, manufacturers began producing smaller and more sophisticated alarming dosimeters, adding features such as dose-recording histograms, electronic checks and dose rate alarms. These alarming dosimeters are now being used by personnel in radiologically controlled areas as incremental dosimetry and have become better known as electronic personnel dosimeters, or EPDs.

The advent of these new dosimeters has significantly improved radiological protection programs and employee safety. The industry had previously relied on self-indicating pocket ion chambers for incremental dosimetry. These dosimeters were very limited in their capabilities, susceptible to shock, and devoid of alarm capabilities. The model of pocket dosimeter fixed the range of exposure measurements. In addition, workers were required to routinely stop and read the dosimeter while in radiation areas, which was difficult to do and contrary to as low as reasonably achievable (ALARA) principles in some environments.

The benefits of EPDs as incremental dosimeters are numerous and their advanced features add convenience and safety to all work within radiologically controlled areas. For example, EPDs can alarm at preset doses and dose rates in the field. This feature alone could have prevented many radiological events. It is my firm belief that the overexposures that have occurred within the radiography industry in the past few years would not have happened if the companies involved had embraced and routinely utilized these new EPDs in their programs. EPD displays can be set to automatically show both dose and dose rate; the worker only needs to glance at the dosimeter. The range of dose measurements is dramatically extended with EPDs. Pocket ion chambers are useless above their range, say 200 mR or 500 mR, and provide very poor accuracy below 5 mR. The new EPDs can read up to 1000 R and down to less than 1 mR. Therefore, one EPD can replace at least three pocket dosimeters on a worker. The advent of the histogram feature on EPDs means that personnel doses during a job can be stored in 200 to 800 retrievable time sequences for job recreations (0.1, 1 or 10 minute increments). This can be an integral part of an ALARA program. I cannot fathom how a site could perform in-depth ALARA reviews without electronic dosimetry. Some electronic dosimeters have a telemetry feature so that dose and dose rate measurements can be transmitted to technicians on a computer terminal outside of the work environment; also a significant ALARA tool. Because these dosimeters can also display dose rate, more information is available to workers than ever before. Pocket dosimeters required manual entries and exits, along with the staff to record the dose readings. EPDs are routinely interfaced to computer systems allowing easy automatic entry and exit recording of individual job exposures.

With the addition of these features, EPDs have become an integral part of radiological protection programs as incremental dosimetry. For many sites, as stated above, it was a natural progression to evaluate additional uses for EPDs.

With that in mind, EPDs were proposed as instruments to replace the passive dosimeters (TLDs and film) in use as record dosimetry at most sites across the U.S. As a technical expert in dosimetry for the National Voluntary Laboratory Accreditation Program (NVLAP) and while implementing Palo Verde Nuclear Generating Station's passive and electronic dosimetry programs, I've had the opportunity to gather information through shared experiences with industry peers. The problems discovered during this process have led me to believe that, for users who might pursue substitution of current passive dosimetry systems with electronic dosimeters, careful evaluation is necessary.

The goal of this article was to list issues of EPD quality, technical considerations, and other practical information that will generate further discussions relative to the general application of these instruments and to the question: Are EPDs ready for the role of primary dosimetry?

6.4 DOSIMETER WEAKNESSES

6.4.1 Variable pin diode energy response curves

Several EPD manufacturers use semiconductor pin diodes as the radiation detectors, while others use GM detectors. Individual pin diode detectors have exhibited highly variable photon energy response curves, observed primarily at lower energies. At least one manufacturer performs a separate 60 keV test to identify the unacceptable detectors. However, data from other EPD users indicates that not all manufacturers are aware of or address the problem with the detectors, or cannot do the tests. The results of some dosimeters irradiated to equal doses of low energy photons vary up to 60%, while those same dosimeters read properly when irradiated to ^{137}Cs. Low-energy photon irradiations should be part of routine acceptance testing for all EPDs if the manufacturers don't provide the tests.

This first problem noted here was a subtle aspect of EPD pin diode detectors initially discovered in an older model of electronic dosimeter. While response-checking EPDs to ^{60}Cs in a heavily shielded source jig, some dosimeters overresponded dramatically compared to the expected exposure. This led to concerns about dosimeter reproducibility, but further tests revealed that the same dosimeters always overresponded. And when exposed to ^{137}Cs in free air, they read correctly. Studies were begun of the "good" and "bad" dosimeters and the problem became readily apparent. When the pin diode detectors of bad dosimeters and were removed and placed onto the boards of the properly responding dosimeters, the problem followed the detectors. Utilizing the spectral measurement properties of TLDs, the spectrum of the heavily shielded source jig was shown to be dominated by scattered, low-energy photons. Some of the pin diodes reacted normally to this change in photon spectrum, but some did not. Thus, within the population of those pin diodes, there were individual ones with very different energy absorption characteristics. It would be analogous to TLDs with inconsistent amounts of dopant material absorbing photons differently. This problem was confirmed to exist within today's population of pin diodes through a review of vendor testing and in discussions with several vendors. At least one manufacturer currently separates out the poor performers with an acceptance criterion. Even now, this has never been published.

6.4.2 Some types of EPDs have been noted to mechanically fail (i.e., in need of repair or replacement) at a significant rate: up to 30% per year

EPDs are electronic devices and as such will have failure rates much higher than passive dosimeters. In addition, some case designs are inadequate and have not matured, and dropping the EPD has been shown over time to cause failures of the detectors, displays, speakers, microprocessors, circuit boards, etc. In addition, water can easily infiltrate some cases.

This issue presented the initial concerns over product durability. In the first year of use, we lost a considerable portion of our inventory, as we have continued to do over the past five years. Primarily, we lost displays, but speakers and complete dosimeter failures were more than we had expected. As instruments, this might not have led to too much concern - simply increase the repair and placement program. But viewed as primary dosimeters, we had concern for the litigation aspects of EPD failures. It did not seem appropriate to go to court and submit that 85% of the dosimeters that a person had worn at our facility had failed and had been repaired or replaced. In fact, after six years of use, we have replaced the entire inventory with new electronic dosimeters. It's interesting to note that TLDs purchased in 1984 for personnel dosimetry are still in use. We do have high hopes for the latest models of electronic dosimeters. The manufacturers have learned to make the EPDs water-resistant, sturdier, and much smaller. But again, as originally pointed out, the devices must mature and there is still little experience with long-term use.

6.4.3 The data loss rate for electronic dosimeters is higher than for passive dosimeters

It has been stated that EPD failures represent only a single entry while TLD failures represent many entries. However, TLDs rarely fail completely. If a single TLD element fails, other elements in the same dosimeter can be used. If a TLD is lost, incremental dosimetry records can be used. In the current programs, very few workers ever encounter a dosimetry failure and workers have high confidence in their reliable passive dosimetry programs. Because the EPD failure rate is higher than for passive dosimeters, and EPDs are processed and recorded after every RCA entry, the number of workers who would encounter a dosimetry failure during the year would increase with the use of EPDs as primary dosimeters. And if EPDs are used as "dose of record" dosimeters, as they would be if the EPD were the only device issued, the number of worker concerns may increase.

In the past five years, as the population of our electronic dosimeters has aged, the failure rate has increased to the point that complete dosimeter failure is a common occurrence. Practically every day, someone wears an EPD that fails during use. During the busy periods of a nuclear power station refueling, broken EPDs pile up in drawers waiting to be repaired or replaced. Over a year, thousands of TLDs are processed with few failures and no worker concerns. As stated before, this problem was one of the reasons for our complete replacement of the population of EPDs, in addition to the major advances that have come about in five years of EPD design.

Just recently, we underwent NVLAP Proficiency testing with our EPD's. We failed the Accident Category because one dosimeter exposed to approximately 330 Rem read over 550 Rem. Below is a portion of the histogram showing two large spikes of 83,500 mrem. During NVLAP testing, this dosimeter was sitting gently on a phantom in a 25 R/hr ^{137}Cs field:

Sat 16 Dec 2000 20:26:32 :	426 mrem
Sat 16 Dec 2000 20:25:32 :	425 mrem
Sat 16 Dec 2000 20:24:32 :	83500 mrem
(the EPD histogram only stores the 1st rate flag)	
Sat 16 Dec 2000 18:13:32 :	428 mrem
Sat 16 Dec 2000 18:12:32 :	425 mrem
Sat 16 Dec 2000 18:11:32 :	83500 mrem
Sat 16 Dec 2000 18:11:16 : rate saturation !	
Sat 16 Dec 2000 18:10:32 :	424 mrem
Sat 16 Dec 2000 18:09:32 :	425 mrem

Our conclusion is that this type of problem is part of the normal aging process of electronic dosimeters. Another theory is that an interaction occurred between the electronic dosimeter next to it and this one, causing these spikes. This effect has been demonstrated in some electronic dosimeters.

6.4.4 Dose rate measurements from one type of EPD have been found to be inaccurate below 200 mR/h

EPDs should be tested in a constant 60 mR/h field. The maximum dose rate (a standard recorded feature on some EPDs) recorded after only about five minutes in a constant 60 mR/h field has been found to be as high as 120 mR/h. This is related to the very short EPD sampling times for the dose rate measurements. Imagine trying to explain to an inspector who was in a radiation area that the general area dose rate was not greater than 100 mR/h, the dosimeter just measured it poorly.

This issue of inaccurate dose rate measurements has been a tricky one. What I was trying to explain above was a situation where an entry is made into an area that is posted as a radiation area, but when the individual leaves the area, his EPD indicates that he entered a high radiation area (>100 mR/hr). This would have people concerned about a potential violation of 10 CFR 20. Further, several organizations were interested in evaluating any worker entries into unauthorized areas. In fact, they wanted to see entries into areas as low as those greater than twice the area background radiation dose rate. EPDs are simply not capable of recording accurate low-dose rate measurements.

6.4.5 EPDs are generally much larger and heavier than passive dosimeters

This may seem minor, but it may lead to significant problems. TLDs and film badges are comfortably worn on a collar or front pocket. Because most EPDs are bulky, they may be worn on worker's hip, which can lead to underresponse due to self-shielding (by the hip or by the arms). They have also been found clipped backwards in worker's front shirt pockets where they can read up to 30% low because of pin diode angular dependence problems.

Even with the best training, workers are not watching their dosimetry every moment. These large dosimeters have been found flipping around on their clips every day. TLDs do not exhibit this problem and generally are designed to be able to flip over and measure dose correctly.

6.4.6 Certain EPD histogram processes need improvement before they can be considered as a strong feature in dose recreation

For example, with one-minute histograms, a reading of 0.5 mR may have been caused by a false, one-second electronic spike equivalent to 1800 mR/h or a real radiation field of 30 mR/hr for the entire duration. Even using a histogram update this short, many events cannot be clearly resolved.

When I wrote this, I had hoped that the manufacturers would take this comment to heart and continue to develop their histograms. They generally have not. RFI spikes remain difficult to prove, but more importantly, with such limited data storage bins, EPDs cannot be the impressive ALARA tool that they should be. One vendor is developing an EPD with 20,000 histogram bins, but it is not ready yet. To date, the best resolution achievable for most site radiation entries is a one-minute bin of exposure and this is not enough to detail individual exposures on specific jobs.

6.4.7 Very little is published in peer-reviewed journals about EPD response and battery voltage

In fact, very little has been published relative to any EPD information. And in this highly competitive electronic dosimeter market, should we expect that manufacturers' literature would detail the problems known about any particular dosimeter?

Well, five years later and nothing new to report. The vendors have performed some excellent studies of their latest dosimeters and are making those available – especially to potential clients. Users have lost the desire to write anything about EPDs. The reason may be that the electronic dosimeters have been put back into the position of secondary devices, or instruments, and there are relatively few reports written about portable instrument responses. This remains disappointing but understandable. However, as long as the personnel dosimetry of the U.S. remains a combination of passive TLD monitoring plus incremental EPD monitoring, excellent dosimetry will be performed.

6.4.8 The resources of the manufacturers are being challenged by the efforts to have EPDs accepted as dose of record

Our industry might be better served if EPD manufacturers were directed to improve the EPDs as incremental and alarming dosimeters in support of passive dosimetry. For example, some EPD alarms sound almost exactly alike. Wouldn't we be providing more radiological safety with distinctive dose and dose rate alarms? If we had louder speakers or vibrating alarms, wouldn't that be better? Wouldn't it be an improvement if the EPDs were half their current size? Because manufacturers have limited resources as they strive to improve them, shouldn't we keep them focused on the most important issues?

In the past five years, manufacturers of EPDs have focused their talents on design changes to better serve the industry. The latest EPDs on the market are less than half of the weight of the previous ones. The cases are clearly much sturdier and water-resistant. The manufacturers have worked diligently to eliminate the dose rate spiking related to radio-frequency interference (RFI). In the future, I remain hopeful that the next generation of EPDs will incorporate built-in vibrating alarms and variable tones, similar to state-of-the-art cellular phones.

6.4.9 Most EPDs are large, single-element dosimeters

The industry generally moved away from one- and two-element TLDs over ten years ago because of their inadequacies.

At the time of this original discussion, some RP staffs did not see the importance of this statement. As mentioned earlier, two issues caused them not be aware of potential problems. First, the only EPD field tests that had been conducted were in environments dominated by high-energy photons. In fields such as those, almost any dosimeter will work. Secondly, the value of a complex dose-measuring tool like a TLD system is rarely understood by many. However, as soon as the RP staffs tested the EPDs in lower-energy environments, problems were discovered. Many electronic dosimeters have demonstrated dramatic underresponse to photons below about 70 keV. Moreover, significant underresponse occurs at angles other than perpendicular (because of the size of the cases, transmitters, and batteries) and non-linear response at high dose rates is exhibited, so very poor personnel measurements do occur with the EPDs in field use. One accident case in particular involved a person who entered a 1000 R/hr field of high energy photons and worked for a short period. His passive monitoring provided a good measure of his true dose of 34 rem. The electronic dosimeter that was worn indicated only 8 rem. A review of our own technical studies showed that this model of EPD underresponds by approximately 75% when exposed to 1000 R/hr of ^{137}Cs photons, or about 8 Rem in this case.

The latest crop of EPDs has been redesigned to improve all three of these issues. They are much smaller (generally 50 to 100 grams) and therefore the angular dependence can be reduced. The general energy response to photons has been improved significantly by the use of either multiple pin diodes or with new pin diode/filtration modifications. Lastly, the dose rate characteristics have been much improved through redesigned EPD circuitry.

6.4.10 Most EPDs do not provide any indication of beta or neutron exposure

Almost all current TLDs and film badges will indicate that conditions have changed and that the workers at a facility are now being exposed to an unexpected beta or neutron environment. This information could be important to identify that a radiological change has occurred and was missed by the operations staff.

Facilities that had considered dropping TLDs had proposed that beta dosimetry was unnecessary because the shallow dose was never limiting at their facilities. In fact, the shallow to deep ratio of total site exposure was routinely near 1.00 (less than 1.10). I continued to argue however that beta dosimetry was necessary for individual workers who do encounter significant beta radiation fields (e.g., after any fuel damage). It is questionable whether the electronic dosimeter manufacturers will ever be successful with a beta dosimeter in the field because to perform beta measurements, the detector has to be near the surface and would be very susceptible to radio-frequency interference. Recently, a new electronic dosimeter has been developed to perform neutron measurements which may prove to be an excellent incremental dosimeter to assist with passive neutron monitoring.

6.4.11 The population of EPDs in use today is very new

EPD problems are being discovered in these first few years of operation. Will they fail more often as the population of these dosimeters ages, as detectors and circuit boards become loose, as wires fray, as detectors become damaged? Are we willing to take this chance? In a few years, practically every electronic dosimeter that a worker has worn will have been replaced because it has failed. How might that be presented in court?

As stated above, in the past five years, we have had to deal with many dosimeter failures while we continue to utilize TLDs that are over fifteen years old.

6.4.12 Could a new dosimeter on the horizon (i.e., an EPD/TLD) be the answer to all of our problems

Data need to be generated and the dosimeter needs to be certified before we make any decisions.

What I was referring to was the Direct Ion Storage (DIS) dosimeter. It was being touted as the new dosimetry answer. Five years later, documented in-field personnel measurements are few and far between.

6.5 UNUSUAL OR ANOMALOUS READINGS

6.5.1 Magnetic Fields have been known to turn some EPDs off

EPDs have turned off completely after exposure to a magnetic field. Others go into a "latent" mode while in a magnetic field and the dosimeter is totally unresponsive to radiation. They resume operational status after removal from that environment. If a worker is in a high dose rate area near a magnetic field (electric motor), the dosimeter may not measure the dose. No one would be aware of the problem if the worker weren't also wearing a thermoluminescent dosimeter (and then only after the fact).

In the past five years, the actual occurrence of this problem at nuclear power stations was probably not significant. But we may never truly know. Recently, one of the latest crop of electronic dosimeters exhibited a similar problem and it has been documented as occurring at several facilities. The dosimeters went into a latent mode, unresponsive to radiation, when exposed to the 125 kHz electro-magnetic field surrounding some computer monitors. As more and more computers are being used in the radiation areas of the plants, such as in containment, these latent mode events could become significant.

Just recently, the following problem was identified about a new EPD in industry event notifications: *"these electronic dosimeters are susceptible to* ***complete loss of accumulated dose*** *when subjected to static discharge. Further testing showed that the loss of dose could be initiated by subjecting the electronic dosimeters to static discharge. Test personnel built up a static charge on themselves, and the dosimeter, which would then discharge when touching the dosimeter to the dosimeter reader or other conductive surface. Any accumulated dose would be lost and a "Power Interrupt" error would appear. However, the "Power Interrupt" error is not limited to the static discharge incidents and therefore is not always an indicator of lost dose."*

Any loss of personnel dose should be a serious event.

6.5.2 EPDs are spiking or scrolling to high and low doses

Anomalous dosimetry events are occurring within the industry where the electronic dosimeters spike to as high as 100 rem, while the associated passive dosimeters (TLDs/Film) read 0 mrem. Radiation protection staffs may not be adequately investigating or documenting the evaluation of these irregular readings. These EPD events would be more noticeable to the workers than the problem of pocket dosimeters going off-scale because, with pocket dosimeter failures, the wearer was presented with no measurement. Anomalous EPD readings inform the individuals that they may have received 8 mrem, or 400 mrem, or 2 rem, or 10 rem. If a worker were also not wearing a TLD, the EPD result would be the only direct record of that individual's dose.

6.5.3 High dose rate events are occurring on EPDs throughout the industry.

This may not be a serious problem to any particular worker's dose determination because many can be short in duration, but it speaks to the lack of stability in some dosimeters. A common statement is that these dose rate spikes, as high as 100 R/h, are probably related to radio-frequency interference and that the EPD manufacturers are aware and have resolved them. However, many other causes of dose rate spikes exist and have not been published to date. They need to be compiled and this article is an initial attempt to begin that process. Yet also, just recently, it was discovered that new 3-watt, 800-MHz hand-held portable radios could saturate pin diode type EPDs to 999 R/h when the antenna is brought within 4 inches of the dosimeter.

6.5.4 Simple occurrences, such as keys rattling in front of the dosimeters, pens tapping on the case, dosimeters being dropped or squeezed, RFI, magnets and microwaves have disrupted EPD measurements.

Measurable doses and dose rates as high as 20 R/h have been generated by placing an EPD near a microwave. The list of causes of high false readings or unknown low responses should be developed before they are used for dose of record. Consider lasers, microphonics, common chemicals, sweat, arc welding, internal electronic processes. . . there are many causes which could be added to this list.

I combined the discussion of these three issues because they related to the same problem: readings of non-radiation induced doses and dose rates in EPDs. Whether the false response was caused by rattling keys, the static charge created when pulling EPDs out of plastic bags on a dry day, sweat dripping onto the case, or by transmitting radios, false high readings are generated within EPDs. At our site, we have had reasonable success identifying and correcting these spikes by using access control software that will download the EPD histogram if a preset maximum dose rate is exceeded.

6.5.5 Unusual performance test failures have been discovered without a determination of the cause

Dosimeters which were irradiated to 120 mR read 76 mR on the first test. A subsequent irradiation showed a reading of 118 mR. Several more runs showed the pattern of approximately 120 mR and then below 80 mR. The answer to this has not been found, but it may be related to either the microprocessor or the associated electronics.

The manufacturer, in five years, has never resolved this issue. It seemed to be related to the manner in which the dose rate is converted to dose, but the problem still exists. Since the dose algorithms of most EPDs remain unknown by their users, these types of problems may continue for many years.

6.5.6 EPD users may not be investigating the high rates of "AS FOUND" failures in their regular performance tests

Users have had EPD failure rates as high as 14% on the six-month performance tests. Are standard follow-up investigations being performed? These investigations should be done if the dosimeters were used as dose of record.

Actually, our studies have shown very good performance testing results from the EPDs that have been evaluated every six months – at least for those that still work. As stated earlier, the repair/replacement rates have been high.

6.5.7 Most facilities are using the EPDs in a pooled environment

It will be a monumental task to investigate EPD problems, who used them previously, and what their true doses were.

I have not heard of anyone issuing EPDs individually so they are all very likely being used in a pooled format. Therefore, if an EPD were discovered operating at minus 60%, the exposures of over one hundred people would need to be re-evaluated at our site each monitoring period if they only wore EPDs.

6.5.8 Increasing the chirp rate on one type of EPD caused them to underrespond by 75%

6.5.9 One version of an EPD, after encountering a 120 mR/h field, continued recording at that rate even though the EPD had been removed from that field

6.5.10 One type of EPD registered high doses when exposed to sunlight

These three items were also combined, as they are part of the larger list of undiagnosed problems with the electronic dosimetry devices. New problems continue to be acknowledged every month. However, some problems are slow to be recognized by the vendors or the users. The identification of a problem may remain within the confines of the facility that discovered it because identifying it to the world sometimes requires a great deal of extra effort.

6.5.11 Failures of the EPD control software at the RCA entrance have been found

An error in software allowed EPDs to be issued without the associated worker's ID being recorded. Because of that, the EPD dose measurement upon exit was not recorded in the computer. Another software error allowed EPDs at the RCA entry point to be removed from the computer interface and worn in the RCA without the dosimeter being turned on. If EPDs are used as the primary dosimetry without a TLD, then the software recording the doses would have to be flawless or it could not be proved that worker doses were always stored.

In the past five years, a few EPD access control software problems have been identified. In addition, numerous events have been recorded where the workers left their EPDs at the access control point in the EPD reader. In response to this, one manufacturer developed software to ensure that the EPDs were worn and turned on at the entrance to the radiologically controlled areas. The software had an error that randomly turned off EPDs – some of the ones that had been on before reaching that point!

6.6 ANGULAR DEPENDENCE ISSUES

6.6.1 Overall EPD angular dependence can be significant

This is due to the detectors, the large cases and the batteries. This can be observed especially as the photon energy is decreased below ^{137}Cs; e.g., an EPD that can read too low by 93%, or a factor of 14 too low, when irradiated with 60 keV photons at an incident angle of 90°. Why do some manufacturers only show the horizontal angular dependence to ^{137}Cs in their technical information? Why don't they also provide the vertical angular dependence to ^{137}Cs, and to lower energy photons? And why don't the users ask for this information? Although EPD performance standards have been drafted, they do not test beyond 60°. It is inappropriate for the industry to be unaware of the results past 60° because this knowledge has to be used to understand the dosimeter responses in the work environment, especially for these large dosimeters. One particular type of EPD reads a factor of 3 low when irradiated to high-energy photons from the bottom with its telemetry device and nine-volt battery attached. The user of one large EPD with a GM detector found that the dosimeter underresponded by 3.3 (response = -

70%) to the deep dose equivalent (DDE) from photons of a submersion ^{133}Xe *environment (81 keV). It is interesting to note that the published energy response data for these EPDs indicates that this dosimeter reads about 5% high at this energy, at 0°. The underresponse observed at that facility is due to the angular response of this submersion irradiation; i.e., the integral of the response of this dosimeter at all angles. What is not routinely published is that electronic dosimeters may respond as low as -80% to -90% to 81 keV photons from 90°. The sum of the weighted response of the dosimeter at this energy through all angles should be near the 0.30 value. The problem with large dosimeters is not energy dependence alone, but angular dependence, which is enhanced and more clearly identified at lower energies. Large pin-diode dosimeters should underrespond in these environments and in any work environment that presents a planar or submersion exposure to the dosimeter. It is my opinion that most work environments, or the environments as integrated by most dosimeters, do represent a plane or submersion as workers move around in them. Most passive dosimeters, such as TLDs, will exhibit an underresponse of only a few percent because they are small.*

Fuel failures! That's one way to radically modify the radiation spectra encountered by a nuclear power plant dosimetry program. The routine radiological spectra at nuclear power stations are high-energy photons. With fuel failures, low energy photons and low- and high-energy betas become significant. In the past five years since this was written, several sites have had to deal with significant energy spectral changes, and, in each case, the EPDs underresponded dramatically to the deep and shallow doses.

6.6.2 Are EPDs being used by Fluoroscopists

These individuals work with low energy photons at large angles to the dosimeter. EPDs will measure the deep doses very poorly because of the information noted earlier.

This comment was a request for someone to put a stop to EPDs as primary dosimetry at hospitals or X-ray labs where they could reveal seriously low and varied personnel dose measurements.

6.7 MICROPROCESSOR PROBLEMS

6.7.1 The EPD algorithms may not be known by the users

Most electronic dosimeters use very complex algorithms to process the dose and dose rate measurements. The manufacturers should be able to provide the complete dose calculation to the users especially as they may change over time. Not having these algorithms is inconsistent with the NVLAP requirements of a record dosimetry program.

Obtaining the EPD algorithms for the users will probably never happen. EPD algorithms have been treated as proprietary information to protect the manufacturers' investments.

6.7.2 EPDs may be asked to process too much

In a constant 8 R/h field, for example, some EPDs have been shown to display 8 R/h, then 12 R/h, then back to 8 R/h, and so on. These irregularities have also been documented at lower exposure rates: 600 mR/h and 75 mR/h. It seems to be related to a problem within the microprocessor and the frequency of the failures may be directly related to the number of features being asked of the dosimeter. For example, if the EPD is set to display dose and dose rate or if the histogram is asked to update every minute, then the rate of these anomalies will increase. Are the users asking too much of these instruments?

This item came about after a discussion with a representative of the electronic dosimeter that we use. He indicated that some of the problems within the EPDs related directly to the complexity of what the dosimeter is trying to do all of the time. Electronic dosimeters are counting instruments, but they are processing this information almost every second. Dose rates and doses are converted from counts, the information is then stored in several places and checked against preset alarms, regular tests of the electronics package, battery, and detector(s) are simultaneously performed, and all of the time, the information is displayed to the workers. Most users don't want to think about the complexity of what is being executed by these electronic dosimeters live time, they just hope that the EPD's work. Occasionally, they don't.

6.7.3 Some EPDs can have a significant dose rate dependence, depending on the type of detector, circuitry, and processing algorithm

This could be a significant problem in high dose rate event recreations.

As I pointed out earlier, EPDs can underrespond dramatically to very high dose rates. Now add that this underresponse is quite variable (-25% at 200 R/hr, …., -75% at 1000 R/hr) and you can appreciate that accident recreation would be almost impossible if the dose rates involved were high *and* changing during the event. The latest crop of electronic dosimeters has shown a vast improvement in this problem through rather complicated rate adjustment algorithms.

6.8 REGULATORY ISSUES

6.8.1 There is no Certification Program for new dosimeter designs

Whether it be for Al_2O_3, EPDs, or any new dosimete, a certification process needs to be available. The NVLAP program has been a testing program directed at TLDs and film badges that had 50 years of published information about their problems, advantages, and disadvantages. Hundreds of reports have been published with regard to these dosimeters detailing angular dependence, energy dependence, fade, light response, filter design, minimum detectable levels, response checks, dose rate dependence, algorithms, and responses to unusual environments. For the EPDs, what has been published? No separate EPD record dose standard has been completed. A certification program, or a detailed type-testing process, would go a long way in identifying and resolving problems before there is widespread use of new dosimeters for dose of record for personnel.

ANSI N13.11 can only do so much. It can test some of the factors of energy and angular response in EPDs. But still, long-term study and publication by users remain the best process to identify and sort out the problems of any dosimetry system.

6.9 CONCLUSIONS

In a 1992 memorandum, the NRC staff concluded that: "since the EPD requires no processing, it does not require a NVLAP processor." Essentially, the staff seemed to determine at that time that 10CFR20.1501 did not apply. It is my opinion that the NRC should reexamine this position and consider other points of view such as those presented here. The NRC also stated in the same memorandum: "The staff believes that an electronic dosimeter which has proven itself reliable in field trials and is on a par with current dosimetry used for permanent records in terms of precision, accuracy and reliability, would be an acceptable alternative to TLD or film." At that time, they stated that one EPD "appeared to," but they did not say that all EPDs had been proven to do so.

Our challenge should be to work with the manufacturers to build smaller, more reliable electronic dosimeters that match or exceed the capabilities of current passive dosimetry. We have worked very hard to provide quality measurements of record doses using passive dosimetry and worker confidence in those measurements is high. Although most EPD measurements have been shown to be generally good, the issues and questions raised in this article may make people stop and consider whether their EPDs have proven themselves to be on a par with passive dosimetry. It may be that some of these issues have been resolved, but a primary dosimeter used for a worker's dose of record should be able to answer all of the above issues and many more, and not just "pass Category IV of NVLAP."

Five years later, there has been no change in the NRC position that EPDs do not require NVLAP accreditation. Several sites have gone forward with the accreditation process anyway because they believed it was important to do so. When a worker's TLD is lost or damaged, the EPD totals would likely be used for the record dose under 10 CFR 20, so those sites believed that this was a positive position to take. But it is not without difficulty because many of the EPDs in use today exhibit problems with these NVLAP proficiency tests.

The manufacturers of the latest EPDs have made enormous strides in energy, angular and dose rate dependence. It is believed that they have been successful in eliminating radio-frequency interference spiking of the dosimeters and the EPDs are rapidly approaching the size of some TLDs. This is important if they are to be utilized for multi-badging on personnel. Problems still exist because they are electronic devices being asked to do many tasks at the same time, but as incremental dosimetry and radiological safety devices, no radiation worker should be without one.

6.10 REFERENCS

ANSI HPS N13.11-1993. American National Standard for Dosimetry, Personnel dosimetry Performance, Criteria for testing, 28 September, 1993.

Lantz, M.W. Should Electronic Dosimeters be Used as Primary dosimetry, radiation Protection management, 13:5, 28-34, 1996.

Chapter 7

AN OVERVIEW OF SOME ELECTRONICS USED IN HEALTH PHYSICS INSTRUMENTS

F. Morgan Cox and Mark D. Hoover

This work is dedicated to the memory of
John S. Handloser (1921-2000), friend and mentor

7.1 INTRODUCTION

This presentation is intended to offer the health physicist an overview of some of the electronics used for various applications in health physics instruments. The emphasis here is on the generic nature of circuitry used in these instruments. Not all types of health physics instruments, detectors or electronics can be covered in such a short paper. The exact circuitry and electronic components used in a specific type instrument will vary with the design from one manufacturer to another. Most instrument suppliers are somewhat sensitive about sharing the exact nature of their particular designs for recognized competitive reasons. One way to view nearly exact circuitry design is to purchase a particular instrument. Even then, some components may be treated by the manufacturer as "proprietary." In defense of the instrument manufacturers, each of the suppliers of health physics instruments may invest hundreds of thousands of dollars and considerable time in the design, testing and manufacturing engineering of a particular instrument. Some instruments may take years of development effort from marketing concept to production of a manufactureable and customer-accepted device, instrument or system.

Some emphasis is given in the presentation to the measurement of tritium for several reasons. Tritium is very difficult to measure because of its physical nature as a gas and as a form of water, and because it has an extremely low energy beta particle which pushes the limits of design and detection.

The authors have recently contributed to the development of the International Electrotechnical Commission (IEC) and American National Standards Institute (ANSI) standards for the detection and measurement of tritium. These standards are IEC Document 710 being revised as IEC Document 60710, IEC Document 761 being revised as IEC Document 60761, and ANSI N42.30 being developed.

7.2 DETECTOR TECHNOLOGIES

Since human beings have no physical sense to detect or measure ionizing radiation, we are dependent on the interaction of radiation through ionization effects in special gases, liquids or solids. Let's review the response of various radiation detection technologies in gaseous media.

Fig. 7.1 shows the classical response curve depicted in Handloser (1959) and Turner (1986) (with permission), of the ideal counter as a function of voltage. The four regions of interest for radiation detection and measurement are:

- saturated ionization chamber
- proportional counter operation
- limited proportionality
- Geiger, Geiger-Mueller or G-M operation

Let's now review the regions of application for health physics instruments.

7.2.1 Ionization chambers

The basic ionization chamber, free air, sealed or flow-through is an enclosed volume, with a central ion current collecting electrode. A potential or voltage is applied between the outer shell electrode and the central electrode in order to establish an electric field, which attracts or drives the charged electrons and ions inside the chamber to their respective electrodes. Charges inside the volume of the ionization chamber are generated either from the radioactive decay of the gas inside, or from ionization of the contained gas or air from external sources such as alpha, beta and/or gamma radiation. Price (1964) and Knoll (1989) provide a great deal of detail in explaining the operation of ion chambers.

A free air ionization chamber is shown schematically in Fig. 7.2 (Handloser 1959, with permission).

Electrons can also be generated directly at the surface elements of the ionization chamber, chiefly the wall material of the chamber by the classical phenomena of the photoelectric effect, Compton scattering and pair production.

The effective measurement volume of an ionization chamber is not always simply defined by its physical dimensions.

At first glance, if an enclosed known volume contains a gas with a particular specific concentration, then the ion current should be simply the product of the volume and the activity of the gas. This only holds true for infinitely large chambers at low specific concentrations and a more than minimum electric field throughout the entire volume.

Several factors must be considered.

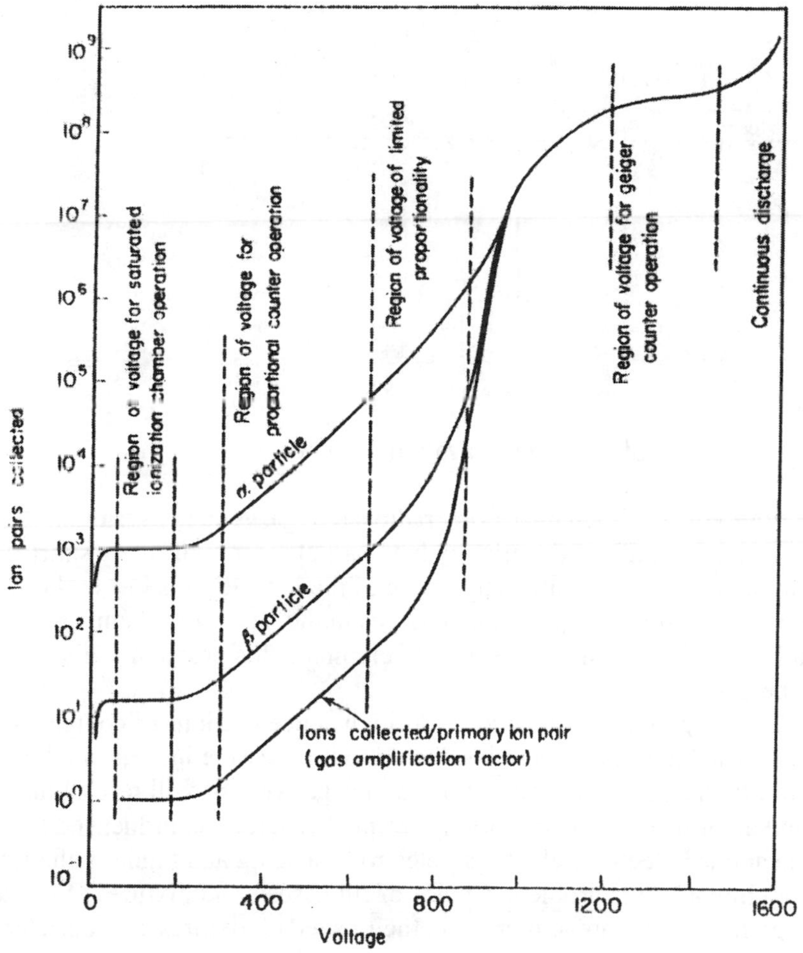

Fig. 7.1. Characteristics of an idealized counter as a function of voltage.

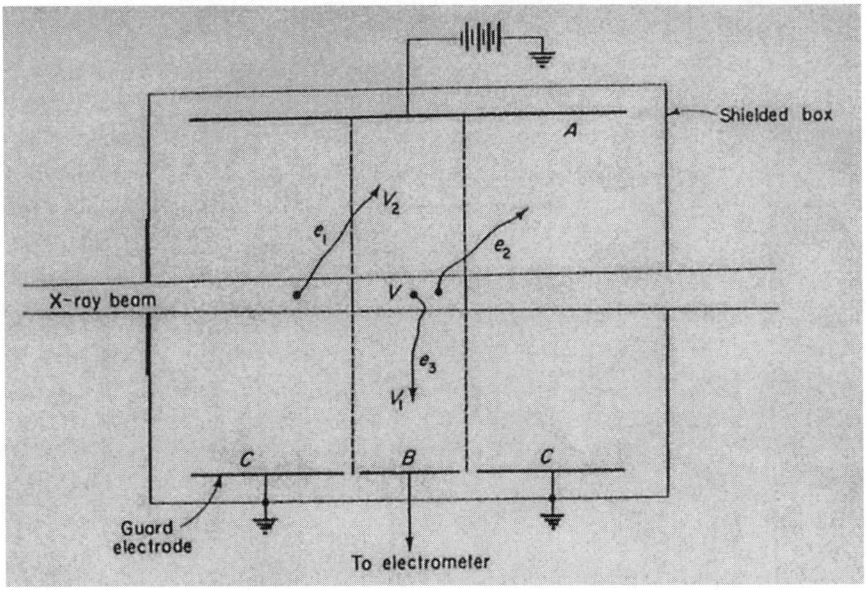

Fig 7.2. Schematic diagram of a free air ionization chamber.

7.2.1.1 Wall effect. When measuring radioactive gases such as tritium, ^{14}C or noble gases, the mean free path of the primary beta particle is not infinitely short. Thus, for small ionization chambers a significant portion of the ionization is lost to the walls of the ionization chamber instead of generating charges in the gas in the chamber.

Phantom wall ionization chambers are used to eliminate this problem. This type of instrument will be described later.

7.2.1.2 Linearity and recombination. At high concentrations of contained radioactive gases, or in high radiation fields, there is a progressive increase in the statistical probability that the generated positive and negative ions will recombine to form neutral atoms and are lost from the ionization current. This effect is influenced by the magnitude of the applied electric field. Small electrode spacing and higher voltages can help to reduce or eliminate this problem. Choice of fill gases or admixtures of fill gases can also help. Again, some manufacturers hold their exact admixtures as proprietary information.

Angular shaped ion chamber with sharp corners can lead to field stagnation, which promotes non-linearity.

Wall effect and recombination are illustrated graphically in Fig. 7.3 and Fig. 7.4 depicts ion chamber designs with square and rounded tops. Figs. 7.3, 7.4, 7.5, 7.6, 7.10 and 7.11 are all courtesy of Overhoff Technology Corporation, manufacturer of modern tritium monitoring instruments.

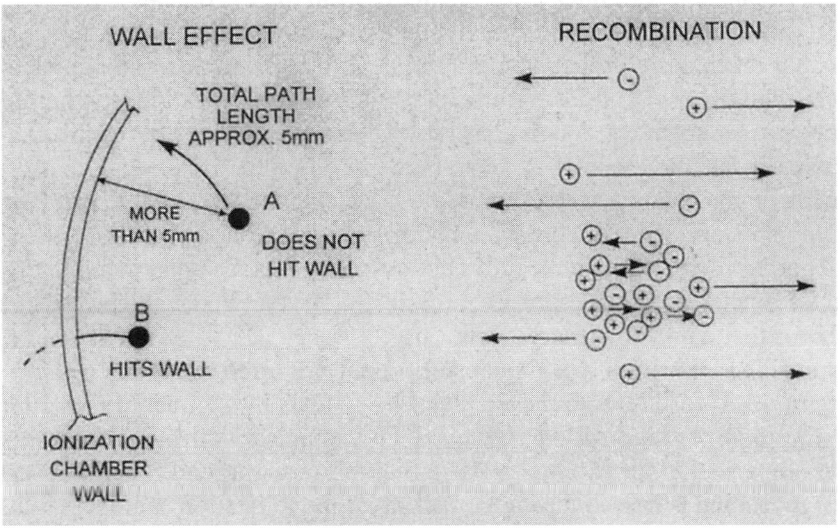

Fig. 7.3. Wall effect and recombination.

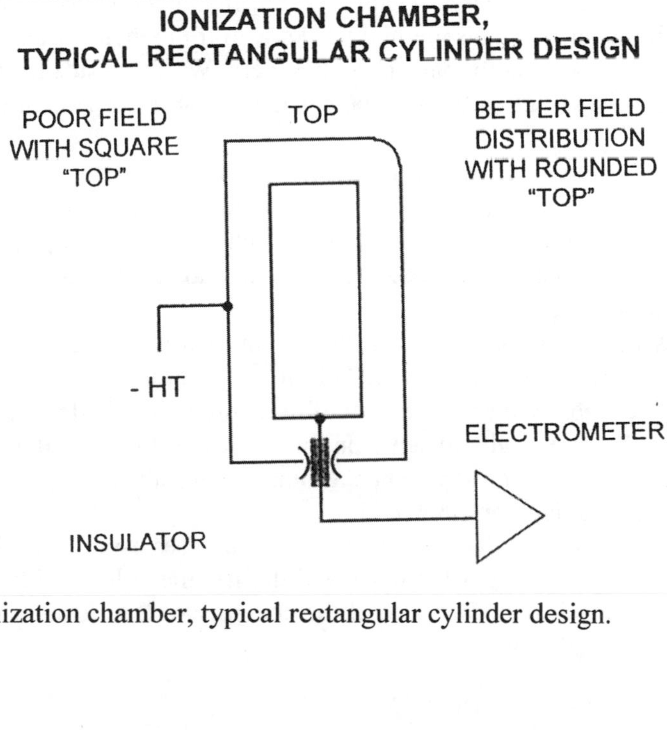

Fig. 7.4. - Ionization chamber, typical rectangular cylinder design.

7.2.1.3 Other types of ionization chambers. For the ionization chambers used for Geiger-Mueller counters, wall interactions actually form the major source of ionization within the chamber. Here, the external beta or gamma radiation can interact with the fill gas, but much more probably will interact with the wall material, which is much more dense than the gas. Thus, a significant portion of the ionization is generated in the wall of the counter chamber, and the gas molecules act as a medium for the Geiger-Mueller avalanche process.

Some neutron ion chamber detectors also use special coatings on the inside of the walls of the chamber. These are termed fission chambers and are coated with fissile materials that generate fission fragments, more neutrons, gamma rays and much energy when irradiated with neutrons. The response of these fission chambers is highest when coated with ^{235}U for thermal or low energy neutrons.

Some gas-filled ionization chambers use Boron Trifluoride (BF_3) or ^{3}Helium for neutron measurements. Please refer to the excellent basic texts by Cember (1976, 1996.)

7.2.1.4 Compensated ionization chambers Two identical ionization chambers can be used for compensation for external gamma radiation background. The radioactive gas sample to be measured is passed through only one of the ionization chambers, while the second is used for background compensation. The two ionization chambers are oppositely polarized and attached to a common electrometer. The effect of the external gamma background is cancelled, leaving a response only to the gaseous radioactivity.

7.2.1.5 Dual chamber systems. Using two ion chambers for background compensation has shortcomings. This method works well if the incident external gamma radiation interacts equally with both ionization chambers. With two side by side ion chambers this requires the radiation beam to be perpendicular to the axis of the two chambers.

If the radiation beam or field is in line with the two chambers, then the foremost chamber will shield the second chamber and compensation will be impaired. To alleviate this potential problem, quadruple ionization chambers have been developed. These work very well in almost every situation except where the external gamma radiation is close, such as in the form of a point source.

7.2.1.6 Plate out, Phantom wall chambers. Another wall effect to consider is termed plate out, where with the exception of Teflon™, most materials from which ionization chamber are made will absorb a film of moisture on the surface. For tritium monitors, tritium is still water, and will settle on the surface of the wall of an ionization chamber. This means emission of a steady background even if the ionization chamber is otherwise free from radioactive constituents.

Phantom wall ionization chambers consist of structures using fine wires as a "wall." Here, the wall is now an invisible fence and defines the volume of the ionization chamber. But, now the physical surface area of the chamber wall is now very small and any absorbed radioactive contaminant no longer plays a significant role. Fig. 7.5 shows the design of a phantom wall ionization chamber.

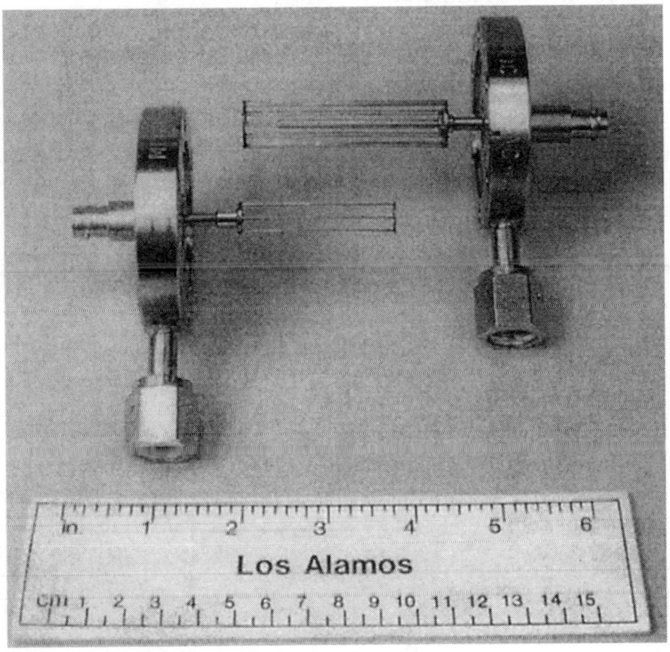

Fig. 7.5 - Phantom wall ionization chamber.

7.2.1.7 Calibration standard chamber. In phantom wall wire cage ionization chambers, ionizing primary particles can pass freely in any direction past the wire walls of the ionization chamber. There is no wall effect and this design now constitutes a true mathematically defined ionization chamber. Such a chamber no longer needs "calibration", and on the contrary, it is itself used to "calibrate" radioactive gases for their true activity.

7.2.1.8 Kanne chamber. The famous Kanne tri-axial ionization chamber developed at the Savannah River Laboratory, circa 1955, for measuring radioactive gases is shown graphically in Fig. 7.6. An early application of the Kanne chamber is shown in Figs. 7.7, 7.8 and 7.9 (Westinghouse 1962). In this case, the large volume ionization chamber was used at the Westinghouse Testing Reactor (WTR), near Waltz Mills, PA, circa 1960. Some specifications for that system were:

- flow rate- 280 Liters/minute (10 cfm)
- ionization chamber volume- 130 Liters

- efficiency- ~26% for fission gases and ^{41}Argon
- sensitivity- 2.48 x 10^{-6} Ampere μCurie^{-1} cm^{-3}
- range- 10^{-11} to 10^{-5} Amperes
- -Keithley micromicroammeter

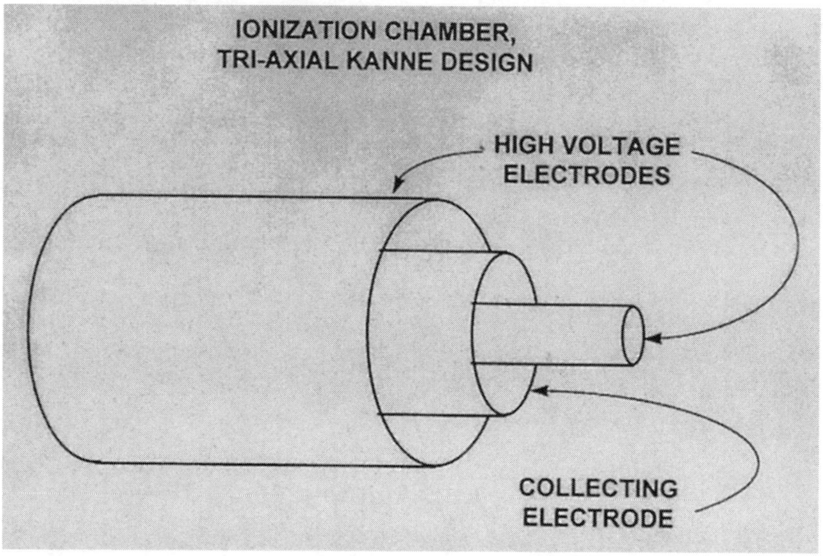

Fig. 7.6 - Tri-axial Kanne chamber.

7.2.2 Proportional Counter Chambers

Proportional counters and Geiger-Mueller counters are quite similar in nature, but the counter geometries are vastly different. In a proportional counter, the electron-collecting electrode is formed from a very thin wire in order that almost all of the electric field is concentrated in the immediate vicinity of the anode wire. The proportional counter works on the principle that a carefully controlled multiplication takes place right at the anode. This relationship is shown graphically in Fig. 7.10. Cember (1976 and 1996) has excellent basic treatments of proportional counters.

The voltage required to obtain gas multiplication, through particle collisions, is now related to the size of the wire used. That is, the thicker the wire, the higher the required applied voltage. In reality, one may visualize the proportional counter as consisting of two regions, one in which the ions are formed from nuclear interaction, then these ions drift toward the center where they are then drawn into the final region for massive amplification.

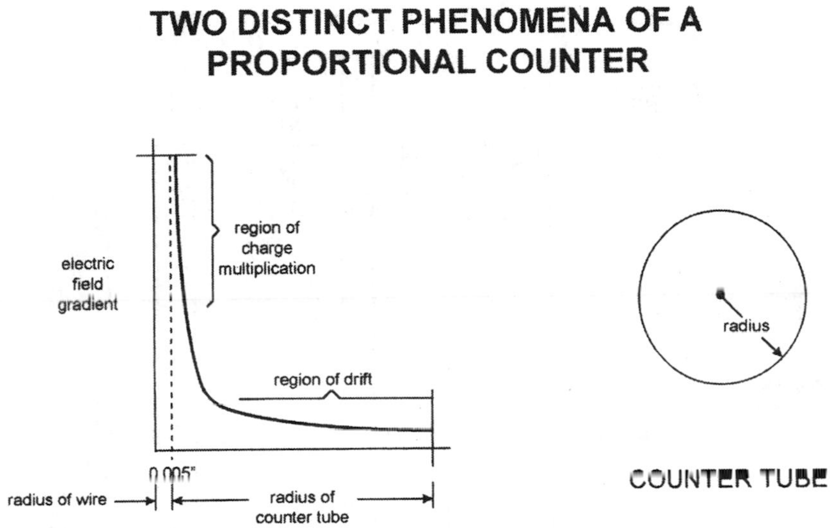

Fig. 7.10 - Distinct phenomena of a proportional counter

The filling gas must be kept free from electronegative gases such as oxygen. Otherwise the desired gas multiplication will suffer. Some people try to operate proportional counters with admixtures of air (samples) and counter gas, typically in a 4 or 5 to one admixture. The results of these are usually poor, as such counters have poor efficiency and are somewhat unstable.

7.2.3 Geiger-Mueller Counters

These detectors are usually built with a fairly massive ionization collection electrode in order to establish an electric field with a gentle gradient so that avalanche ionization will occur throughout the volume of the chamber. These devices are filled with gases which exclude electronegative components such as oxygen which tend to quench ionization because they promote recombination and thereby reduce the free ion population. Trace gases such as halogens are sometimes added to enhance ionization at low radiation levels.

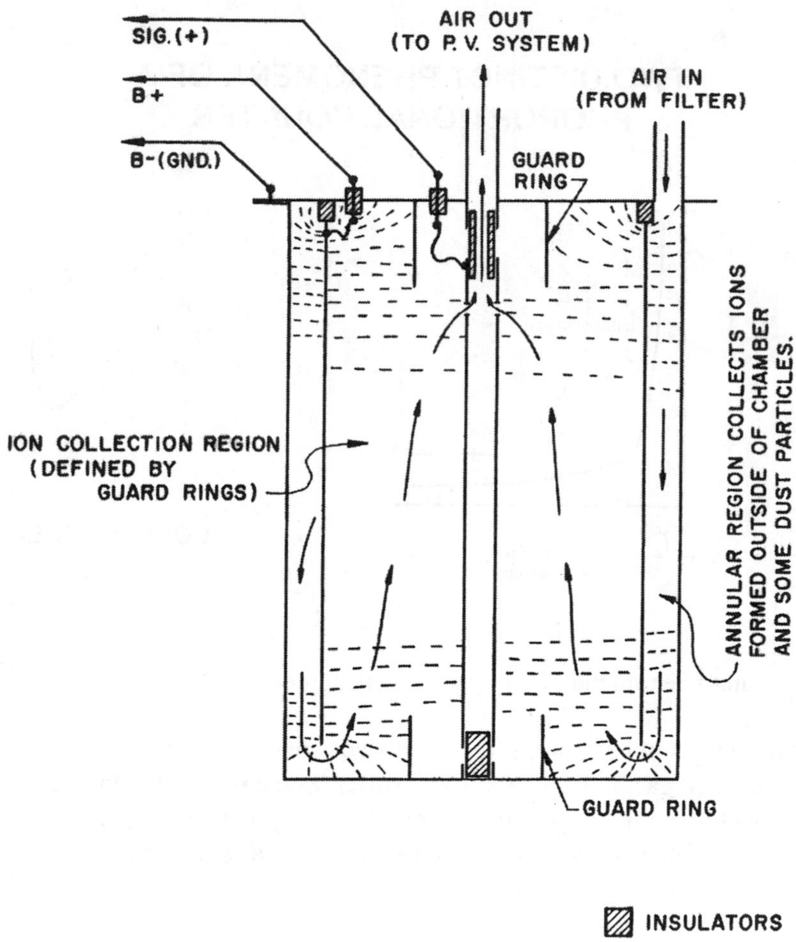

Fig. 7.7 - Simplified Kanne chamber.

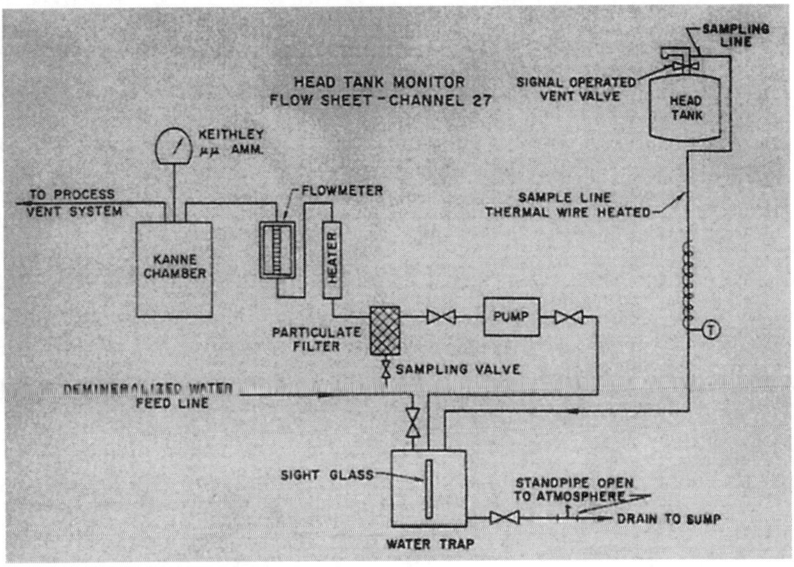

Fig. 7.8 - Head tank monitor.

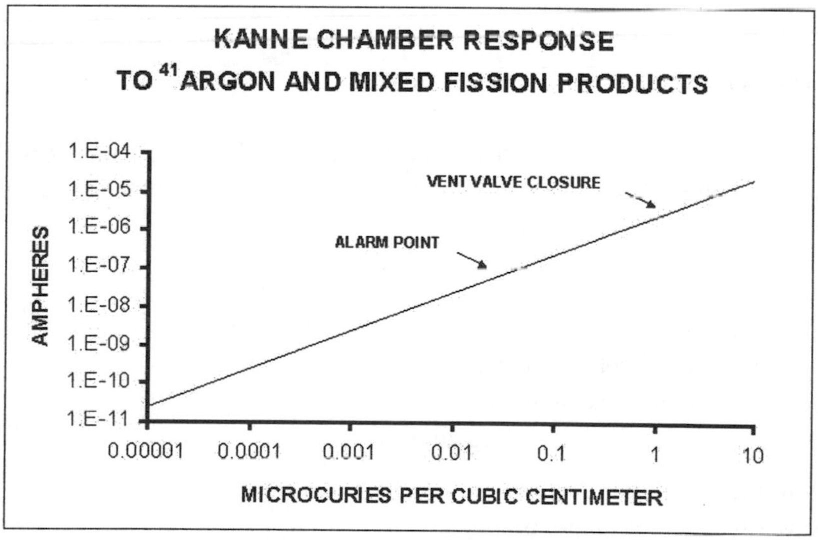

Fig. 7.9 - Kanne chamber response.

7.2.4 Scintillation Counters

The classical scintillation counter consists of a solid, liquid or gaseous scintillator coupled to a photomultiplier (PM) tube. The scintillator is a passive device, with some low efficiency, and the high gain or amplification of the PM tube is required to make up for the low scintillation efficiency . Scintillation materials include the inorganic phosphors such as NaI (Tl), CsI(Tl, CsI(Na), etc. Harshaw (1984) provides and illustrates the principles of scintillation counting in great detail, including 1) the characteristics of the various scintillation phosphors, 2) photomultiplier tubes and photodiodes used in scintillation counting, 3) temperature effects, 4) basic interactions, etc. This text of about 60 pages describes the scintillation technology including some electronics in extreme detail. Knoll (1989) also provides an excellent and detailed treatment of scintillation detection technology.

There are also some organic scintillators such as anthracene and stilbene used in health physics instruments. These will be generically discussed later.

7.3 ELECTRONICS

Let's review some of the basic electronics used by the various radiation detection technologies.

7.3.1 Ionization chambers

The heart of the electronic systems associated with ionization chambers is the electrometer. As with any electronic device there are several requirements for acceptable performance. These include:

- zero stability with fast warm up and long term drift
- linearity
- noise level
- input bias and offset requirements
- speed of response and slew rate
- ability to withstand abuse (ruggedness)

Secondary requirements can involve

- power supply requirements
- output drive capability
- packaging, size, availability, special handling requirements

Electrometer design is greatly influenced by the particular ionization chamber with which it is used. We shall consider electrometers for use in the measurement of tritium or other radioactive gases where ultra-low detection is needed. Such electrometers must be ultra-sensitive, drift free, stable with temperature and relatively noise free.

A number of semiconductor manufacturers fabricate operational amplifiers which are designed for electrometer applications. A typical specification sheet gives the following values:

Burr-Brown OPA 128LM

- Input bias current, typical ≅75 x 10^{-15} Amps, increasing by a factor of 2/11°C
- Offset current, typical 30 x 10^{-15} Amps
- Drift, 5μV per °C

These amplifiers are used in the conventional voltage feedback arrangement, also termed a trans-impedance amplifier since it converts current into voltage. Fig. 7.11 shows some properties of a modern, advanced electrometer.

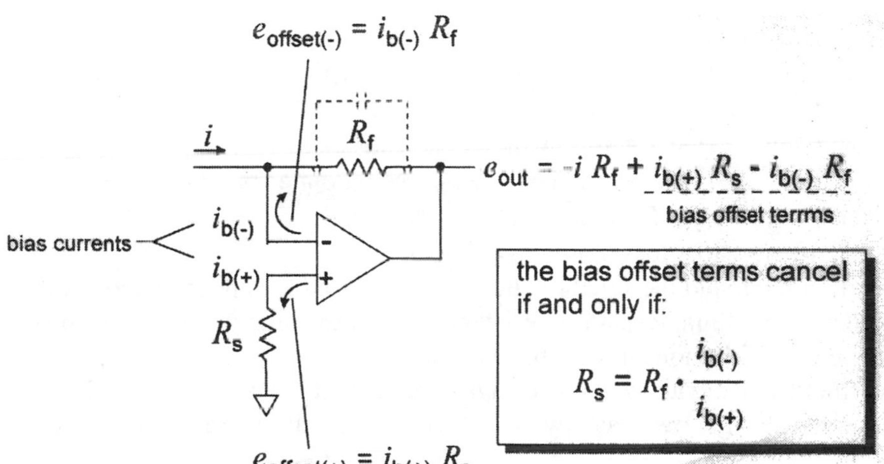

Fig. 7.11 - Electrometer properties.

If we concern ourselves with measurements of tritium at low levels, such as in the workplace, we can examine the electronics requirements for a modest measurement of 3.7 x 10^4 Bq per m^3(1 μCi m^{-3}).

Let's assume that an ionization chamber with a volume of one liter containing air with 3.7 x 10^4 Bq m^3(1 μCi m^{-3}) of tritium will generate an ion current of 1 x 10^{-15} Amperes. To obtain a suitable output voltage from the amplifier, the feedback resistor R_l

must be large. A resistor with the value of 10^{12} Ohms will then give an electrometer response of 1 milliVolt per 3.7 x 10^4 Bq (1 μCurie $^{-3}$) of tritium for a one liter chamber.

However, the bias and offset currents cannot be neglected. The bias currents are the internally generated currents appearing at each of the two inputs of the amplifier. These currents are also measured by the feedback resistor and produce a voltage response.

We note that a bias current of 75 femtoamps will show up as a zero error of 2.78 x 10^6 Bq (75 μCi m^{-3})!

This is an untenable situation. To alleviate this situation, we can balance out the bias currents by introducing a resistor R_s into the positive input of the amplifier. If the bias currents are the same, and the resistors R_f and R_s have the same values, then the effects of the bias current would exactly cancel. But we would need a precision of 1% or better, with temperature stability. This is tough nut to crack!

The manufacturer quotes an input offset current, implying that there is a difference between the two input currents. Now design is necessary to compensate for the situation.

There are several possibilities:

- the quoted value of 75 fA is actually too conservative, and is only about 10 fA at room temperature
- hand-select devices with low offset currents
- live with this situation and balance out the output error
- find another manufacturer with better devices
- trim the resistors for optimum voltage balance

To further confound the situation, the insulators used to support the ion collection electrode have not only finite insulation resistance, but occasionally possess piezoelectric properties, generating false ion currents themselves.

Nevertheless, using all available design strategies, it is possible to build electrometers with offset currents as low as 10^{-16} Amperes that are free from temperature drifts.

Some earlier efforts in electrometer design used insulated gate MOSFETs, but these devices not only exhibit poor warm up drifts, but also have significant bias (leakage) currents.

Most instrument manufacturers are reluctant to specify electrometer design in detail. The best estimate is that matched junction FETs or JFETs provide the lowest noise and lowest offset input currents in the current generation of electrometers.

7.3.2 Proportional counters

Here is a review of amplifiers and signal processing for proportional counters, primarily for measurement of tritium and other gases requiring high detectability.

For linear ionization chambers, the previously discussed amplifier is used to measure essentially "d.c." current. It speed of response is typically such that it exhibits a rise time of only the order of milliseconds. Proportional counters produce extremely fast, small pulses and the same type of feedback amplifier is used here. There is a slight but significant difference, and in the application, the amplifier is termed a charge amplifier.

The instantaneous output voltage upon detecting a pulse is

$$V=Q/C, \text{ where } Q=\int i\,dt.$$

The feedback resistor imparts an exponential decay to the initial voltage. In essence, the charge amplifier produces an output voltage in proportion to the charge from the proportional counter, hence the name.

Here the requirement is very high speed, coupled with reasonable amplifier characteristics. A good choice for an amplifier for this application is the Model CLC440 with the following salient characteristics:

- Rise time- 2 nanoseconds
- Unity gain bandwidth- 750 MHz
- Input bias current – 10 μA
- Input offset current- 0.5 μA
- Output current- ±90 mA

The amplifier output feeds processing electronics via a coaxial line.
For simple applications, the processing can consist of just a window comparator. This is a device which generates a digital pulse in response to a pulse whose amplitude is higher than a lower threshold value, but less than a maximum value. Hence the origin of the name window comparator. One can also construct comparators with stacked windows. A two or three window comparator can be built using similar architecture.

One can, for example, use a proportional counter to simultaneously measure alpha and beta particles simply on the basis of the difference in generated charge by using a few windows. By way of example, the decay of tritium produces a beta particle of 18 keV maximum and 5.6 keV average. In contrast, an alpha particle has energies between 4 and 8 MeV. This is a ratio of 1,000 to 1, so one can sort out radon progeny when measuring tritium. A similar technique can also be used for higher energy beta particles from noble gases, ^{14}C, ^{99}Tc, or such in combination with alpha particles.

For high sensitivity measurements, such as 100 Bq (2.7 nCi) tritium, it is necessary to be able to distinguish between the tritium pulses and background beta particles generated by Compton scattering. A combination of pulse height and pulse duration analysis is used since the tritium beta tracks are very short, of the order of 10 mm or less, whereas tracks from gamma rays can be the length of the counter and take longer to appear and show at the counter anode. This phenomenon still takes of the order of 100 nanoseconds or less, and requires very fast signal processing electronics.

7.3.3 Geiger-Mueller (G-M) Counters

One can use a charge amplifier for measuring the pulse from a G-M tube. However, the output pulse is so large that it can be transferred directly to a CMOS digital counter integrated circuit. In other cases a junction FET is used to amplify the pulse.

As previously mentioned, the G-M counter is an extremely insensitive device. It is therefore suitable for use in measurements of high radiation fields.

The G-M tube has a characteristic slow response rate. When the gas in a G-M tube is ionized by incoming radiation, a cloud of positive ions and electrons is formed. The electrons are collected quickly and result in a pulse from the device, but the ions take some time to drift to the cathode to be neutralized. This is known as dead time, during which time the G-M tube is incapable of more response. Dead time is well characterized in Knoll (1989).

In order to extend the dynamic range of G-M counters, one may make use of an ingenious method involving pulsing the high voltage to the detector tube for very short periods of time. During the short time that voltage is applied one observes the statistical probability that a radiation event occurs. This technique is termed "time to count" and is used in high range G-M counters.

7.3.4 Scintillation counters

The scintillation counter consists of a solid, liquid or gaseous scintillator coupled to a photomultiplier or PM tube. The scintillator is a passive device with relatively low efficiency, and the gain of the photomultiplier tube is needed to make up for the low efficiency.

The photomultiplier tube is truly a unique device. Not only does it possess extremely high gain, up to 10^7 and as much as 10^8, but it also has a fast response, in the nanosecond range, and it has little noise in the conventional Johnson or thermal noise sense. In fact, the PM tube does exhibit some thermal noise in the form of tiny random pulses known as "dark current." Dark current can be suppressed in magnitude and reduced in root mean square (rms) variation by the use of Peltier cooling. Photomultiplier tubes are well described in Handloser (1959), Price (1964), Cember (1976 and 1996), Harshaw (1984) and Knoll (1989).

In scintillation counters, incoming radiation produces photons or light flashes in the scintillation material, which are converted into charges or pulses by the PM tube. Again, a classical feedback charge amplifier is used.

There are also some interesting variations in this electronics as the pulse response of scintillation materials can vary widely. For example, plastic scintillators used for beta, gamma and neutron detection are extremely fast, on the order of 1 nanosecond or less. On the other hand, inorganic scintillators, such a sodium iodide (NaI(Tl)), are slow with decay time constants in the μ second region.

A technique can be used to employ one PM tube with time resolving electronics or pulse shape detection electronics to simultaneously measure two distinct types of ionizing radiation. This type of scintillation detector is called a "phoswich" scintillation counter where, for examples, one can use a combination of a thin sheet of plastic scintillator optically coupled to a NaI scintillator for beta and gamma detection, or a sheet of Zinc Sulfide (ZnS) coupled to a plastic scintillator for alpha and gamma detection.

Again, the amplifier used is the classical voltage feedback amplifier as a "charge" amplifier. The voltage output is inversely proportional to the size of the feedback capacitor, and there is value in making it small.

7.3.5 Pulse detectors

Solid state crystal detectors, including Zinc Sulfide (ZnS), are usually totally passive, without intrinsic gain. There are some new developments, involving avalanche photo-detectors, but there are exceptions. Thus, in general, their uses are limited to the detection of heavy ionizing particles which can generate substantial ionization pulses for each nuclear event.

Still, these signals are small and weak, requiring the utmost limits in low-level amplifier design. Here the simple charge amplifier circuit is not sufficient, and a finite band amplifier is required. Here there are both upper and lower frequency limits, in order to reduce background thermal noise. Peltier coolers can be used to reduce noise, but this is cumbersome, especially when using small multi-stage Peltier coolers.

The low-level bandwidth limited amplifiers are used as preamplifiers, and are usually located close to the detector itself to reduce noise from flexing coaxial cables. The preamplifiers are then followed by post-amplifiers and an assortment of window comparators or multi-channel analyzers.

Typical applications include aerosol detectors for transuranic measurements needed in the USDOE weapons laboratories, Uranium mining and milling, etc. They are also used for other alpha spectroscopy and for photonics applications.

7.3.6 Digital electronics

Digital electronics is Digital Signal Processing (DSP), and affects nearly every form of communications and instrumentation in worldwide use today. From cellular phones and modems to cardiac monitors and compact disk (CD) players, DSP is having an impact that those of us raised on 78's, 45's and "long playing" records never imagined. Over the past ten years, significant progress has been made in applying DSP to radiation energy spectroscopy, providing new benefits and higher performance in some systems. At the same time, DSP has had little impact on the health physics instrument business. Digital technology has long been employed in interfaces such as liquid crystal display (LCD) readouts and keypads, while microprocessors have replaced and/or augmented analog circuitry. The real benefits of DSP remain to be applied, and for that reason we use a crystal ball to peer into the future. Here we also use a generic crystal ball plus a combined several hundred years of experience by the authors and other listed contributors.

7.3.6.1 ***Challenges*** There are specific aspects of health physics (HP) instrumentation that could see substantial changes in the near term. The main question is whether these changes are technically or economically feasible, but rather when the marketplace will be prepared to accept them. Among the challenges the instrument manufacturers face are:

1) "We have always done it this way" is so true and has been for decades. Change is slow to come into this business as evidenced by the slow rate of acceptance in the spectroscopy market. Basic HP instrumentation is still quite similar to what it was forty years ago-- and is quite often adequate to meet the current survey requirements. One can see the consistency in comparing radiation detection technology in Handloser (1959) to Knoll (1989), a thirty year span.

2) The selling prices of most HP instruments are quite low. So the buying influence is not inclined toward higher priced instruments, even with superior capabilities.

3) The current user interfaces are generally simple and effective. Digital interfaces must well designed to compete, particularly with respect to cost.

What factors might make DSP technology more attractive, even at higher costs. New and evolving regulatory requirements, decreasing training budgets and turnover among HP technicians are among the major factors.

7.3.6.2 Potential advances and improvements. What can DSP do to warrant the investment on the part of manufacturers and the user community? The advantages of DSP begin at the interface to the radiation detector and extend all the way to the presentation of results to the user. Analyzing the basic advantages, the range of improvements will affect at least these aspects:

1) Detector interfacing- Although DSP can do impressive things with traditional interface waveforms, as other have pointed out, much valuable information is lost when a detector pulse is integrated as in pulse height analysis. The details of the pulse shape are lost, as is the potential to process the pulse very rapidly in real time. Direct signal sampling of the detector pulses allows the pulse shape characteristics to become part of the surveying process. Why is that important? Different radiation interactions with detectors often produce different pulse shapes. It is possible to separately count alpha, beta, gamma and neutron radiation simultaneously. One can also obtain energy information such as with beta radiation surveys containing multiple maximum beta energies. Pulse shape details and digital pulse processing by DSP hardware and software permit instruments to provide much more information without requiring more from the user.

2) Diagnostics- Although digital circuitry has been in use for some time, HP instruments have yet to fully benefit from the power of DSP processors. Compared to conventional microprocessors, DSP processors such as the Analog Devices Model 218L, offer 100 times the processing power and interrupt responses in the tens of nanoseconds rather than microseconds. The capability to shut down the detector input to the instrument and run diagnostics and circuit calibration checks in a millisecond is here now, establishing a new level of confidence in the operation of the instrument at any time while it is in use. Internally generated calibration pulses coupled with statistical analysis of the current performance of the instrument, allow QA/QC requirements to be met and easily exceeded. Most importantly, the user has greater confidence in his instrument while he's using it.

3) Count rate range and linearity- Because of the digital pulse interface, pulse processing can be performed in real time so that count rates from count per minute to millions of counts per minute can readily be handled. With on-board correction, the user gets a straightforward output in real time, already corrected.

4) Dose information- Careful selection of detectors can allow limited energy determination of the incoming pulses. If the detector is capable of responding with some degree of energy information, the DSP-based instrument can dynamically better estimate what the real dose to the worker is. Not only is it important for survey results which affect radiation work permits, but it can be a true, live "stay time" readout for the technician while he is performing his survey. By analyzing the incoming pulse energies in real time, the instrument can dynamically alert the technician at appropriate intervals, even multiple times per second, ensuring awareness of current radiological conditions.

5) User interface and ease of use- Perhaps some of the greatest challenges of DSP are in this realm. The range of possibilities is immense, while the question of what is most useful must be answered. Audible feedback to the user is traditional, with clicks or beeps. Visual output has been restricted to meters, analog or digital. More creative and advanced video and audio methods will surely evolve, perhaps with speech-like audio forms. The extensive use of DSP's in cell phones, pagers, modems, etc., includes everything from voice recognition to speech synthesis. The question to be answered is what helps the user most with least effort on his part, with safety and productivity in mind. Acceptance by the trained user is extremely important.

7.4 CONCLUSION

Digital signal processing holds tremendous promise for making more and more accurate measurements, and reducing the number of surveys required for measurements. By allowing pulse shape capabilities to determine detector design and selection, instruments will be able to provide much more and better information with fewer surveys and much less data manipulation by the technician. The user can take advantage of a selection of detectors via user interfaces. The real advantages of existing instruments should not be lost as the advantages of DSP are incorporated.

7.5 ACKNOWLEDGEMENTS

The authors are very grateful to the following friends and colleagues who contributed to this effort: Louis Costrell of NIST, John S. Handloser, Jr, of Health Physics Instruments, Ken O'Dell of Westinghouse-Savannah River Site, Mario Overhoff of Overhoff Technology Inc, John Umbarger of Flour-Daniel-Hanford and Mike Wolf of Cell Robotics.

7.6 REFERENCES

Cember, H; Introduction to health physics. Second edition. Oxford: Pergamon Press, 1976.

Cember, H; Introduction to health physics. Third edition. New York: McGraw-Hill, 1996.

Handloser, J.S.; Health physics instrumentation. New York: Pergamon Press, 1959.

Harshaw Chemical Company. Harshaw Radiation Detectors, Cleveland, Ohio, 1984.

Knoll, G.F.; Radiation detection and measurement. Second edition. New York: John Wiley, 1989.

Price, W.J.; Nuclear radiation detection. Second edition New York: McGraw-Hill, 1964.

Turner, J.E.; Atoms, radiation and radiation protection. First Printing. New York: Pergamon Press, 1986.

Westinghouse Electric Corporation. Health physics instrumentation and protective equipment used at the Westinghouse Testing Reactor, WTR-64, 1962.

Chapter 8

AIR SAMPLING INSTRUMENTS

Douglas Draper and Mark Hoover

8.1 INTRODUCTION

Air sampling comprises an important aspect of protection: of personnel, of the environment, and of facilities and equipment. Air sampling is conducted to meet regulatory requirements, assess the need for a monitoring program, measure known releases, investigate personnel exposures, monitor plant processes and effectiveness of engineering controls, and to assess non-routine releases. An effective air sampling program is based on an Air Sampling Quality Assurance Program, a site-specific plan that forms the technical basis that assures quality objectives are met.

8.2 AIR SAMPLING IN RADIOLOGICAL PROTECTION-WHAT IT IS AND ISN'T

An effective radiological control program includes contamination control, source control, dose optimization, and monitoring to assure that program requirements are met. Monitoring includes measurements of various media, including biological samples (bioassay), surfaces such as floors and walls, materials such as soils and liquids for volume contamination, and the air for radioactive contamination. In general, these media are not monitored in their entirety, rather, samples of each are collected for subsequent analysis. Air sampling is the collection of a portion of the total air volume over a period of time, which then is analyzed by various methods. The results of the analyses are then compared to criteria to meet project goals. Air monitoring is a part of a comprehensive radiation safety program. Air sampling provides the samples that are subsequently analyzed in the air monitoring program. There are a number of methods used to provide an adequate air sample that can be used to meet the safety and project engineering goals. Air sampling is not air monitoring; rather, air sampling is a component of an air monitoring program.

8.2.1 Radiation and Radioactive Decay

Aerosols of radioactive materials have all the physical and chemical forms of nonradioactive aerosols. They can range from ultrafine metal fumes to large liquid droplets. Their physical form has the usual influence on their aerodynamic behavior and on the choice of measurement technique. Their chemical form has a similarly important influence on their biological or environmental behavior and on the technique that will be used to confirm their chemical form after collection. It is their radioactive properties that can make them both easier to detect, yet, in many cases, may require special techniques for collection and handling.

8.2.2 Types of Radiation

Three types of radiation are of concern in studies of airborne radioactive materials: alpha, beta, and gamma. Neutrons and positrons are also of concern in some special circumstances. Details of the origin and characteristics of all types of radiation as well as the radioactive decay schemes for radionuclides are presented in references such as the Table of Isotopes (Lederer 1978) and the Health Physics and Radiological Health Handbook (Schleien 1998).

8.2.3 Half Life for Radioactive Decay

The rate at which a radioactive material decays is described by its half-life: the time for half of the atoms present to undergo a spontaneous nuclear transformation. Half-life is an important consideration in determining how to collect, handle, and quantify samples of a radioactive material. For example, samples of short-lived materials may need to be analyzed immediately after collection, or even in real time. Conversely, samples of long-lived radionuclides may remain radioactive for years, centuries, or even millennia. Experimental determination of half-life is also one way of helping to identify a given radionuclide.

The equation describing radioactive decay is:

$$A(t) = A_0 \, e^{-0.693t/t_{1/2}} \tag{8.1}$$

where:

$t_{1/2}$ is the decay half life for the radionuclide (in time units),

t is the elapsed time (in the same time units as $t_{1/2}$),

e is the base of the naperian logarithm system (2.718),

0.693 is the natural logarithm of 2,
A_0 is the activity of the sample at time t=0 (in disintegrations per second), and
A(t) is the activity of the sample after an elapsed time of t.

Another way of writing the relationship involves the decay constant, λ, where:

$$\lambda = 0.693/t_{1/2} \text{ , and } A(t) = A_0e^{-\lambda t} \quad (8.2)$$

In terms of the number of elapsed half-lives, n, the amount of activity remaining can be determined as:

$$A = A_0(½)^n \quad (8.3)$$

Thus, after, two half lives, only ¼ (25%) of the original activity is present, and after seven half lives only 1/128 (0.8%) of the original activity remains. Radioisotopes with long half lives continue to accumulate in activity throughout the sample collection process. Radioisotopes with short half lives can reach an equilibrium activity during sample collection. If the aerosol sampling rate is Q (l/min) and the airborne concentration of the radionuclides is C (Bq/M^3) then the equation for the buildup of radioactivity on the collection filer is:

$$A(t) = (QC(\lambda) (L-e^{-\lambda t}) \quad (8.4)$$

If the sampling period is long compared to the half life of the radionuclide, then equilibrium will be established and A(t) = QC/λ when sampling is stopped, the radioactivity in the sample decays according to the relationship $A(t) = A_0e^{-\lambda t}$

8.2.4 Specific Activity

The specific activity of a radioactive material is the activity, or amount of decay per unit mass. Materials with a short half-life have a high specific activity. The traditional unit for activity, the curie (Ci) was defined in terms of the specific activity of radium (1 Ci/g) 3.7×10^{10} disintegrations per second (2.22×10^{10} disintegrations per minute). The international system (SI) unit for activity is the Becquerel, which is 1 disintegration per second. Thus, 1 nanocurie is 37 Bq. Specific activity is of biological concern for radioactive materials in the body because it determines the rate at which energy will be deposited and damage will occur in tissues per unit of mass incorporated into the body. Specific activity is of practical concern for aerosol measurement because it determines the amount of radioactivity that will be present in a sample of a given mass, and the mass or number of particles that will be associated with a sample of a given activity. The dependence of airborne mass concentration, C_m (g/m^3), on activity concentration, C_a (Bq/m^3), involves the specific activity of the material, A (Bq/g), in the following way:

$$C_m = C_a / A \tag{8.5}$$

The number of particles per cubic meter, C_n (particles/m^3), required to provide a given activity air concentration, C_a, involves a similar relationship of specific activity, A, particle density, ρ (g/m^3), and the volume of the airborne particles. For a simple example in which particles are assumed to be monodisperse in size and spherical in shape (volume = $\pi d_p^3/6$, with d_p in m), this relationship is:

$$C_n = C_a / \left(A \rho \left(\pi d_p^3 / 6 \right) \right) \tag{8.6}$$

The DAC is the concentration to which a worker could be exposed for an entire work year of 2000 hours without exceeding the annual limit on intake for the radionuclide of concern. The 8-DAC-h level is equivalent to the integrated exposure over a single workday at the DAC. The general calculation to convert a derived air concentration (DAC) in Bq/m^3 to a radioactivity accumulation, A, in counts per minute on a collection substrate is:

$$A = \frac{DAC \cdot Q \cdot T}{E_R \cdot E_S} \tag{8.7}$$

Where Q is the air sampling flowrate, T is the duration of sampling, E_R is the radioactivity detection efficiency for the system, and E_S is the aerosol collection efficiency. For example, if the integration period is 1 hour, and the concentration of concern is 1 DAC, then A will be the activity associated with 1 DAC-h of integrated air concentration. See Table 8.1.

Table 8.1 Number of particles per cubic meter as a function of monodisperse particle size for selected toxic materials at their derived air concentration (DAC).

Particle Diameter (μm)	$^{238}PuO_2$	$^{239}PuO_2$	Enriched Uranium	Beryllium Metal
10	0.0001	0.02	54	2065
5	0.0008	0.15	433	16,518
3	0.004	0.7	2007	76,471
1	0.1	19	54,180	2,064,715
0.5	0.8	150	433,443	16,517,716

[a]Insoluble ^{238}Pu has a specific activity of $6.44 \cdot 10^{11}$ Bq/g and a DAC of 0.3 Bq/m^3.
Insoluble ^{239}Pu has a specific activity of $2.26 \cdot 10^9$ Bq/g and a DAC of 0.2 Bq/m^3.
For enriched uranium, the specific activity is $2.35 \cdot 10^6$ Bq/g (dominated by the contribution from ^{234}U, which is present at 1% by mass), and the DAC is 0.6 Bq/m^3.
The effective density of beryllium metal aerosol particles with a slight oxide coating is 2 g/cm^3 (Hoover *et al.* 1989) and the occupational exposure limit for beryllium is 2 μg/m^3.

8.3 AIR SAMPLING FUNDAMENTALS

8.3.1 Types Of Air Sampling

Air samples are taken by means of a "train" consisting of a sample collector, an air flow meter, and an air mover (pump) for drawing the air through the sample collector. The entire train must be calibrated for air flow in order to accurately determine the concentration of the contaminant inn the air. Various types of air sampling include personnel sampling, normally with personnel air samplers, or breathing zone (BZ) samplers, continuous area sampling, grab sampling, or special sampling such as duct or stack monitoring. Sampling methods include volumetric, filtration, adsorption, absorption, aerodynamic, and electrostatic sampling.

8.3.2 Volumetric Grab Samples, Impingers, Cold Traps, and Adsorbers

Sampling techniques such as evacuated volumes, liquid bath impingers, cold traps, activated charcoal and zeolite adsorbers, and various absorbers such as propylene glycol work equally well for capturing radioactive and nonradioactive materials. These methods are used to collect samples of vapors and gasses, but also collect particles. Filter media is normally used to collect particulate forms of airborne contaminants. Standard radioactivity counting techniques can be applied, depending on sample geometry, to any of these samples.

8.3.3 Placement of Samplers

Concentrations of airborne contaminants can vary considerably in the workplace and depend on a number of factors including work processes, airflow patterns, contamination levels, chemical and physical nature of the contaminants, and personnel activities. For example, high levels of fixed contamination normally would not create an airborne hazard from walking; however, grinding, cutting, or brazing of surfaces with fixed contamination can create localized areas of very high airborne contamination. Powders and other finely divided contaminants rapidly become airborne from mechanical stresses, whereas wetted materials generally require more energy to become airborne. For these reasons, to determine placement of samplers requires an understanding of the activities that are occurring in the workplace, the nature of the materials, and airflow patterns. Consideration of stratification and stagnation is important in determination of exact locations for samplers. Also, airflow patterns may change with elevation, so airflows at worker breathing level may not be the same as airflows at ceiling or floor height. In general, Table 8.2 shows the relationship between placement of samplers and the purpose for the sampling. NUREG-1400 presents an excellent discussion on the positioning of samplers.

Table 8.2 Sampling Purpose and Sampler Location

Purpose of sampler	Placement of sampler
Estimate worker exposure for dose estimate	In worker's breathing zone
Identify requirement for confinement control	In airflow pathway near release point
Verify engineered controls (confinement)	Downstream of confinement
Estimate from many potential release points	Downstream at exhaust point
Determine posting requirements	At worker locations and boundaries
Special cases, such as airflow studies	Case by case evaluation

8.3.4 Statistical Approach To Air Sampling

Normally distributed sizes arise when a multitude of small additive or subtractive factors influence a process or when the particles are nearly all the same size. Examples are diameters of ball bearings or latex spheres of the same nominal diameter. A lognormal distribution is a distribution in which the logarithms of a variable such as particle size, are normally distributed. Lognormally distributed sizes arise when multiplicative factors act on a variable that would otherwise have a single magnitude. Examples include the breakup of large objects into smaller objects, or certain types of agglomeration of smaller objects into larger objects. Lognormal distributions are more common than normal distributions in aerosol science and technology. Note that particle sizes and inhalation exposures can never be smaller than zero.

8.3.4.1 Probits and probabilities A probit is 1 standard deviation unit. Probit units are another name for the "normal standard deviate" or "normal standard variable". Z is frequently used to denote the normal standard variable. The relationship for the normal and lognormal distributions follow the normal and logarithmic mathematical patterns.

Normal Distribution	Lognormal Distribution
$z = \dfrac{x - \mu}{\sigma}$	$z = \dfrac{\ln x - \ln \mu}{\ln \sigma}$
$D_{0.135} = \mu - 3\sigma$ $D_{2.27} = \mu - 2\sigma$ $D_{15.9} = \mu - 1\sigma$ $D_{50} = \mu$ $D_{84.1} = \mu + 1\sigma$ $D_{97.7} = \mu + 2\sigma$ $D_{99.865} = \mu + 3\sigma$	$D_{0.135} = D_{50} / \sigma^3$ $D_{2.27} = D_{50} / \sigma^2$ $D_{15.9} = D_{50} / \sigma$ $D_{50} = D_{50}$ $D_{84.1} = D_{50} \bullet \sigma$ $D_{97.7} = D_{50} \bullet \sigma^2$ $D_{99.865} = D_{50} \bullet \sigma^3$

Specific relationships of interest for the lognormal distribution are:
$D_{50} / D_{16} = D_{84} / D_{50} = 1\ \sigma_g$
$D_{50} / D_5 = D_{95} / D_{50} = \sigma_g^{1.64}$
$D_{50} / D_{2.5} = D_{97.5} / D_{50} = \sigma_g^{1.96}$

These relationships hold because:
$\pm 1\ \sigma_g = \pm 1$ probit = range from 16% to 84% = 68% of the distribution
$\pm 1.64\ \sigma_g = \pm 1.64$ probit = range from 5% to 95% = 90% of the distribution
$\pm 1.96\ \sigma_g = \pm 1.96$ probits = range from 2.5% to 97.5% = 95% of the distribution

8.3.4.2 Radioactivity counting statistics. The net count rate for the sample is determined by the gross count rate minus the background count rate:

$$R_s = R_g - R_b = C_g / T_g - C_b / T_b \tag{8.8}$$

For example, an aerosol filter sample counted for 10 minutes results in 1000 counts. The background count of an unexposed filter was 250 counts in a 10-minute period. Assuming negligible radioactive decay of the sample during the counting period, the net count rate in counts per minute and the standard deviation are approximately:

(1000 counts/10 min - 250 counts/10 min) = 100 cpm - 25 cpm = 75 counts per minute.

The standard deviation is found from the following equation for propagation of errors for sums or differences:

$$\sigma = \sqrt{\frac{R_g}{T_g} + \frac{R_b}{T_b}} = \sqrt{\frac{100}{10} + \frac{25}{10}} = \sqrt{12.5} = 3.5 \text{ cpm} \qquad (8.9)$$

For example, an aerosol filter sample has been split into two unequal parts and analyzed. The results are 75 ± 6 Bq for one portion and 39 ± 7 Bq for the other. The total activity should be reported as the sum of the toal activities:
W = x+y = 75+39=114Bq.
The standard deviation for sums or differences is the root mean square of the individual standard deviations is calculated:

$$\begin{aligned} \sigma_w &= \sqrt{\sigma_x^2 + \sigma_y^2} \\ &= \sqrt{6^2 + 7^2} \\ &= \sqrt{85} \\ &= 9 \end{aligned}$$

Thus, the correct answer for the total activity is 114±9Bq.

Note that σ for a sum is neither the sum of the individual σ's or weighted sum of the individual σ's.

8.3.4.3 Minimum Detectable Activity The decision level (DL) and critical level (CL) are two terms are used to describe the same "false alarm" or Type I error concept for deciding if a sample is above background. The decision level, L_c is defined so that there is less than a certain probability (called an α error, or a Type I error) that count rates greater than DL will occur by random chance for a clean sample. A Type I error can be thought of as "jumping to a conclusion." Note that the type I error rate, α is usually selected to be 0.05. The standard normal variable, z or k value associated with α=0.05 is 1.64.

The DL is a fundamental performance indicator for the counting system and can be used an "alarm setpoint." Thus, an alarm setpoint of DL for a fixed counting system will give a false alarm rate of 5% when no activity is actually present. The minimum detectable activity (MDA) is the true activity for which there is only a chance of (1-β) that the observed count rate will be less than the decision level DL. The MDA is has been historically called the Detection Level or Limit (L_D), and is also sometimes referred to as the Lower Limit of Detection. MDA accounts for the "false negative" or Type II error. The value of β is usually taken to be 0.05. MDA is an indicator of what might be accomplished with the counting system. Thus, the MDA is a value large enough that there is only a 5% chance that it will NOT set off an alarm.

Traditional Definition of Minimum Detectable Activity

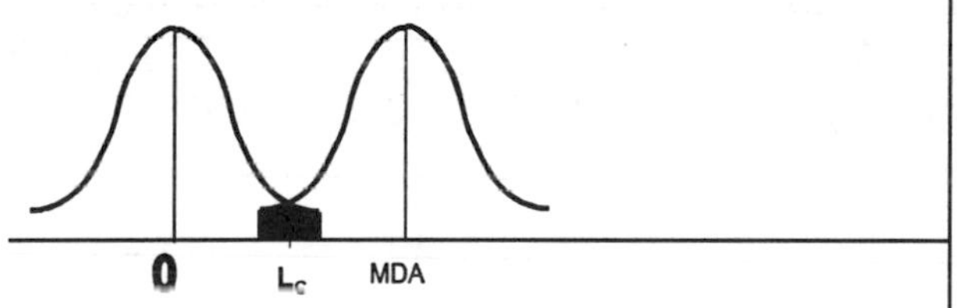

5% probability of Type I (false positive) and Type II (false negative) Errors

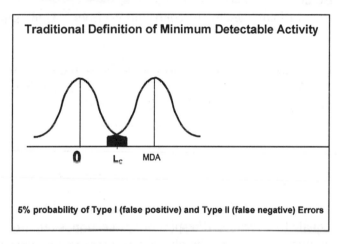

The DL is a basic performance indicator for the counting system, and MDA is an indicator of what can be done with the counting system. After these parameters are determined they can be used in combination with the characteristics of the aerosol sampling system to calculate the MDC. The relationship between the net count rate of radioactivity on an aerosol-sampling filter R_n cpm and airborne radioactivity concentration C (Bq/m^3) is:

$$C = \frac{R_n}{E_C E_f Q T} \qquad (8.10)$$

where:
E_C is the counting efficiency of the detection system (cpm/dpm),
E_f is the collection efficiency of the filter or sampler,
Q *is* the sampling flowrate (such as m^3/min), and
T is the duration of aerosol sample collection.

Thus,

$$MDC = \frac{MDA}{E_C E_f QT} \tag{8.11}$$

And, to have a low MDC, it is desirable to have a low MDA (cpm), high collection efficiency, a high sampling rate, and a long sampling time. The Derived Air Concentrations (DACs) for the insoluble class of these radionuclides are:

^{239}Pu	0.2 Bq/m^3
^{233}Pu	0.3 Bq/m^3
Enriched U	0.6 Bq/m^3 (dominated by the contributions from ^{234}U, present at 1% by mass)

The ratios of DAC/MDC for a system with an MDC of 0.065 Bq m^{-3} corresponding to an MDA of 10 cpm, a counting efficiency of 20%, a filter collection efficiency of 95%, and a sample flow of 1 cfm for 8 hours are:

3 for ^{239}Pu
4.6 for ^{238}Pu
9.2 for Enriched U

Therefore, in this case the MDC is adequate to determine if the DAC has been exceeded, but not adequate to determine if respiratory protection is required (typically at 0.1 DAC) for the plutonium aerosols, and barely adequate for the uranium aerosols.

8.4 APPLIED AIR SAMPLING

The mass median aerodynamic diameter (MMAD) is defined as the diameter for which "half of the mass of the aerosol is associated with particles having a smaller aerodynamic diameter, and half of the mass is associated with particles having a larger aerodynamic diameter." The MMAD is usually determined with a particle size "spectrometer," such as cascade impactor. If the aerosol is not homogeneous with respect to its composition, that is, if it includes several different materials, then the MMAD of the total aerosol may not be a useful indicator of radiation risk.

The activity median diameter (AMD) is the diameter for which "half of the radioactivity is associated with particles having a smaller physical diameter, and half of the radioactivity is associated with particles having a larger physical diameter." The

AMD is not normally determined because it requires simultaneous knowledge of the amount of radioactivity per particle (perhaps from autoradiography), and the physical size of the particle (perhaps from electron microscopy). It can sometimes be determined if the aerosol is homogeneous in composition and morphology, because of the relationship between aerodynamic and physical diameters.

The count median aerodynamic activity diameters is the diameter for which "half of the total number of particles in the aerosol are smaller than the given aerodynamic diameter, and half of the number of particles in the aerosol are larger than the given aerodynamic diameter." The CMAD and the CMD (count median diameter) are of interest when optical techniques are used to determine particle size distribution in the sampled air. The ability of an instrument to determine CMD or CMAD depends on its method of measuring diameter. Time-of-flight instruments that accelerate particles through a nozzle and measure their velocity by detecting the interruption of two light sequential beams can measure CMAD. Devices that sense a particle size by light scattering measure CMD. Other diameters of interest are related to surface area and volume.

For a typical workplace aerosol of homogeneous composition, the AMAD will be related to the MMAD in that the condition of homogeneity requires that AMAD = MMAD. If the activity is distributed uniformly with in the particle mass, then the diameters are equal. If the workplace aerosol is a nonhomogeneous mixture of radioactivity and other carrier materials, the AMAD is not related to the MMAD. The radioactivity could preferentially reside in any portion of the size distribution. This is especially true for aerosols that may result from (a) physical processes like welding or torch cutting that form ultrafine aerosols by vaporization-condensation as well as larger droplets by physical dispersion of molten metals; (b) situations where different processes are occurring at the same time such as rough machining and fine polishing; or (c) situations where a more volatile radionuclide may attach to the available surface of a preexisting aerosol. In the condensation case, the radioactivity would be concentrated in the small particle fraction where the surface-to-volume ratio is largest. Attachment of radon progeny occurs preferentially on smaller particles, typically on the surfaces of background aerosol particles with diameters of about 0.2 μm.

In a typical workplace aerosols have density greater than 1 g/cm^3. This means that the aerodynamic diameter will always be larger than the physical diameter. Particles that have the same terminal settling velocity are said to be aerodynamically equivalent, i.e., they have the same aerodynamic diameter. Particles of size d_1 and density P_1, are aerodynamically equivalent to particles of size d_2 and density P_2 when:

$$d_1^2 P_1 = d_2^2 P_2 \tag{8.12}$$

The CMD is usually much smaller than the AMAD because of the D^3 relationship between the size of a particle and its mass. It takes many small particles to equal the mass of a single larger particle. The mass median diameter, MMD, is related to the count median diameter, CMD, by

$$\text{Log MMD} = \log \text{CMD} + 6.908 (\log \Sigma_g)^2 \tag{8.13}$$

where Σ_g = geometric standard deviation of the particle size distribution.

8.4.1 Aerosol Settling Behavior

The terminal settling velocity, V_T, of a spherical particle diameter and density P is:

$$v_t = \frac{d_2 Pg}{18\eta} \tag{8.14}$$

where g = gravitational acceleration
η= viscosity of air (= 185 μ poise at room temperature)

If aerosol particles are relatively large (>~1μ), the air behaves as if it were a continuous fluid medium. When the aerosol particle sizes approach the mean free path of the vibrating air molecules however, the air ceases to behave as a continuous fluid medium, and the aerosol particles "slip" downwards between the air molecules. The resulting increased terminal settling velocity is calculated with the Cunningham correction factor, f_c:

$$v_t = \frac{d^2 Pg}{18\eta} \times f_c = \frac{d^2 Pg}{18\eta} \times \left(1 + \frac{\lambda}{d}\right) \tag{8.15}$$

where λ = mean free path of air molecules (=~7 x 10^{-6}cm at 25°C)

For many practical purposes including air sampling, small particles have such slow terminal settling velocities that they may be considered to be suspended in the air, and their movement is controlled solely by the movement of the air in which they are suspended.

8.4.2 Corrections for Temperature and Pressure

The operation of an air flow meter, such as a variable area flowmeter (Rotameter) is affected by the viscosity and density of the air, which in turn depends on the temperature and pressure. Temperature and pressure corrections therefore must be applied to metered values when measurements are made under environmental conditions that differ from calibration conditions. The actual volumetric airflow, Q (a), is related to the metered airflow, Q(m) by:

$$Q(a) = Q(m) \sqrt{\frac{P(cal) \times T(m)}{P(m) \quad T(cal)}} \qquad (8.16)$$

Where p(cal) = atmospheric pressure at calibration
P(m) = atmospheric pressure at measurement
T(cal) = absolute temperature at calibration
T(m) = absolute temperature at measurement

8.4.3 Sampling Of Gaseous Contaminants

Gaseous materials are often sampled for Industrial Hygiene purposes. Many common chemicals are gaseous at normal temperature and pressure, including many organic and organo-metallic compounds. These compounds are less common as radiological hazards, but some notable exception include the noble gasses, many ^{14}C compounds, many tritiated compounds and especially HTO, volatized radioiodine and technicium compounds, many ^{32}P compounds, etc. Some gaseous or particulate radioactive compounds can interact with air when released, and thereby increase their hazard potential. For example, one mole of UF_6 gas, when released to the air is hydrolyzed to one mole uranyl fluoride, UO_2F_2, and four moles fluoride (F), a highly toxic gas whose OSHA limit is 0.1ppm. Radionuclides not considered as gaseous hazards can become gaseous in nature if subjected to certain conditions. For example, plutonium contaminated metals that are welded, cut with a torch, or otherwise heated or abraded can become airborne as such small particles that they are gaseous rather than particulate in nature.

8.4.4 Sampling Of Particulate Materials

Filtration is the most widely used method for collecting samples of radioactive aerosols. The methods and equipment range from high volume samplers (sampling rates up to about 60 m^3/h) for environmental or short-term workplace sampling, to low-volume, miniature lapel samplers (1 L/min or less) for collecting aerosols in the breathing zone of individual workers (U.S. DOE 1988). Low-pressure-drop, cellulose filters are commonly used; the sample filters can be easily reduced to ash or dissolved for analysis by analytical chemistry or radiochemistry.

Concerns for penetration of particles into the filter matrix are a function of the type of filter, the type of radiation, and the radiation counting method being used. Membrane filters with their superior front-surface collecting characteristics are preferred over fiber-type filters when alpha particle spectroscopy is used for counting. Shielding by the filter media is seldom a concern for detection of gamma radiation. Although energy degradation concerns are greatest for alpha-emitting radionuclides, even glass microfiber filters such as the Gelman A/E glass filter collect particles near enough to the filter surface that radioactivity counting results from the ZnS method are as accurate as radiochemical results.

Long-term storage of filter samples for archival purposes is not always feasible for high specific activity, alpha-emitting radionuclides such as ^{241}Am. Radiation damage to the filter, packaging tape, or plastic container may result in release of radioactivity. Care should be taken when handling samples that have aged.

The presence of airborne dust can lead to burial and underestimation of plutonium or uranium deposited on the sample collection filter. The accumulation of ambient dust on the collection filter leads to attenuation of alpha energy. Such burial of plutonium leads to underreporting of air concentrations, ranging from 10% to 100% when airborne dust concentrations are greater than 1 mg/m^3 (Hoover et al. 1988b and 1990). Alpha particles from plutonium that is buried by 2 mg/cm^2 of salt on a filter are prevented from reaching the detector of an analyzer. In the case of a real time monitor, this does not prevent the monitor from responding to large puff releases of radioactivity that will be deposited on the surface of the collection substrate and detected before it can be attenuated by additional collection of dust. However, it does limit the ability to directly count the sample for the detection of ambient conditions. The sample can be chemically processed to eliminate the interfering material, and the radioisotope redeposited and counted.

Continuous air monitoring for low airborne concentrations of high-specific-activity radionuclides, such as ^{238}Pu, will suffer from statistical limitations in achieving a representative sample (Birchall 1991, Cember 1996, Scott 1997, DOE 1998b, and Scott 1999). This arises from the fact that a small number of particles per unit volume can account for a concentration of concern. For example, a spherical particle of plutonium-238 dioxide (of nominal specific activity 560 GBq/g, and nominal density of 10 g/cm^3) with a physical diameter of 5 μm and a corresponding aerodynamic equivalent diameter of 15.8 μm (based on the fact that the aerodynamic diameter is approximately equal to the physical diameter times the square root of the density) has an activity of 365 Bq, (1 x 10^{-4} μCi/particle). The DAC for $^{238}PuO_2$ is 0.3 Bq/m^3. Thus, the DAC corresponds to a particle concentration of 0.0008 particles/m^3, or 1 particle in 1250m^3! There is a very low probability that a CAM operating at 56 L/min (2 cfm) will collect even a single particle in a grab sample during a short sampling of time. However, if it did collect even a single particle of pure $^{238}PuO_2$, during an 8 hour sampling period, the result would show a ^{238}Pu concentration of 13.6 Bq/m^3, or 45 times the DAC. Such concerns do not exist for isotopes with very low specific activity where the aerosol cloud can be expected to have a larger number concentration and be more homogenous at concentrations of concern. In any case, the DOE Standard for Internal Dosimetry (DOE-STD-1121-98) notes that air sampling data should not be used to assign radiation dose to workers except in cases (such as stable tritiated particulate, for example) where no approved bioassay method is available.

When a sample is collected for particulate contamination, the isotope of interest is collected along with some amount of radon progeny. The radon progeny often have a shorter half life, so the sample is allowed to decay for a period, often a week, before counting for the isotope of interest. Currently, many facilities desire to have fast turnaround on particulate samples. This can be achieved by counting the sample by alpha spectroscopy rather than gross alpha. In this way, the alpha peaks from the isotope of interest can be separated from radon progeny. An active area of recent work has related to the selection of appropriate filter media for alpha constant air monitors (CAM) which use alpha detectors. The quality of the alpha energy spectrum reported by a semi-conductor detector system is strongly dependent on the type of filter substrate used for particle collection. Membrane filters have long been accepted as having both the excellent particle collection efficiency and the excellent surface collection characteristics needed for alpha spectroscopy. Lindeken et al. (1964) demonstrated that, compared to submicrometer pore filters, large-pore membrane filters show no serious sacrifice in collection efficiency until the pore diameters exceed 5 μm. Thus, larger pore filters, with their lower pressure drop, are preferred. A filter widely recommended by CAM manufacturers is the 5-μm-pore-size, membrane filter (mixed cellulose ester). However, these filters are very fragile and gravimetric confirmation of particle mass collection were unreliable because of filter breakage under field conditions. Results of evaluations to find a more rugged membrane filter that would have a reasonably low pressure drop and excellent surface collection efficiency are summarized in Annex D of ANSI N13.1 (1999). Because of the extremely poor spectral quality obtained when a conventional fiber filter such as the Whatman 41 is used in an alpha CAM, the use of fiber-type filters must be avoided if the filter is to be counted by alpha spectroscopy.

An alternate approach to handling radon progeny background interference was developed at the Department of Energy Savannah River Site (Collins 1956; Tait 1956; Alexander 1966; Chen 1999). It uses an impactor jet to deposit aerosol particles directly onto a greased planchet that is coated with a thin layer of ZnS(Ag). Flashes of light from the interaction of alpha emissions with the ZnS(Ag) scintillator are detected by a photomultiplier tube. This provides a realtime capability for detecting plutonium particles deposited on the collection substrate. Collecting the aerosol by inertial impaction eliminates most of the radon progeny, which are usually attached to the small particle fraction of ambient aerosol (diameter less than 0.3 μm) and are smaller than the effective cutoff diameter of the impactor. Smaller plutonium particles are also undetected, but most releases of alpha-emitting radionuclides in the workplace are dominated by larger particles with an activity median aerodynamic diameter of about 5 μm and a geometric standard deviation of about 2.5 (Dorrian 1995). Radioactivity counting efficiency for the collected plutonium is approximately 50% (nearly 100% of the emissions that occur in the direction of the ZnS layer) because the light emissions from the ZnS(Ag) are emitted isotropically, and all emissions can be seen with nearly equal likelihood by a photomultiplier or photodiode detector.

A variation of the Savannah River approach uses an impactor jet to deposit the aerosol directly onto the lightly greased surface of a semi-conductor detector (Kurz Model 8300, Kurz Instruments, Inc., Monterey, CA). Detection efficiency was excellent for the collected material, and direct detection of the alpha particles led to good separation of the plutonium and radon progeny peaks. However, particle bounce off the detector surface degraded the collection efficiency of large particles (more than 90% of the 10 μm aerodynamic diameter particles are lost to bounce off the lightly greased surface) (Hoover 1993). This problem may be solved by the use of a virtual impactor or particle trap approach (see Biswas and Flagan (1988) for a description of the particle trap.)

Other instruments have used coincidence-counting measurements of beta or gamma emissions from radon progeny to correct for alpha emission interference. However, these indirect measurements are susceptible to errors when the composition or concentration of the radon progeny background is altered. The multiple windows and peak stripping methods are the most widely used algorithms in current alpha CAMs.

8.4.5 Particle Size Considerations

Inertial sampling using cascade impactors, spiral duct centrifuges, and cyclones has been the major approach for characterizing the aerodynamic particle size distribution of radioactive aerosols. A number of specialized versions of these instruments have been developed specifically for use with radioactive aerosols (Mercer 1970 and Kotrappa 1972). Special requirements for instruments used in handling radionuclides generally include being compact and easy to assemble and disassemble while wearing protective gloves in a confined space such as a glovebox enclosure. They also need to be easy to clean. Low sample collection rates are usually adequate for collection of small samples for particle size determination. Analytical methods for collected samples are straightforward by radioactive counting. The spiral duct centrifuge has been widely used to estimate the density or shape factor (the shape factor accounts for the non-spherical shape of most particles) of individual particles. This works equally well for radioactive and nonradioactive particles. Because the aerodynamic diameter associated with each particle deposition location is known, electron microscopy can be used to determine the physical size and shape of particles found at those locations, and density or shape factor can be calculated (Stoeber and Flachsbart 1969).

Real-time inertial techniques, such as time-of-flight measurements of particles accelerated through a nozzle, are also useful for radioactive aerosols. This assumes a willingness to purchase dedicated instruments for use with radioactive aerosols, because decontamination of equipment for return to unrestricted use is not always practical.

8.5 SPECIAL TOPICS IN AIR SAMPLING

8.5.1 Decommissioning and Decontamination

Decommissioning and Decontamination (D&D) activities often involve operations where significant dust is generated. The accumulation of ambient dust on the collection filter of a sampler leads to attenuation of alpha energy as discussed earlier. Also, because activities generally change day to day, it is often difficult to obtain meaningful data for subsequent trend analysis. This is complicated by the progress of the project; as one area is cleaned, the job moves, and therefore so do the monitoring locations. D&D tasks may be completed in a few days or may take months. It is often difficult to provide engineering controls that are adequate to eliminate the need for personnel respiratory protection.

8.5.2 Sampling from Stacks and Ducts

Because the Clean Air Act sets limits on allowable releases of radioactive materials to the environment (U.S. EPA 1991), effluent monitoring or stack sampling may be required for radiological facilities. This must be done routinely to verify that environmental releases are within limits and ALARA. Measurements are often made both on site and off site. Because concentration limits for environmental releases are generally lower than for the work place, greater sensitivity is usually required for environmental measurements than for workplace measurements. This is achieved by using higher sampling flow rates, sampling for longer periods of time, or using more sensitive analytical techniques.

The original version of the American National Standard Guide to Sampling Airborne Radioactive Materials in Nuclear Facilities (ANSI N13.1-1969) has been revised and reissued as the American National Standard for Sampling and Monitoring for Releases of Airborne Radioactive Substances from the Stacks and Ducts of Nuclear Facilities (ANSI N13.1-1999). Both versions of the standard emphasize the importance of obtaining a representative sample. However, the original standard included faulty assumptions about aerosol mixing in exhaust ducts and faulty technical guidance on the use of multipoint sampling to obtain a representative sample. The new standard minimizes reliance on simplistic concepts such as isokinetic sampling (considered by many to be broadly misapplied in turbulent sampling conditions) and emphasizes "qualification" of a well-mixed sampling location, which enables the more efficient collection of samples from a single point. Compact and cost effective mixer designs have been developed to create a well-mixed condition in an exhaust stack (McFarland 1999a and 1999b). The revised standard also contains information about using the new and more efficient shrouded probes for sample extraction (McFarland 1989 and Rodgers 1996). Significant improvements have also been made in modeling the deposition of radioactive aerosols in sampling transport lines. Modern codes such as DEPOSITION 4.0 (Anand 1996 and Riehl 1996) can now be used to predict particle deposition as a function transport line. An overview of the methodology for sampling effluent air from stacks and ducts of the nuclear industry is available (McFarland 1998).

8.5.3 Radon And Its Progeny

Two naturally occurring radon isotopes are commonly encountered while sampling for other radionuclides. These include ^{222}Rn, or thoron, half life = 55 sec, a progeny of ^{232}Th, and ^{222}Rn, or "radon," half life 3.82 days, a progeny of ^{226}Ra. Each of these noble gas isotopes is the parent of progeny that emit alpha, beta and gamma radiation. Virtually all dose from exposure to radon and its progeny comes from the progeny. The progeny are commonly attached to particulate material in the air, generally of 0.3 micron or smaller size. Radon is adsorbed by charcoal filters, and is soluble in water. Radon progeny are absorbed, adsorbed, or captured by most air sample media. Sometimes it is necessary to monitor the environment for radon, generally for regulatory compliance, sometimes it is necessary to monitor for radon progeny, generally for worker exposure control. Sometimes radon or its progeny interfere with the analysis of samples for other radionuclides.

Radon is sometimes captured in a Lucas cell, an interesting variation of a grab sampling. Filtered air, which contains only radon gas but no dust particles, is drawn into a chamber whose interior walls are coated with a layer of crystalline ZnS(Ag) or other scintillator that is sensitive to alpha radiation. Nearly 100% of the alpha rays reaching the scintillator will result in a flash of light. Flashes of light that occur within the chamber are observed by a detector. This method can be applied to radon gas, radon progeny, or other alpha-emitting radionuclides that can be drawn into the chamber. Progeny are typically excluded by drawing the sample into an evacuated Lucas cell through a particulate filter. The half-life of the radionuclides being sampled influences the delay time or cleaning requirements before the cells can be reused.

Radon is often monitored in real time by the use of a flow through ionization chamber. Other gaseous radionuclides may interfere with this method of measurement. Ionization chambers are also sensitive to water, smoke and particles from welding, etc. that give rise to ionized particles.

Radon progeny are typically captured on a filter media and analyzed real time with a surface barrier alpha detector or equivalent. The inhalation exposure to radon progeny is typically expressed in working level- months, which is a measure of the inhaled alpha energy. (see chapter 16)

The alpha emissions of naturally occurring radon progeny such as ^{218}Po and ^{212}Bi with alpha energies of 6.0 MeV and 6.08 MeV, respectively, are similar enough in energy to the alpha emissions of ^{239}Pu (alpha energy 5.2 MeV) and ^{238}Pu (alpha energy 5.5 MeV), to cause interference or false positive reports of plutonium air concentrations. Early alpha CAM designs did not include spectrometry, but used a single channel analyzer to detect radioactivity in a plutonium energy region of interest (ROI), and a second analyzer to both detect radon progeny activity in a second energy region and to allow a simple background correction for the counts seen in the plutonium ROI. The correction method was crude and the limit of detection high, but it provided a useful, real-time means for detecting relatively large airborne releases of plutonium. The first successful algorithm for subtraction of radon progeny interference in the plutonium energy region was developed by Unruh (1986) at Los Alamos National Laboratory which utilized four ROIs.

Recent advances in microcomputers have enabled a major development effort to include alpha spectroscopy in semi-conductor detector CAMs for real-time monitoring of alpha-emitting radionuclides. Again, a semi-conductor detector is placed just above the surface of a filter onto which alpha-emitting aerosols are drawn, but the detector output goes to an embedded spectrometer in the CAM, rather than to single-channel analyzers.

In a batch operation, the air gap between the filter and the detector can be evacuated. This prevents attenuation of the alpha particle energies as the alphas pass through the air gap and minimizes any spectral overlap of the plutonium and radon progeny alpha energy peaks. For real-time sampling of the aerosol, however, the region between the filter and the detector cannot be evacuated, and attenuation of the alpha particle energies occurs in the air gap.

8.5.4 Air Sampling Quality Assurance Program

An ever-increasing emphasis is being placed on having a "technically defensible basis" for measuring radioactive aerosols in the work place and environment. New terms such as "Conduct of Operations," "Total Quality Management," and "Integrated Safety Management" have been coined to provide new paradigms for management and worker attitudes and responsibilities, record keeping and quality assurance requirements, demonstration of compliance, and proper integration of concerns for industrial hygiene, health physics, and traditional safety hazards. Some of the notable deficiencies in the technical basis for radioactive aerosol measurement are quality and handling of radioactive calibration sources, selection of appropriate instrumentation for specific tasks, placement of instrumentation in the work place or environment, and criteria for collection of an adequate sample. Improvements in the limit of detection for work place and environmental releases are also needed. This might come through wider application and improvements in virtual impaction techniques to concentrate airborne particles prior to collection.

Pressure for improved quality and capabilities can be expected to be the greatest in four classes of nuclear applications: environmental restoration of radioactively contaminated facilities such as those located throughout the Department of Energy weapons production complex; design, construction, and operation of facilities for safe disposal of low-level and high-level radioactive waste; monitoring and handling of radioactive materials in medical settings; and continued operation or design of new nuclear facilities for generation of electricity, including those that would use mixed uranium-plutonium oxide fuels from disposition of nuclear weapons.

Opportunities are available for improving the quality of regulations and standards and for preparing guidance documents for state of the art sampling and monitoring. The American National Standard for Radiation Protection Instrumentation Test and Calibration (ANSI N323-1983) is being revised to include a specific new standard to be issued as ANSI-N323C for air monitoring instruments. International standards for air monitoring and radiation protection instrumentation are under continuous development and improvement. However, continued work will be needed to fully understand the basic behavior of radioactive aerosols, and to synthesize total measurement approaches that are technically defensible and cost-effective.

8.6 REFERENCES

Alexander, J.M. A Continuous Monitor for Prompt Detection of Airborne Plutonium. Health Phys, 12: 553-556. 1966.

Anand, N.K.; McFarland, A.R; Dileep, V.R.; Riehl, J.D. Deposition: Software to Calculate Particle Penetration through Aerosol Transport Lines. NUREG/GR-0006, Washington, DC: U.S. Nuclear Regulatory Commission. 1966.

American National Standard Guide to Sampling Airborne Radioactive Materials in Nuclear Facilities. ANSI N13.1-1969. New York: American National Standards Institute. 1969.

Radiation Protection Instrumentation Test and Calibration. ANSI-N323-1978, reaffirmed 1983. New York: American National Standards Institute. 1983.

American National Standard for Sampling and Monitoring for Releases of Airborne Radioactive Substances from the Stacks and Ducts of Nuclear Facilities. ANSI N13.1-1999. New York: American National Standards Institute. 1999.

Biswas, P; Flagan, R.C. The particle trap impactor. J. Aerosol Sci. 19: 113-121. 1988.

Cember, H. Introduction to Health Physics, 3rd ed. McGraw-Hill, New York, 1996.

Chen, B.T.; Hoover, M.D.; Newton, G.J.; Montano, S.J.; Gregory, D.S Performance Evaluation of the Sampling Head and Annular Kinetic Impactor in the Savannah River Site Alpha Continuous Air Monitor. Aerosol Sci. Technol. 31: 24-38. 1999.

Dorrian, M.D.; Bailey, M.R. Particle Size Distributions of Radioactive Aerosols Measured in Workplaces. Radiation Protection Dosimetry, 60: 119-133; 1995.

Hickey, E.E.; Stoetzel, G.A.; Olsen, P.C. Air Sampling in the Workplace. NUREG-1400, Washington, DC.: U.S. Nuclear Regulatory Commission.1991.

Hoover, M.D.; Eidson, A.F.; Mewhinney, J.A.; Finch, G.L.; Greenspan, B.J.; Cornell, C.A. Generation and characterization of respirable beryllium oxide aerosols for toxicity studies. Aerosol Sci. Tech. 9:83-92; 1988a.

Hoover, M.D.; Newton, G.J.; Yeh, H.C.; Seiler, F.A.; Boecker, B.B. Evaluation of the Eberline Alpha-6 Continuous Air Monitor for Use in the Waste Isolation Pilot Plant: Phase I Report. 21 December 1988. Albuquerque, NM: Inhalation Toxicology Research Institute; 1988b.

Hoover, M.D.; Castorina, B.T.; Finch, G.L.; Rothenberg, S.J. Determination of the Oxide Layer Thickness on Beryllium Metal Particles. Am. Ind. Hyg. Assoc. J. 50: 550-553; 1989.

Hoover, M.D.; Newton, G.J.; Yeh, H.C.; Seiler, F.A.; Boecker, B.B. Evaluation of the

Eberline Alpha-6 Continuous Air Monitor for Use in the Waste Isolation Pilot Plant: Report for Phase II. 31 January 1990. Albuquerque, NM: Inhalation Toxicology Research Institute; 1990.

Kotrappa, P.; Light, M.E. Design and performance of the Lovelace aerosol particle separator. Rev. Sci. Instrum. 43: 1106-1112; 1972.

Lederer, C.M.; Hollander, J.M. Perlman, I. editors. Table of Isotopes. 7th Edition. New York: John Wiley and Sons; 1978.

Lindekin, C.L.; Petrock, F.K.; Phillips, W.A; Taylor, R.D. Surface collection efficiency of large-pore membrane filters. Health Phys. 10: 495-499; 1964.

McFarland, A.R.; Ortiz, C.A.; Moore, M.E.; DeOtte, Jr. R.E.; Somasundaram, A. A Shrouded Aerosol Sampling Probe. Environ. Sci. Technol. 23:1487-1492; 1989.

McFarland, A R. Methodology for Sampling Effluent Air from Stacks and Ducts of the Nuclear Industry. LA-UR-96-2958, Revised July 1998, Los Alamos, NM: Los Alamos National Laboratory; 1998.

McFarland, A.R.; Anand, N.K.; Ortiz, C.A.; Gupta, R.; Chandra, S.; McManigle, A.P. A Generic Mixing System for Achieving Conditions Suitable for Single Point Representative Effluent Air Sampling. Health Phys. 76: 17-26; 1999a.

McFarland, A. R., R. Gupta, and N. K. Anand. Suitability of Air Sampling Locations Downstream of Bends and Static Mixing Elements. Health Phys. 77: 703-712; 1999b.

Mercer, T.T.; Tillery, M.I.; Newton, G.J. A multi-stage low flow rate cascade impactor. J. Aerosol Sci. 1: 9-15; 1970.

Riehl, J.R.; Dileep, V.R.; Anand, N.K.; McFarland, A.R. DEPOSITION 4.0: An Illustrated User's Guide. Aerosol Technology Laboratory Report 8838/7/96, Department of Mechanical Engineering, College Station, Texas: Texas A&M University; 1996.

Rodgers, J.C.; Fairchild, C.I.; Wood, G.O.; Ortiz, C.A.; Muyshondt, A.; McFarland, A.R. Single Point Aerosol Sampling: Evaluation of Mixing and Probe Performance in a Nuclear Stack. Health Phys, 70:25-35; 1996.

Scott, B.R.; Hoover, M.D.; Newton, G.J. On Evaluating Respiratory Tract Intake of High-Specific Activity Emitting Particles for Brief Occupational Exposure, Radiat. Prot. Dosim., 69(1): 43-50; 1997.

Scott, B.R.; Fencl, A.F. Variability in PuO_2 Intake by Inhalation: Implications for Worker Protection at the US Department of Energy. Radiat. Prot. Dosim. 83: 221-232; 1999.

Schleien, B.S.; Slabeck, L.A., Jr.; Kent, B.K. editors. Handbook of Health Physics and Radiological Health. Third Edition. Baltimore, MD: Williams and Wilkens; 1998.

Stoeber, W.; Flachsbart, H. 1969. Size-separating precipitation of aerosols in a spinning

spiral duct. Environ. Sci. Techol. 3: 1280-1296; 1969.
Tait, G.W.C. Determining Concentration of Airborne Plutonium Dust. Nucleonics, 14: 53-55; 1956.
U.S. Department of Energy. Operational Health Physics Training. ANL-88-26. Prepared by Argonne National Laboratory, Argonne, IL, for the U.S. Department of Energy Assistant Secretary for Environment, Safety, and Health. Oak Ridge, TN: National Technical Information Service; 1998.
U.S. Department of Energy. The Department of Energy Laboratory Accreditation Program for Radiobioassay, DOE STD 1112-98, December 1998, Washington, DC, U.S. Department of Energy; 1998.
U.S. Environmental Protection Agency. National Emission Standards for Hazardous Air Pollutants. Title 40, Code of Federal Regulations, Part 61. 1 July 1991. Washington, DC: U.S. Government Printing Office; 1991.
Unruh, W.P. Development of a Prototype Plutonium CAM at Los Alamos. LA-UR-90-2281. 15 December 1986. Los Alamos, NM: Los Alamos National Laboratory; 1986.

8.7 OTHER READING

Allen, M. D.; G. Raabe. Slip correction measurements of spherical solid particles in an improved Millikan apparatus. Aerosol Sci. Technol. 4: 269-286; 1985.
American National Standard for Radiation Protection Instrumentation Test and Calibration. ANSI N323-1978. New York: American National Standards Institute; 1978.
American National Standard for Performance Specifications for Reactor Emergency Radiological Monitoring Instrumentation. ANSI N320-1979. New York: American National Standards Institute; 1979.
American National Standard for Specification and Performance of On-Site Instrumentation for Continuously Monitoring Radioactivity in Effluents. ANSI N42.18-1980. New York: American National Standards Institute; 1980a.
American National Standard for Performance Criteria for Instrumentation Used for Inplant Plutonium Monitoring. ANSI N317-1980. New York: American National Standards Institute; 1980b.
American National Standard on Performance Specifications for Health Physics Instrumentation—Portable Instrumentation for Use in Normal Environmental Conditions. ANSI N42.17A-1989. New York: American National Standards Institute; 1989a.
American National Standard on Performance Specifications for Health Physics

Instrumentation—Occupational Airborne Radioactivity Monitoring Instrumentation. ANSI N42.17B-1989. New York: American National Standards Institute; 1980b.
American National Standard on Performance Specifications for Health Physics Instrumentation—Portable Instrumentation for Use in Extreme Environmental Conditions. ANSI N42.17C-1989. New York: American National Standards Institute; 1980c.
Cember, H. Introduction to Health Physics. Third Edition. New York: McGraw Hill. 1996.
Cohen, B. S.; M. Eisenbud; Harley, N.H.; Measurement of the α-radioactivity on the mucosal surface of the human bronchial tree. Health Phys. 39: 619-632; 1980.
Cohen, B. Sampling Airborne Radioactivity. In Air Sampling Instruments for Evaluation of Atmospheric Contaminants. 8th Edition. B. S. Cohen and S. V. Hering, editors. Cincinnati: American Conference of Governmental Industrial Hygienists; 1995.
Radiation risks from plutonium recycle. Environ. Sci. Tech. 11: 1160-1165; 1995.
Eicholz, C.G.; Poston, J.S.; Principles of Nuclear Radiation Detection. Ann Arbor, MI: Ann Arbor Science; 1979.
Eisenbud, M. Environmental Radioactivity from Natural, Industrial, and Military Sources. San Diego: Academic Press; 1987.
Evans, R. D. The Atomic Nucleus. New York: McGraw-Hill; 1955.
Finch, G. L.; Hoover, M.D.; Mewhinney, J.A.; Eidson, A.F. Respirable particle density measurements using isopycnic density gradient ultracentrifugation. J. Aerosol Sci. 20: 29-36; 1989.
Grivaud, L.; Fauvel, S.; Chemtob, M. Measurement of Performances of Aerosol Type Radioactive Contamination Monitors, Radiat. Prot. Dosim., 79(1-4): 495-498; 1998.
Hobbs, C.H.;. McClellan, R.O. Toxic effects of radiation and radioactive materials. In Casarette and Doull's Toxicology: The Basic Science of Poisons. 3rd edition. C. D. Klaassen, M. O. Amdur, and J. Doull, editors. New York: MacMillan Publishing Co; 1986.
Hoover, M.D. Newton, G.J.; Yeh, H.C.; Eidson, A.F. Characterization of aerosols from industrial fabrication of mixed-oxide nuclear reactor fuels. In Aerosols in the Mining and Industrial Work Environments. V. A. Marple and B. Y. H. Liu, editors. Ann Arbor: Ann Arbor Science; 1983.
Hoover, M.D., Finch, G.L.; Castorina, B.T. Sodium metatungstate as a medium for measuring particle density using isopycnic density gradient ultracentrifugation. J. Aerosol Sci. 22: 215-221; 1991.
Hoover, M.D.; Newton,G.J. Preliminary Evaluation of Optical Monitoring for Real-Time

Detection of Radioactive Aerosol Releases. Albuquerque, NM: Inhalation Toxicology Research Institute; 1991.

Hoover, M.D.; Newton, G.J. Radioactive Aerosols. In: Aerosol Measurement: Principles, Techniques, and Applications, 1st Edition, K. Willeke and P. A. Baron, editors, New York: Van Nostrand Reinhold; 1993.

Hoover, M.D.; Newton, G.J.; Guilmette, R.A.; Howard, R.J.; Ortiz, R.N.; Thomas, J.M.; Trotter, S.M.; Ansoborlo, E. Characterisation of Enriched Uranium Dioxide Particles from a Uranium Handling Facility. Radiat. Prot. Dosim. 79(1-4): 57-62; 1998.

Hoover, M.D.; Newton, G.J. Performance Testing of Continuous Air Monitors for Alpha-Emitting Radionuclides. Radiat. Prot. Dosim. 79(1-4): 499-504; 1998.

Hoover, M.D.; Mewhinney, C.J.; Newton, J. Modular Glovebox Connector and Associated Good Practices for Control of Radioactive and Chemically Toxic Materials. Health Phys. 76: 66-72; 1999.

International Atomic Energy Agency. Particle Size Analysis in Estimating the Significance of Airborne Contamination. Technical Report Series No. 179. Vienna: International Atomic Energy Agency; 1978.

International Commission on Radiological Protection. Principles of Environmental Monitoring Related to the Handling of Radioactive Materials. International Commission on Radiological Protection Publication 7. Oxford: Pergamon Press; 1965.

International Commission on Radiological Protection. General Principles of Monitoring for Radiation Protection of Workers. International Commission on Radiological Protection Publications 12 and 35. Oxford: Pergamon Press; 1968 and 1982.

International Commission on Radiological Protection. Implications of Commission Recommendations That Doses Be Kept as Low as Reasonably Achievable. International Commission on Radiological Protection Publication 22. Oxford: Pergamon Press; 1973.

International Commission on Radiological Protection. Reference Man: Anatomical, Physiological and Metabolic Characteristics. International Commission on Radiological Protection Publication 23. Oxford: Pergamon Press; 1975.

International Commission on Radiological Protection. Limits on Intakes of Radionuclides by Workers. International Commission on Radiological Protection Publication 30 and addendums. Oxford: Pergamon Press; 1979.

International Commission on Radiological Protection. Principles of Monitoring for the Radiation Protection of the Population. International Commission on Radiological Protection Publication 43. Oxford: Pergamon Press; 1985.

International Commission on Radiological Protection. 1990 Recommendations of the

International Commission on Radiation Protection. International Commission on Radiological Protection Publication 60. Ann. ICRP 21 (1/3). Oxford: Elsevier Science Ltd.; 1994a.
International Commission on Radiological Protection. Human Respiratory Tract Model for Radiological Protection. International Commission on Radiological Protection Publication 66. Ann. ICRP 24 (4). Oxford: Elsevier Science Ltd; 1994b.
International Commission on Radiological Protection. Dose Coefficients for Intakes of Radionuclides by Workers. International Commission on Radiological Protection Publication 68. Ann. ICRP 24 (4). Oxford: Elsevier Science Ltd; 1994c.
International Commission on Radiation Units and Measurement. Radiation Protection Instrumentation and its Application. ICRU Report 20. Washington, DC:; 1971.
International Electrotechnical Commission Radiation Protection Instrumentation. Calibration and Verification of the Effectiveness of Radon Compensation for Alpha and/or Beta Measuring Instruments, IEC Standard 61578, Geneva, Switzerland: 1997.
International Electrotechnical Commission Radiation Protection Instrumentation. Equipment for Continuously Monitoring Radioactivity in Gaseous Effluents, Part 1: General Requirements, IEC Standard 61761-1, Geneva, Switzerland: 2000a.
International Electrotechnical Commission. Radiation Protection Instrumentation. Equipment for Continuously Monitoring Radioactivity in Gaseous Effluents, Part 2: Specific Requirements for Radioactive Aerosol Monitors including Transuranic Aerosols, IEC Standard 61761-2, Geneva, Switzerland; 2000b.
International Electrotechnical Commission. Radiation Protection Instrumentation. Equipment for Continuously Monitoring Radioactivity in Gaseous Effluents, Part 3: Specific Requirements for Radioactive Noble Gas Monitors, IEC Standard 61761-3, Geneva, Switzerland; 2000c.
International Electrotechnical Commission. Radiation Protection Instrumentation. Equipment for Continuously Monitoring Radioactivity in Gaseous Effluents, Part 4: Specific Requirements for Radioactive Iodine Monitors, IEC Standard 61761-4, Geneva, Switzerland; 2000d.
International Electrotechnical Commission. Radiation Protection Instrumentation. Equipment for Continuously Monitoring Radioactivity in Gaseous Effluents, Part 5: Specific Requirements for Tritium Effluent Monitors, IEC Standard 61761-5, Geneva, Switzerland; 2000e.
Kanapilly, G.M.; Raabe, O.G.; Goh, C.H.T.; Chimenti, R.A. Measurement of the in vitro dissolution of aerosol particles for comparison to in vivo dissolution in the respiratory tract after inhalation. Health Phys. 24: 497-507; 1973.
MacArthur, D.W.; Allander. K.S. Long-Range Alpha Detectors. LA-12073-MS. Los

Alamos, NM: Los Alamos National Laboratory; 1991.

Mewhinney, J.A. Radiation Exposure and Risk Estimates for Inhaled Airborne Radioactive Pollutants including Hot Particles, Annual Progress Report July 1, 1976 - June 30, 1977. NUREG/CR-0010, Albuquerque, NM: Inhalation Toxicology Research Institute; 1978.

Mewhinney, J.A.; Eidson, A.F.; Wong, V.A. Effect of wet and dry cycles on dissolution of relatively insoluble particles containing Pu. Health Phys. 53:337-384; 1987a.

Mewhinney, J.A.; Rothenberg, S.J.; Eidson, A.F.; Newton, G.J.; Scripsick, R.. Specific surface area determination of U and Pu particles. J. Colloid Interface Sci. 116: 555-562; 1987b.

Moore, M.E.; McFarland, A.R.; Rodgers, J.C. Factors that Affect Alpha Particle Detection in Continuous Air Monitor Applications. Health Phys. 65: 69-81; 1993.

Morrow, P.E.; Mercer. T.T. A point-to-plane electrostatic precipitator for particle size sampling. Am. Ind. Hyg. Assoc. J. 25: 8-14; 1964.

National Council on Radiation Protection and Measurements. Instrumentation and Monitoring Methods for Radiation Protection. NCRP Report No. 57. Bethesda, MD: NCRP. 1978b. A Handbook of Radiation Protection Measurements Procedures. NCRP Report No. 58. Bethesda, MD; 1978a

National Council on Radiation Protection and Measurements. Deposition, Retention and Dosimetry of Inhaled Radioactive Substances. NCRP Report No. 125. Bethesda, MD; 1997.

Newton, G.J.; Hoover, M.D.; Barr, E.B.; Wong, B.A.; Ritter, P.D. Collection and characterization of aerosols from metal cutting techniques typically used in decommissioning nuclear facilities. Am. Ind. Hyg. Assoc. J. 48: 922-932; 1987.

Price, W.S. Nuclear Radiation Detection. New York: McGraw-Hill; 1965.

Raabe, O.G. Instruments and methods for characterizing radioactive aerosols. IEEE Transactions on Nuclear Science. NS-19(1): 64-75; 1972

Raabe, O..G.; Newton, G.J.; Wilkenson, C.J.; Teague, S.V. Plutonium aerosol characterization inside safety enclosures at a demonstration mixed-oxide fuel fabrication facility. Health Phys. 35: 649-661; 1978.

Rasetti, F. Elements of Nuclear Physics. New York: Prentice-Hall; 1947.

Rothenberg, S.J.; Denee, P.B.; Cheng, Y.S.; Hanson, R.L.; Yeh, H.C.; Eidson, A.F. Methods for the measurement of surface areas of aerosols by adsorption. Adv. Colloid Interface Sci. 15: 223-249; 1982.

Rothenberg, S.J.; Flynn, D.K.; Eidson, A.F.; Mewhinney, J.A.; Newton, G.J. Determination of specific surface area by krypton adsorption, comparison of three different methods of determining surface area, and evaluation of different specific surface area standards. J. Colloid Interface Sci. 116: 541-554; 1987.

Rutherford, E..; Chadwick, J.;. Ellis, C.D. Radiation from Radioactive Substances. New York: Macmillan; 1930.
Shapiro, J. Radiation Protection, A Guide for Scientists and Physicians. Cambridge: Harvard University Press; 1981.
Teague, S.V.; Yeh, H.C.; Newton, G.J. Fabrication and use of krypton-85 aerosol discharge devices. Health Phys. 35: 392-395; 1978.
Turner, J.E.; Bogard, J.S.; Hunt, J.B.; Rhea, T.A. Problems and Solutions in Radiation Protection. New York: Pergamon; 1988.
Turner, J.E. Atoms, Radiation, and Radiation Protection. Second Edition. New York: John Wiley and Sons; 1996.
U.S. Department of Defense. Narrative Summaries of Accidents Involving Nuclear Weapons 1950-1980. Washington, DC: U.S. Department of Defense; 1981.
U.S. Department of Energy. Occupational Radiation Protection, Title 10, Code of Federal Regulation, Part 835 (10 CFR 835), 14 December 1993. Washington, DC: U.S. Government Printing Office; 1993.
U.S. Department of Energy. Airborne Release Fractions/Rates and Respirable Fractions for Nonreactor Nuclear Facilities, DOE-HDBK-3010-94, Vol. 1 and 2, December 1994, Washington, DC: U.S. Department of Energy; 1994.
U.S. Department of Energy. Implementation Guide for Air Monitoring for use with Title 10 Code of Federal Regulations Part 835 Occupational Radiation Protection, DOE G 441.1-8. 17 March 1999. Washington, DC: U.S. Government Printing Office; 1999.
U.S. Department of Transportation. Regulations for Transportation of Hazardous and Radioactive Materials. Title 49, Code of Federal Regulations, Parts 171-177. 1 October 1990, Washington, DC: U.S. Government Printing Office; 1990.
U.S. Environmental Protection Agency. Hazardous Waste Management Regulations. Title 40, Code of Federal Regulations, Parts 260 through 268. 1 July 1990. Washington, DC: U.S. Government Printing Office; 1990.
U.S. Nuclear Regulatory Commission. Proceedings of the CSNI Specialists Meeting on Nuclear Aerosols in Reactor Safety. NUREG/CR-1724. Washington, DC: U.S. Nuclear Regulatory Commission; 1980.
U.S. Nuclear Regulatory Commission. Standards for Protection Against Radiation. Title 10, Code of Federal Regulations, Part 20 et al., 21 May 1991. Washington, DC: U.S. Government Printing Office; 1991.
U.S. Nuclear Regulatory Commission. Calibration and Error Limits of Air Sampling Instruments for Total Volume of Air Sampled. Regulatory Guide 8.25. Washington, DC: U.S. Nuclear Regulatory Commission; 1992.
Voigts, Chr.; Siegmon, G.; Berndt, M.; Enge, W. Single alpha-emitting aerosol particles.

In AEROSOLS, Formation and Reactivity, 2nd Int. Aerosol Conf. Berlin, p. 1153. Oxford: Pergamon Press; 1986.

Yeh, H.C.; Newton, G.J.; Raabe, O.G.; Boor, D.R. Self-charging of ^{198}Au-labeled monodisperse gold aerosols studied with a miniature electrical mobility spectrometer. J. Aerosol Sci. 7: 245-253; 1976.

Yeh, H.C.; Newton, G.C.; Teague, S.V. Charge distribution on plutonium-containing aerosols produced in mixed-oxide reactor fuel fabrication and the laboratory. Health Phys. 35: 500-503.; 1978.

Chapter 9

CALIBRATION OF HEALTH PHYSICS INSTRUMENTS

Alan Justus

9.1 INTRODUCTION

In this presentation, calibration is that set of all mechanical, electronic, and response adjustments necessary for an accurate measurement. It will be distinct from documentation and record keeping requirements and methods, and from repair (although some form or extent of which is often required at or before the time of recalibration). It will also be distinct from detailed discussions of traceability, systematic, and stochastic errors and their modeling and propagation, and from detailed discussions of ICRU/ICRP concepts regarding external dosimetry. It will also be distinct from user operational/ functional tests. However, such tests are so important a component of the overall calibration (or measurement quality) program that a brief discussion is included as section 9.4.

The reason for the initial calibration of a health physics instrument is obvious. The reason for recalibration might be less obvious, the driving force behind recalibration being the compensation for long-term drifts or for minor repair, both caused or necessitated by so-called 'component aging.' The 'component' could be actual electronic components, the detector, cabling, or mechanical components or connectors. The 'aging' could be oxidation or drying of components, detector leakage (e.g., PIC gas), detector degradation (e.g., water in hygroscopic NaI(Tl), gas of TE (tissue equivalent) multiplying ion chamber), detector window degradation (e.g., pokes, etc., although replacement would be a better option), detector contamination (although D&D or replacement might be a better option), normal wear and tear, the effects of mechanical shock, human tampering, sabotage, etc. It should be noted here that in the past, vacuum tubes, electrolytic capacitors, and un(temperature)compensated circuit designs were notorious culprits as regards drift. However, vacuum tubes are no longer used within the electronics, electrolytic capacitors are hermetically sealed, and most good circuit designs incorporate temperature-compensating feedback design techniques. Hence, it is typically the detector probe rather than the electronics (as long as they are functioning!) that is now the primary concern regarding long-term drift.

The primary goals of health physics instrument (re)calibration are (1) energy (i.e., gain) calibration, e.g., through HV and discrimination level(s) adjustment,

(2) detector efficiency check (i.e., sensitive volume and /or spectral region) and, if applicable, crosstalk check, and (3) the determination of the response or the adjustment of readout with respect to an appropriate standard. The secondary goals of (re)calibration relate to the general appearance and to the general working condition of the instrument. These primary and secondary goals are accomplished through acts of so-called general maintenance, precalibration adjustment, and primary calibration.

There are numerous health physics instrument types and specific models, such as Portable α/β Contamination Survey Meters, Hand & Shoe/ Personnel Contamination Monitors, α/β Laboratory Assay Instruments, Portable xγ Contamination Survey Meters, Portable βγ Field Survey Meters, Portable neutron Field Survey Meters, Area Monitors, Electronic Personnel Dosimeters, Air Monitors, Air Samplers, Portable and Laboratory HPGe Systems, Stack Monitoring Systems, Transfer Instruments, etc. Due to this vast variety, there is neither the space nor the time here to discuss calibration particulars of each type or model. However, common to all these particulars would be the use of just a finite number of 'calibration tools', i.e., the electronic standards, radiation field/ radioactivity standards, air flow (and other parameter) standards, as well as the necessary statistical tools for determination of appropriate alarm set points for the required sensitivities.

9.2 CALIBRATION TOOLS

9.2.1 Electronic Standards

Pulse generators are electronic instruments, available commercially from Eberline and Ludlum, used in the calibration (and troubleshooting) of pulse counting health physics instruments. They are essentially electronic pulse height and pulse rate standards. The Eberline MP-2 generator, for example, has a repetitive pulse rate that is variable from 0.1 cpm to 1.6×10^6 cpm. Its pulse shape is designed to simulate many health physics detectors, being negative with a rise time of 0.2 μsec, a width of 2 μsec, and a fall time of 3 μsec. Its pulse heights are variable from 0.2 mV to 3 V. These pulse generators are capacitively decoupled, providing isolation from the DC high voltages utilized in health physics pulse counting instruments.

Electrostatic kilovoltmeters are instruments designed to measure high voltages. They are characterized by input resistance values of 1×10^{15} ohms. On DC high voltages, the instrument draws a momentary charging current which instantly drops to a negligible value. Because of the very high input impedance, circuit loading is insignificant and the voltage measurement is quite accurate. Models ranging from 500 to 5000 V full scale are available (e.g., from Sensitive Research).

Digital multimeters (DMM), electrometers, and oscilloscopes are often used in the calibration (and troubleshooting) of health physics instrumentation. The DMM's and oscilloscopes require no further introduction. The electrometer is a specialized electronic instrument with a greatly extended measurement range relative to that of a typical DMM. It is utilized for the measurement of very low currents or voltages as well as very large resistance values. They are available from Inovision (Keithley and Victoreen).

9.2.2 Radiation Field/Radioactivity Standards

Recommended reference α/β sources for calibration of instruments that measure contamination on surfaces are given in NCRP 112 and ISO 8769. These include ^{241}Am, ^{238}Pu, ^{239}Pu, ^{230}Th α sources and ^{14}C, ^{147}Pm, ^{99}Tc, ^{204}Tl, ^{36}Cl, ^{90}Sr+^{90}Y, and ^{106}Ru+^{106}Rh β sources. These reference sources are utilized as activity and/or energy standards in the calibration of portable α/β contamination survey meters, hand & shoe/ personnel contamination monitors, and α/β laboratory assay instruments. Since most of the detector windows of these instruments are protected by grillwork, the reference α/β sources used for efficiency determinations should have dimensions at least several times greater than the grillwork spacing. In this way, efficiency determinations will be independent of placement on the grillwork. Consideration should be given to the decision if one's reference standards utilize activity electroplated on Ni or Fe backings, sealed into anodized Al micropores, etc. For the calibration of α or β continuous air monitors and stack monitoring systems, the source need not be one of the above recommended by NCRP or ISO, but should be or adequately represent the isotope being monitored on the filter media and must have a physical size that accommodates the filter sample holder.

Gas activity standards are utilized in the calibration of grab air samplers and continuous air and stack monitoring systems. The techniques employed for ^{3}H, ^{85}Kr, ^{14}C, and ^{133}Xe gas can involve the injection of known aliquots into a given closed-loop volume. The techniques employed for ^{222}Rn can use either a NIST emanation standard or a so-called flow-through source. The use of a NIST emanation standard involves the accumulation within a sealed air volume of gas from a standard having a known amount of ^{226}Ra and a known ^{222}Rn emanation fraction. The use of the flow-through sources can involve either the purging of ingrown gas into a given closed-loop volume, or the steady-state emanation into a continuous volumetric flow rate (note: this later technique can also be used for ^{220}Rn).

Xγ reference sources are utilized as activity and/or energy standards in the calibration of portable xγ contamination survey meters, portable or laboratory HPGe and

NaI(Tl) xγ spectrometry and stack monitoring systems. A very large number of single or multiline isotopes or combinations thereof are available. A large number of source geometries is also available or fabricated, such as point, electroplated disc, evaporated planchet, filter media, charcoal cartridge, Marinelli beaker (for soil, water, milk, etc.), and many other volumetric containers. Many sample matrices (i.e., effective Z and ρ) are also available or fabricated.

Recommended reference xγ radiation field standards for calibration of dose measuring instruments are given in NCRP 112 and ISO 4037. These include ^{137}Cs, ^{60}Co, ^{226}Ra, ^{241}Am, and various K fluorescent and filtered bremsstrahlung x-ray sources. These reference sources are utilized as exposure and energy standards in the calibration of portable βxγ field survey meters, xγ area monitors, xγ electronic personnel dosimeters, xγ transfer instruments, and even portable neutron field survey meters and neutron area monitors. For most health physics purposes, ^{137}Cs is normally a good choice due to its reasonably high energy and long half-life. The xγ radiation field standard is exposure in air, determined from a NIST-traceable air-equivalent ion chamber measurement free-in-air. A conversion factor (ICRP 74 1997) to cSv (or rem) for ^{137}Cs of 1.06 (i.e., 1.21 times 0.873) is normally not applied in the USA; exposure rate is the unit of display on the health physics instrument. It is theoretically possible to use a free-in-air source, a collimated source, or a source in a well or just a box. Some of these choices present large scattering and facility design problems. For most health physics purposes, a collimated source is normally a good choice. The collimation typically presents a 30° conical or pyramidal beam, and with the use of attenuator plates, the range is increased beyond that due to inverse-r^2 alone. Wall scatter is normally a negligible concern if the wall eventually intercepting the beam is no closer than twice the usable range. The effects of air attenuation and scatter for ^{137}Cs is typically negligible and ignored.

Recommended reference β radiation field standards for calibration of instruments that measure β absorbed dose are given in NCRP 112 and ISO 6980. These include ^{14}C, ^{147}Pm, ^{99}Tc, ^{36}Cl, ^{204}Tl, ^{90}Sr+^{90}Y, DU, and ^{106}Ru+^{106}Rh. These reference sources could be utilized in the calibration of or the determination of correction factors (CF's) for portable βxγ field survey meters such as the Eberline RO-20 and Bicron RSO-5/50/500 instruments. The β radiation field standard is tissue cGy (or rad) from an extrapolation ion chamber measurement or the PTB secondary calibration assembly with compensation filters. Please note however, that during a health physic instrument's use in the workplace, the β CF's would apply to those β fields that are quite similar to those of the calibration, and of course do not apply to any accompanying xγ components of the workplace field. The goal should therefore be that if one knows that one will be dealing with significant β fields in the workplace, then one should use an instrument that is specially designed for both βx and xγ measurements in the workplace such as the HPI 1075 dual ion chamber instrument. The βx chamber could then be calibrated to a recommended reference β radiation field or, if justified by the manufacturer's or other's type testing, to a reference photon field.

Recommended reference neutron radiation field standards for calibration of instruments that measure neutron fields for radiation protection purposes are given in NCRP 112 and ISO 8529. These include bare and D_2O-sphere-moderated ^{252}Cf spontaneous fission sources, and ^{241}Am:Be, ^{241}Am:B, and ^{239}Pu:Be (α ,n) sources. These reference sources are utilized in the calibration of portable neutron field survey meters, neutron area monitors, neutron electronic personnel dosimeters, and neutron transfer instruments. For most health physics purposes, so-called bare ^{252}Cf or ^{241}Am:Be (α ,n) neutron sources are normally chosen due to their well known spectra, availability, and, for ^{241}Am, its long half-life. The neutron standard is 4π emission rate, determined at NIST through a manganese sulfate bath measurement. Since the sources are always encapsulated and the source fluence rates at 90° utilized, an anisotropic correction factor near 1.04 is often applied. In addition, for the so-called ^{252}Cf sources, the non negligible ^{250}Cf contribution must always be summed with the ^{252}Cf emission rate and for the so-called ^{239}Pu:Be sources the ingrowth of ^{241}Am from ^{241}Pu decay must always be summed with the ^{239,240}Pu. Dose equivalent conversion factors of 3.33×10^{-5} and 3.73×10^{-5} mrem per n per cm^2 (Schwartz et al 1982 and Buckner et al 1992) are typically used in the USA for bare ^{252}Cf and ^{241}Am:Be (α ,n) neutron sources, respectively. The (ICRP 74 1997) ambient dose equivalent, H*(10), conversion factors of 385 and 391 pSv per n per cm^2 (Bohm et al 1999) for bare ^{252}Cf and ^{241}Am:Be (α ,n) neutron sources, respectively, are normally not applied in the USA; mrem/h is the unit of display on the health physics instrument. The neutron source is always used free-in-air, hence large scattering and facility design problems are always present. Either empirically determined scattering corrections or shadow shield sets are employed. For most health physics (i.e., inexpensive) purposes, the empirically determined scattering corrections are normally chosen to account for dosimeter-specific air scatter and room return (Schwartz et al 1982 and IAEA 285 1988).

Simulations via Monte Carlo (MCNP, etc.)/Point Kernel (MicroShield, QAD, etc.) computer codes are often utilized with portable and laboratory HPGe and NaI(Tl) xγ spectrometry and stack monitoring systems. Applications include in-situ HPGe soil measurements, charcoal cartridge (i.e., for ^{131}I) and Marinelli gas beaker (i.e., for ^{41}Ar) measurements, and portable NaI(Tl) detector in-situ response to contaminated or activated soils, metals, concrete, etc.

9.2.3 Air Flow/Pressure/Temperature Standards

Air flow standards consist of both mass flow and volumetric flow standards. Mass flow meters, such as those from Kurz, Omega, or Unit Instruments, utilize a pair of flow and temperature sensors. The flow sensor is heated to a constant temperature and responds to the mass flow by sensing the cooling effect of the air as it passes over the heated flow element. The temperature sensor accurately compensates for a wide range of ambient temperature variations. These meters are referenced to a standard temperature and pressure. If volumetric flow rate is desired, the indicated slpm reading must be multiplied by ρ_s/ρ_a , where ρ_s is the air density at the standard conditions and ρ_a the actual air density during use. Volumetric flow standards can consist of calibrated rotameters, variable-head meters utilizing a Magnehelic gage, soap film flowmeters, etc. A mercury (Hg) barometer/ thermometer is normally utilized as a pressure/temperature standard, accurate to 0.2 mm Hg and 0.5 °C when properly used with manufacturer-recommended corrections applied.

9.2.4 Statistical Tools

Statistical tools are required to determine and set appropriate alarm set points for the required sensitivities. These sensitivities and alarm set points depend on the given set of efficiencies, background count rates and count times, acceptable sample count times, overall false alarm probability, and detection probability. It is normally required to calculate Poisson distributions for essentially all α and neutron counting applications and even for β counting applications when unequal statistical α and β probabilities (type I and type II errors) are involved. When equal statistical α and β probabilities are involved, then the use of simple Normal (Gaussian) distributions is acceptable for β counting applications if the total count exceeds 30 counts (with of course Poisson-based standard deviations). Please note that, to a health physicist, a statistical $\alpha = 0.01$ would correspond to a 'false alarm' probability of 1% and a statistical $\beta = 0.05$ would correspond to a 'detection' probability of 5%.

9.3 REPRESENTATIVE EXAMPLES

As previously mentioned, due to the variety of health physics instrument types and models in existence, the calibration particulars of each type and model will not be discussed here. However, as stated previously, common to these particulars would be the use of just a finite number of the 'calibration tools.' The typical usage requirements of these calibration tools within the calibration process is summarized in Table 9.1 for various types of health physics instrumentation. Due to these common elements involving the use of the 'calibration tools', only a few specific examples of their representative use will be required in order to demonstrate the essential features of their use. Additional guidelines may be found in IAEA 133 (1971) and 285 (1988), NBS 633 (1982), and NCRP 112 (1991).

9.3.1 Hand & Shoe Monitor Calibration

An α/β hand and shoe monitor calibration will demonstrate the typical usage of the pulse generator, the electrostatic kilovoltmeter, wide-area reference α/β activity sources, and the necessary Poisson statistical calculations. The α/β hand and shoe monitor considered here could consist of a dual-channel Aptec OmniTrak-α/β instrument attached to a matched pair of hand and shoe detectors. The calibration begins by inspecting (and repairing as necessary) all detector windows, screens, and cabling, switches and sensors, displays and indicators, and other mechanical. The P-10 gas flow is inspected and adjustments made, if necessary.

The pulse generator is used to check and set a 2 mV input sensitivity, which effectively biases against spurious pulses and other noise. The pulse generator is then used to check midscale cpm readouts on all applicable ranges. With the α threshold raised as high as possible and placing a wide-area reference $^{90}Sr+^{90}Y$ β source on a detector, high voltage is adjusted until count rates level out. Note that this corresponds to gas amplifying the β dE/dx curve of Fig. 3.10 in Chapter 3 over the input sensitivity. The α threshold is then brought down until a very small but reproducible amount of the β dE/dx signals crosstalk over into the upper α channel. Using the electrostatic kilovoltmeter, check and confirm an expected or typical high voltage for the detector model. Determine the β efficiency from the source response in the lower β channel. Place a wide-area reference ^{241}Am α source on the detector and determine the α efficiency from the source response in the upper α channel. Remove sources and determine background count rates. Confirm that the efficiencies and backgrounds are as expected or typical for the detector models.

Table 9.1. Typical usage of the 'calibration tools' within the calibration process for various types of health physics instrumentation.

HP Instrument Type	Electronic Standard	Radiation Standard	Air Flow/etc. Standard	Statistical Tools
Portable α/β Contamination Survey	PG, kVm	αβA		
Hand & Shoe/ Personnel Contamination Monitor	PG, kVm	αβA		P
α/β Lab. Assay (PC, dual scint., LSC)	PG, kVm, D/Em	αβA, αβE		P, N
Portable xγ Contamination Survey	PG	γA, γE		
Portable βγ Field Survey	PG, kVm	γF, βF	(Hg)	
Portable neutron Field Survey	PG, kVm	γF, nF		
Area Monitors	PG, kVm, D/Em	γF, nF	(Hg)	(P)
Electronic Personnel Dosimeters		γF, nF		(P)
Air Monitors	PG	αβE, Gas	MF, VF, Hg	P
Air Samplers			VF, MF, Hg	
Portable and Lab. HPGe	PG, Osc	γA, γE, MC		N
Stack Monitors	PG, kVm, Osc	αβA, αβE, Gas, γA, γE, MC	MF, VF, Hg	(P)
Transfer Instruments	PG, kVm, D/Em	Gas, γF, nF	Hg	

Key to Table 9.1:

- PG=pulse generator, kVm=electrostatic kilovoltmeter,
- D/Em=digital multimeter/electrometer, Osc=oscilloscope
- αβA=α/β wide-area activity standards, αβE=α/β energy standards

- Gas=tritium, krypton, radon, etc. gas activity standards
- γA=xγ activity standards (multi), γE=xγ energy standards (multi)
- γF=γ radiation field standard, βF=β radiation field standard, nF=neutron radiation field standard
- MC= Monte Carlo or other simulations
- MF=mass flow standard, VF=volumetric flow standard, Hg=pressure/temperature standard
- P=Poisson distribution-based analysis, N=Normal distribution-based analysis
- ( )=might be used but normally is not

As a demonstration of the necessary Poisson statistical calculations, let's assume that the α response is 0.2 cts/dis, the α background can be as high as 12 cpm, the sensitivity (MDA) required is 500 dpm α at a reasonably high detection probability, an acceptable false alarm probability for this α channel is 0.5 to 1%, and effective background count times are long enough compared to sample count times that their measurement error can be neglected. Additionally, the sample count time should be less than 10 seconds. An iterative approach is utilized to arrive at a count time and alarm set point that satisfies the requirements. In this case, this is a count time of 7 sec with an alarm set point of 0.85 cps or 510 cpm. At the maximum background of 12 cpm (or 0.2 cps), the Poisson mean would be 1.4 counts. The false alarm probability for 6 or more counts (i.e., 0.85 cps) is 0.32%. The Poisson mean that corresponds to a detection probability of 95% for 6 or more counts is 10.513 counts. The best L_D is therefore 10.513-1.4 net cts per 7 sec = 1.3 cps or 78 cpm. The best MDA is 78 cpm ÷ 0.2 cts/dis = 390 dpm. The worst would correspond to a minimal background such as near 0 cpm. The false alarm probability would be negligible, but the L_D would then be 10.513-0 net cts per 7 sec = 1.5 cps or 90 cpm. The worst MDA is then 90 cpm ÷ 0.2 cts/dis = 450 dpm. As one can see, all the requirements have been met.

9.3.2 Portable xγ Contamination Survey Instrument Calibration

A portable FIDLER xγ contamination survey instrument calibration will be used to demonstrate the typical usage of the pulse generator, the electrostatic kilovoltmeter, and xγ reference energy sources. The mini-FIDLER (field instrument for the detection of low energy radiation) instrument considered here could consist of an Eberline ASP-2e instrument attached to a Ludlum 44-17 2mm by 2"-diameter NaI(Tl) detector. The calibration begins by inspecting (and repairing as necessary) detector window and screens, cabling, batteries, switches, displays and indicators, and other mechanical.

The pulse generator is used to check and set a 5 mV lower threshold, which effectively biases against spurious pulses and other noise. For temporary calibration purposes, the upper threshold is set to 10 mV. The pulse generator is also used to confirm the cpm calibration and that the microprocessor is properly sensing switch changes by checking midscale cpm readouts on all ranges. The rate meter response times are appropriately chosen, as is the scaler time, and the dead time set to the manufacturer recommended value of 12 μsec. With an appropriate ^{241}Am xγ reference source at the detector, the Run Plateau is initiated from the Calibration menu for 500 to 1200 V. The resulting plateau is inspected for the locations of the 59.5 keV γ and the 17 keV L_β x-rays. The HV at the center of the second peak (17 keV L_β x) is re-entered under Channel Parameters of both Main and PHA. The source is then removed and the background count rate in Main mode determined and confirmed to be indicative of the typical mini-FIDLER background in a typical ambient γ background field.

9.3.3 Portable βγ Field Survey Instrument Calibration

A portable energy-compensated GM βγ field survey instrument calibration will be used to demonstrate the usage of the pulse generator, the electrostatic kilovoltmeter, and the reference xγ radiation field standard. The energy-compensated GM survey instrument considered here could consist of an Eberline ASP-1 instrument attached to a high-sensitivity TGM N378S/BNC energy-compensated GM detector. The calibration begins by inspecting (and repairing as necessary) the detector probe, cabling, batteries, switches, displays and indicators, and other mechanical. The rate unit should be set to mR/h and the dial to x1K to x.01. The audio should be set to divide by 1.

The electrostatic kilovoltmeter is used to check and adjust, if necessary, the high voltage to 450 V. The pulse generator is used to check and adjust, through the gain/threshold controls, a 30 mV input sensitivity, which effectively biases against reflected pulses and other noise. The (repetitive) pulse generator is then further utilized to confirm that the microprocessor is properly sensing switch position changes by checking midscale response on all switch range positions. The detector is placed in a reference xγ radiation field of approx. 1 mR/h and the linear constant adjusted to yield a response within 5-10% of the established calibration field value. The detector is then placed in a reference xγ radiation field of approx. 150 mR/h and the dead time constant is increased to yield a response within 10% of the established calibration field value. Responses are then additionally checked for at least three additional intermediate xγ radiation field values.

9.3.4 Portable neutron Field Survey Instrument Calibration

A portable neutron remmeter survey instrument calibration will be employed to demonstrate the typical usage of the pulse generator, the electrostatic kilovoltmeter, the reference xγ radiation field standard, and the reference neutron radiation field standard. The neutron remmeter survey instrument considered here could consist of an Eberline ASP-1 instrument attached to an Ludlum 42-31 (the 9-inch Cd-loaded Hankin's sphere with a N.Wood G-5-1 $^{10}BF_3$ detector). The calibration begins by inspecting (and repairing as necessary) detector, cabling, batteries, switches, displays and indicators, and other mechanical. The rate unit should be set to mrem/h and the dial to x100K to x1. The audio should be set to divide by 1.

The pulse generator is used to check and adjust, through the gain/threshold controls, a 2 mV input sensitivity, which effectively biases against spurious pulses and other noise. The electrostatic kilovoltmeter is used to check and adjust, if necessary, the high voltage to 1600 V. This should correspond to gas amplifying the α plus Li energy deposition portion (at the high pulse heights of Fig. 3.11 of Chapter 3) over the input sensitivity, while discriminating against the secondary electron dE/dx portion (at the low pulse heights of Fig. 3.11). In order to confirm this, a gamma insensitivity test is performed at approximately 10 R/h and, if necessary, the HV reduced. Note that after the linear constant is determined (see next paragraph), the constant is checked to be typically as large as expected for that model of Hankin's sphere remmeter. If it is not, the HV could be increased or the detector tube replaced.

The (repetitive) pulse generator is then further utilized to confirm that the microprocessor is properly sensing switch position changes by checking midscale response on all switch range positions. The remmeter detector is placed in a reference neutron radiation field of approx. 50 mrem/h and the linear constant adjusted to yield a response within 5-10% of the established calibration field value. Since the remmeter can in no way be placed in a reference neutron radiation field near 100 rem/h, the dead time constant is simply adjusted to the manufacturer's recommended value of 10 μsec. Responses are then additionally checked for at least three additional neutron radiation field values.

9.3.5 Portable HPGe System Calibration

A portable (or laboratory) HPGe system calibration will be used to demonstrate the usage of a pulse generator, oscilloscope, xγ reference sources utilized as both energy and activity standards, and possible the results of Monte Carlo or Point Kernel computer code simulations. The portable HPGe xγ spectrometry system considered here could consist of an Ortec Nomad Plus HV/amp/MCA instrument attached to an Ortec n-type coaxial HPGe detector and utilized for in-situ soil measurements. The calibration begins by inspecting (and repairing as necessary) detector tripod and cabling, MCA and laptop PC batteries, switches and sensors, displays and indicators, any other mechanical, and the laptop PC state of operation. The LN_2 (liquid nitrogen) supply is inspected and filled, if necessary. A ^{137}Cs button γ source should be placed at the detector for pole-zero adjustments. If pole-zero adjustments are not automatic, then an oscilloscope will be required. A short count is then made with a γ reference source typically comprised of a combination of single gamma energy and multiple gamma energy isotopes. The system is then energy calibrated by a quadratic fit of peak channel number to calibration library peak energy. The library appropriate for the measurements at hand is then chosen and/or edited. Efficiency values from MCNP or other computer code simulations are then also chosen for the measurements at hand (normally as cts per γ per gram rather than the more typical cts per γ).

9.4 USER FUNCTIONAL TESTS

Recommendations have been made by various commissions and committees regarding the measurement accuracy requirements for health physics instruments. At dose levels higher than those corresponding to one-quarter of the occupational limits, the desired measurement accuracy is ±30% (ANSI N323A 1997). Note that this standard does not address the difficult topic of the subsequent workplace measurement systematic and random errors. This measurement accuracy is to be achieved through (1) the establishment of calibration fields or activities with uncertainties within ±5-10% with respect to a secondary standard's calibration, (2) instrument calibrating errors within ±10-20% with respect to these established calibration fields or activities, and (3) limiting instrument drift with respect to the instrument calibration to within ±20% as determined by the user operational/functional tests. In this way the measurement accuracy of a health physics instrument in a user's hand at any time between recalibrations would be expected to be better than ±30% (i.e., $0.1^2 + 0.2^2 + 0.2^2 = 0.3^2$).

9.5 ACKNOWLEDGMENTS

The author is indebted to Mr. Scott Borkowski for his contribution to Section 9.2.1, to Mr. Jerry Letizia for his contribution to Sections 9.3.1 and 9.3.3, and to Mr. Todd Teel for his contribution to Sections 9.3.2 and 9.3.4. Thanks are also extended to Dr. V. Rao Veluri and Mr. McLouis Robinet for their review and comments.

9.6 REFERENCES

American National Standards Institute. Radiation Protection Instrumentation Test and Calibration, Portable Survey Instruments. ANSI N323A. Institute of Electrical and Electronics Engineers, Inc. New York, 1997.

Bohm, J.; Alberts, W.G.; Swinth, K.L.; Soares, C.G.; McDonald, J.C.; Thompson, I.M.G.; Kramer, H.M. ISO Recommended Reference Radiations for the Calibration and Proficiency Testing of Dosimeters and Dose Rate Meters used in Radiation Protection. Rad. Prot. Dos. 86: 87-105; 1999.

Buckner, M.A.; Sims, C.S.. Neutron Dosimetric Quantities for Several Radioisotopic Neutron Sources. Health Phys. 63: 352-355; 1992.

International Atomic Energy Agency. Handbook on Calibration of Radiation Protection Monitoring Instruments. IAEA Technical Report Series No. 133. IAEA, Vienna, Austria, 1971.

International Atomic Energy Agency. Guidelines on Calibration of Neutron Measuring Devices. IAEA Technical Report Series No. 285. IAEA, Vienna, Austria, 1988.

International Commission on Radiological Protection. Conversion Coefficients for use in Radiological Protection against External Radiation (Adopted by the ICRP and ICRU in September 1995). ICRP Publication 74. Pergamon Press, 1997.

International Organization for Standardization. X and γ reference radiations for calibrating dosemeters and dose ratemeters and for determining their response as a function of photon energy. ISO 4037, current version.

International Organization for Standardization. Reference beta radiations for calibrating dosemeters and dose ratemeters and for determining their response as a function of beta radiation energy. ISO 6980, current version.

International Organization for Standardization. Neutron reference radiations for calibrating neutron-measuring devices used for radiation protection purposes and for determining their response as a function of neutron energy. ISO 8529, current version.

International Organization for Standardization. Reference sources for the calibration of surface contamination monitors - Beta-emitters (maximum beta energy greater than 0.15 MeV) and alpha-emitters. ISO 8769, current version.

National Council on Radiation Protection and Measurements. Calibration of Survey Instruments Used in Radiation Protection for the Assessment of Ionizing Radiation Fields and Radioactive Surface Contamination. NCRP Report No. 112. NCRP, Bethesda, Maryland, 1991.

Schwartz, R.B.; Eisenhauer, C.M. Procedures for Calibrating Neutron Personnel Dosimeters. NBS Special Publication 633, National Bureau of Standards, Washington, D. C., 1982.

Chapter 10

INSTRUMENTATION FOR THE MEASUREMENT OF ELECTROMAGNETIC FIELDS 3 kHz-300GHz

John A. Leonowich

10.1 INTRODUCTION

We are literally surrounded by electromagnetic radiation and fields, from both natural and man-made sources. The identification and control of man-made sources of has become a high priority of radiation safety professionals in recent years. For the purposes of this chapter, we will consider radio frequency radiation (RFR) to cover the frequencies from 3 kHz to 300 MHz, and microwaves from 300 MHz to 300 GHz, and will use the term RFR interchangeably to describe both. Electromagnetic radiation and fields below 3 kHz is considered Extremely Low Frequency (ELF) and will be discussed in the next chapter. Unlike x- and gamma radiation, RFR is *non-ionizing*. The energy of any RFR photon is insufficient to produce ionizations in matter. The measurement and control of RFR hazards is therefore fundamentally different from ionizing radiation. The purpose of this chapter is to acquaint the reader with the fundamental issues involved in measuring RFR fields with currently available instrumentation.

10.2 THE ELECTROMAGNETIC SPECTRUM

Understanding of the physical properties of electromagnetic waves and the components of the electromagnetic spectrum are absolutely essential in order to measure non-ionizing radiation. Electromagnetic energy can have many forms: static electric or magnetic fields; power frequency ELF fields, RFR and microwaves, optical radiations (infrared, light and ultraviolet); and the ionizing radiations. Ionizing radiations, such as x-rays and gamma rays, have extremely high frequencies (i.e. very high energy) and are able to produce ionization (positive and negative electrically charged molecules) by breaking the bonds that hold molecules in cells together. Non-ionizing radiations, on the other hand, have very small amounts of photon energy and are much too weak to be able to break these molecular bonds. Therefore, even high intensity non-ionizing radiation cannot cause ionization or radioactivity in the body. Fig. 10.1 shows the electromagnetic spectrum, along with the frequency and location in the spectrum of each radiation and field.

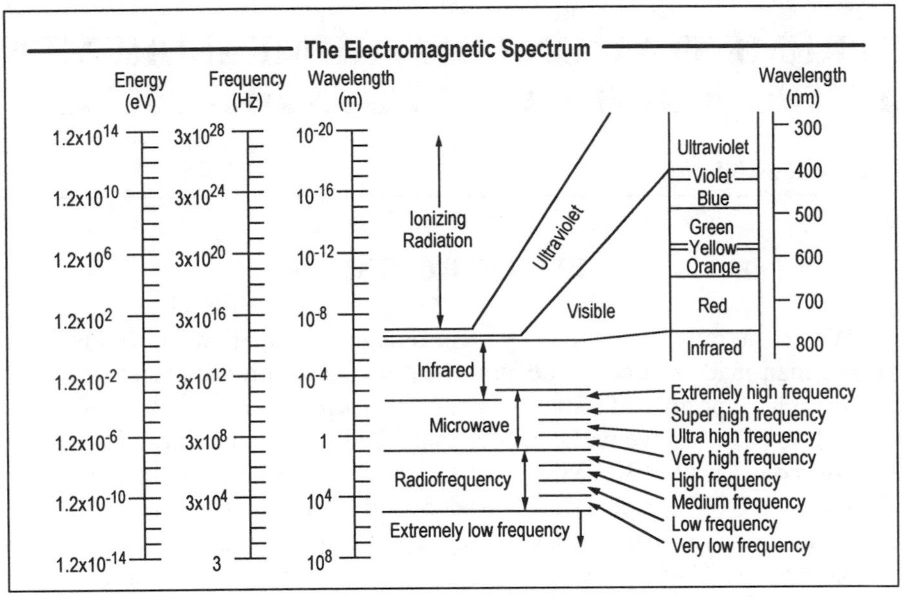

Fig. 10.1 - The electromagnetic spectrum.

10.3 PHYSICS OF ELECTROMAGNETIC WAVES

The information in this section is a summation of the basic information required to understand the basic physics of electromagnetic waves. More in-depth treatment of the subject can be found in a number of other references (Leonowich, 1992, 1997).

Electromagnetic energy is propagated through space in the form of a wave as shown in Fig. 10.2. For plane waves traveling in free space, the electric field vector **E** and the magnetic field vector **H** are mutually orthogonal and are in phase - i.e. maxima and minima occur at the same point in time and space. The units of **E** are volts/meter (V m^{-1}) and for **H** amperes/meter (A m^{-1}).

When RFR is sufficiently far away from a radiation source (the so-called far field), the orthogonal relationship between the **E** and **H** field vectors is taken for granted. The energy carried in an electromagnetic wave is usually expressed in terms of the energy passing through a fixed area per unit time. For an electromagnetic wave this *power density* **S** at any point may be calculated from the vector product of the electric and magnetic field strength vectors, i.e. **E x H** = **P**. **P** is called Poyntings Vector and represents the power density and the direction of energy propagation. The Poynting vector points in the direction in which the electromagnetic wave is traveling, thus transporting energy in that direction. The magnitude of the power density propagated in the wave can be calculated from the vector product:

$$|\mathbf{ExH}| = |E|\,|H| \sin \Theta \qquad (10.1)$$

For $\Theta = 90^{\circ}$, as is the case for a plane wave in free space:

$$|\mathbf{ExH}| = |E|\,|H| \qquad (10.2)$$

Note that if **E** has dimensions of V m^{-1} and **H** is in units of A m^{-1} the dimensions of **P** and **S** are W m^{-2}. More typically in hazard analyses **S** is expressed in milliwatts per square centimeter (mW cm^{-2}). Power density can be equated to **E** and **H** as follows:

$$\mathbf{S} = \mathbf{E}^2 / 377 = 377\,\mathbf{H}^2 \qquad (10.3)$$

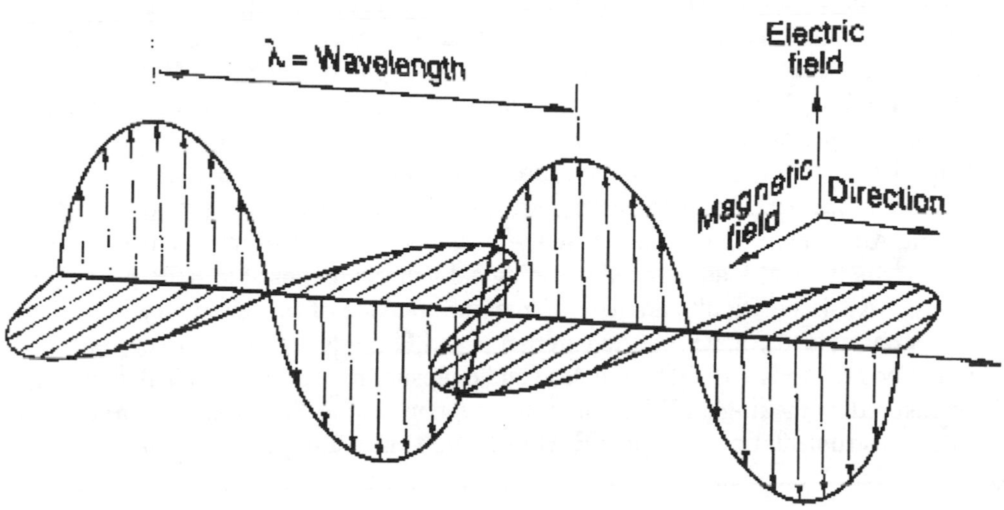

Fig. 10.2 - Typical electromagnetic wave showing the relationship between the **E** and **H** field vectors.

Polarization is another important property of electromagnetic waves and is characterized by the oscillatory behavior and orientation of the electric field vector. A wave referred to as being linearly polarized means that the electric field vector varies in amplitude in only one direction as it travels. It is conventional to describe polarization in terms of the electric field only, not the magnetic field. An electromagnetic wave may exhibit linear, circular, elliptical, or random polarization (such as in a light bulb). A receiver (or absorber) of electromagnetic radiation must have the same sense of polarization as the incident wave for it to be detected or absorb energy from it <u>most efficiently</u>.

Suppose we could "freeze" the motion of an electromagnetic wave traveling in free space that we have previously discussed and take a snapshot of its **E** and **H** fields. The ratio of the electric field strength to the magnetic field strength is known as the characteristic impedance (or wave impedance Z_O) of free space. Therefore:

$$Z_O = |\mathbf{E}/\mathbf{H}| \tag{10.4}$$

Note that Z_O is the ratio of **E** to **H** and is independent of their absolute magnitudes and has a value in free space of 120π or approximately 377 ohms. Therefore, in the far field, if one can measure either field component (**E** or **H**), then it is easy to calculate the other. Unfortunately, in many cases measurements are needed to be made in the so-called "near field", where the orthogonal relationship between **E** and **H** does not hold. In this case, both **E** and **H** fields must be measured. It is usually assumed that near field conditions are present below 300 MHz.

Table 1 lists both the dose and exposure metrics associated with RFR safety standards. As can be seen, for a very large portion of the RFR spectrum (100 kHz – 10 GHz), Specific Absorption Rate (SAR) in W kg^{-1} is the dose metric of importance. SAR is a measure of the thermal load induced in biological material exposed to RFR. Unfortunately, SAR is a difficult parameter to measure outside of the laboratory. However, it is possible to equate SAR to external **E** and **H** fields, as well as to currents (I) induced in the body. Above 10 GHz, the dose and exposure metric are identical, and can easily be measured by making an **E** field measurement and using Eqn. 10.3. Therefore, all RFR safety instrumentation measures **E**, **H** or I. None measure power density directly.

Table 1. Metrics in exposure standards for RFR.

Frequency range	*Biomechanism*	*Dose metric*	*Exposure metrics**
VLF / LF (3-100 kHz)	Neuromuscular stimulation	Current density in excitable tissues	E, H, induced currents & contact currents (I)
Intermediate RFR (MF – SHF) (100 kHz–10 GHz)	Tissue heating	Specific Absorption Rate (SAR) in W/kg	E^2, H^2 (<300 MHz), I^2 from induced currents, & contact currents (<100 MHz)
Millimeter Wavelength Microwaves (10 – 300 GHz)	Surface heating	Power Density In W/m^2	E^2, Power Density In W/m^2

*All RFR exposure metric magnitudes are root-mean-squared (RMS) values.

10.4 DEVELOPMENT OF BROADBAND INSTRUMENTATION FOR MEASURING RFR FIELDS

Characterization of field strengths for RFR safety applications is a demanding task. High level, or *potentially* high level, fields need to be compared against national and international standards. Standards, while providing excellent information about the field limits and their biological basis, typically provide little information about instrumentation used to demonstrate compliance.

Most measurements are performed with wideband instrumentation consisting of a metering instrument and a broadband field probe. The broadband field probe is comprised of isotropically mounted antennas, which are the detector (in the case of thermocouple arrays) or have a detector at their junction (as a diode based system, see Fig. 10.3). The advantages of these types of systems over narrowband (i.e., small frequency range) equipment, is that they are able to measure the total field strength of multiple sources simultaneously. With isotropic sensor designs, wideband sensors can accurately sum these fields, regardless of the polarization employed, or the direction of propagation. This assumes that the detector attached to the antenna is accurately processing the incoming waveform, which is not always the case. Little information is available to the user about how their instrument will fare in other than ideal situations. In most, if not all situations, the detector used will be affected by outside forces. Effects of signal type (frequency, modulation, harmonic content, number of signals) and environment (temperature, humidity, etc.) can seriously degrade measurement accuracy.

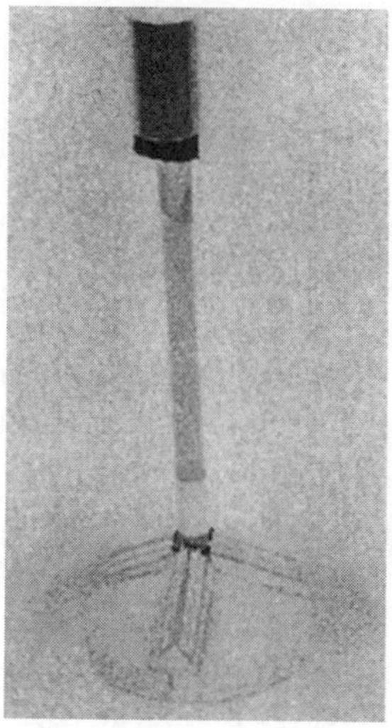

Fig. 10.3 - Typical isotropic RFR field sensor using three orthogonal elements.

Documents such as IEEE C95.3-2001*, as well as its predecessor (IEEE, 1991) provide detailed information about sensor technology, but very little guidance as to when and where a particular sensor will operate with the highest accuracy. In order to perform proper measurements of radiated fields, an operator needs to be aware of the sensor, and its characteristics. This chapter introduces the various types of sensors available and their characteristics. The most important are the diode, compensated diode and thermocouple type of sensors.

10.5 DEVELOPMENT OF RFR INSTRUMENTATION

Measurements of RFR fields for safety purposes have been performed over the last 30 years or so with broadband equipment. The average user is usually unaware of design limitations and compromises that different manufacturers have reached for their particular customer base or measurement philosophy. These compromises will dictate how well an instrument will perform in a given RFR environment. Approximately 35 years ago, the commercialization of microwave ovens generated a need for instrumentation operating at 915 and 2450 MHz in order to obtain leakage information

* Institute of Electrical and Electronics Engineers., Draft recommended practice for measurements and computations with respect to human exposure to radio frequency electromagnetic fields, 3 kHz to 300 GHz, C95.3-200x. New York, NY. Institute of Electrical and Electronics Engineers; 2001.

for manufacturers and repair organizations. Awareness of RFR energy and its possible effects led to developments of broader frequency range monitors that at first were circularly polarized in an attempt to respond to all polarizations. These instruments were therefore not isotropic in their detection capability and their effectiveness was markedly affected by geometric considerations. Without a priori knowledge of the field to be measured it was therefore quite possible to make a totally erroneous conclusion on the amount of RFR energy present. These initial products were very broadband for their time, covering the spectrum from 1 - 14 GHz by the use of thermistor detectors. While the thermistor was very linear in its response, the receiving antenna design was not, necessitating multiple frequency calibrations to overcome polarization and frequency sensitivity errors of up to 10 decibels (dB). The next generation of these circularly polarized monitors incorporated thermocouple detectors and improved antenna designs that reduced frequency sensitivity errors to about 6 dB. About 25 years ago the first isotropic detection probes came on the market. Electric field probes were made available covering the spectrum from 300 MHz to 18 GHz with a frequency sensitivity of only 3 dB, and a measuring range of 30 dB. During the early 1970's there were also many advances in calibration methods and procedures for quantifying RFR fields from then United States National Bureau of Standards (now NIST). Near field calculations and transverse electromagnetic (TEM cell) developments allowed for even higher calibration accuracies over a broad range of frequencies to uncertainties of +/- 0.5 dB. Also in this time period the development of magnetic field probes was accomplished in part to measure the magnetic fields associated with high frequency (HF) communication systems. The impetus for the development of much of this isotropic instrumentation was the United States military, in particular the U.S. Air Force. The development of this broadband instrumentation overcame many of the problems associated with the earlier measurement equipment. Previously tedious measurements with narrowband equipment were replaced with broad sensors covering many decades of frequency. Standards had constant field limits over frequency during this period of time. Approximately 20 years ago, RFR standards became frequency dependent as research determined the need to further limit the fields at the human resonant frequencies, due to increased absorption of RF energy. In addition, fields tested (**E** and/or **H**) changed depending on the particular standard, and frequency of emitters. Sensors have not changed as much as the instrumentation used to display the field detected. The main reason for this is that most limits are still based on the RMS, time averaged field level. Since it is the RMS average power level available to do the work, or create the heating of tissue, thermal limits still dominate field limit considerations. New metering instruments simplify the recording of surveys, but the field probe determines the overall accuracy. Some meters can enhance accuracy by compensating for temperature or measurement level errors, particularly when diodes are employed as sensors.

10.6 DETECTORS FOR RFR: DIODES VERSUS THERMOCOUPLES

All currently available RFR instrumentation relies on either diodes or thermocouples as detectors. A modern broadband RFR survey meter, incorporating probe, cable and meter is shown in Fig. 10.4.

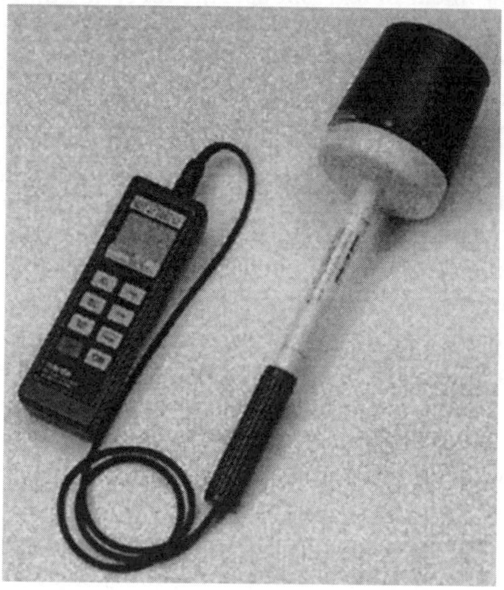

Fig. 10.4 - Modern Broadband RFR Survey Meter (courtesy of NARDA Microwave, Inc.).

Diode sensors have been used for measurements of RF and microwave power for many years. Semiconductor diodes are used routinely to measure absolute RF power. Diodes exhibit many characteristics that make them desirable for measuring RFR, but also do have disadvantages. They are relatively cheap and rugged. Diodes work well at low RFR field levels when the diode is in its "square law" region. A diode is a non-linear device, which when operated at higher levels will detect peak rather than average levels of RFR, as shown in Fig. 10.5. Diodes are therefore not recommended for multiple signals or pulsed environments. They do, however, have intrinsic "zero stability" and seldom drift.

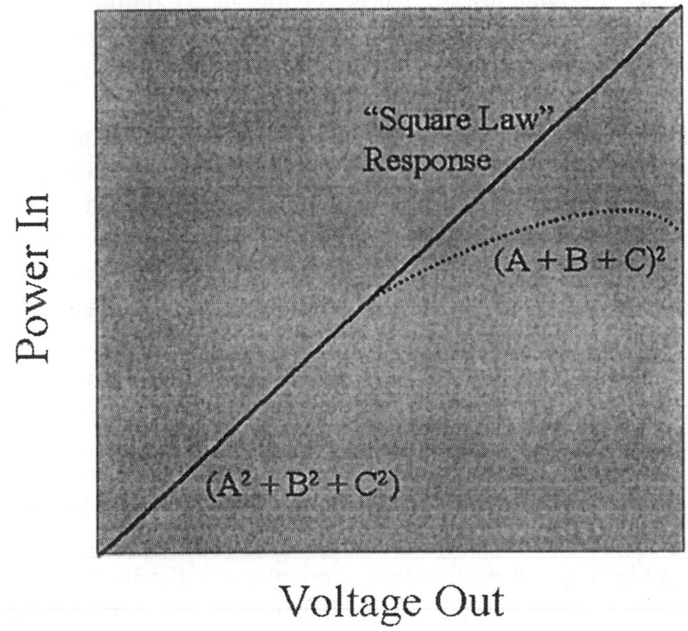

Fig. 10.5 - Diode Linearity Versus Input Level.

The RF induced current in a diode can be represented mathematically as:

$$I = I_s[\exp(V_j / V_t) - 1] \quad (10.5)$$

The characteristics of the diode are reflected in the "thermal-voltage," V_t and the reverse saturation current I_s. This current is a function of device area, materials used to form the junction, and temperature. The temperature dependence is very important because most of the change in detector performance can be related to variations in I_s, which changes by approximately a factor of 2 every 20° C. Therefore, it is imperative that a diode field sensor incorporate at least a basic temperature compensation circuit to correct for I_s variations. There are some improvements to accuracy that metering instruments used with diodes can perform. The most common is to correct for linearity errors in diode detectors. Circuits that compensate diode output will only correct for the diode's response to a particular signal, generally a sinusoid. If the signal is not sinusoidal, the relationship of peak to RMS voltage (power) is different and results in an error for the power measurement. The second accuracy enhancement is to compensate for the diode temperature sensitivity. This sensitivity can be quite large, as was mentioned as much as a 2:1 change for 20 degrees centigrade. Compensated diode sensors can be utilized for multiple-signal environments, but are not recommended for pulse modulated emitter measurements.

Thermocouple detectors have been the detection technology of choice for sensing microwave frequencies since their introduction in the late 1960's. The main reason for this being that they feature, unlike diodes, an inherent square-law detection characteristic (input RF power is proportional to DC voltage out). The energy of the pulse is detected by a series of thermocouples arranged in series to form a resistive dipole antenna. As is well known, energy is deposited at the speed of light. Since thermocouples are heat-based sensors, they are true "averaging detectors." This recommends them for all types of signals formats from CW to complex digital phase modulations.

Thermocouples are based on the fact that dissimilar metals generate a voltage proportional to temperature differences at a hot and cold junction of two metals. Probes that utilize thermocouple detectors function in two different modes. Between 300 MHz and 18 GHz, they operate as a resistive dipole. Above 12 GHz, they utilize the phase delay of a traveling wave to produce additional output. As a resistive dipole, in the 1500 MHz to 18 GHz region, each probe contains three pair of mutually perpendicular elements, as shown in Fig. 10.6. Resistive thermocouples are distributed along the length of the dipole at a spacing that will not permit resonance over the operating range of frequencies. The dipole may be viewed as a group or series connected small resistive dipoles or as a very low Q resonant circuit.

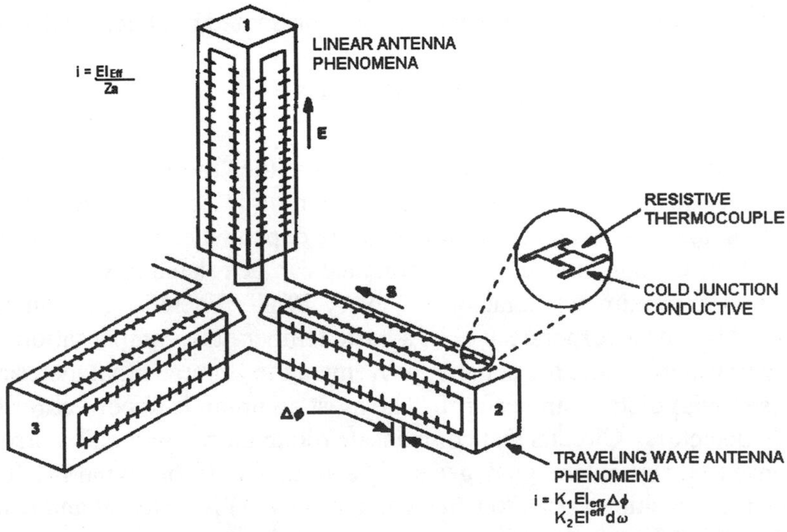

Fig. 10.6 - Typical thermocouple antenna array (courtesy of NARDA Microwave, Inc.).

Negative points associated with thermocouples are inherently lower output voltage than diode causing "zero drift" and lower overload level. Early thermocouple based detectors exhibited excellent accuracy in complex, modulated field environments but were limited by overload specifications that a careless operator could exceed, thereby damaging or destroying these probes, although state of the art thermocouples now have overload protection of at least 2000% of full scale reading or 13 dB.

10.7 MEASUREMENT PROBLEMS AND THE SELECTION OF EQUIPMENT

With any diode-based detector, there will be a point within its dynamic range (or distance from the RFR emitter) where it neither underestimates nor overestimates the average field reading. It may not be accurate except for this one point, or level, in front of the radar. Both diode and thermocouple sensors were recently tested by the authour in front of the ROLAND air defense radar system (Fig. 10.7). The Roland system features a fast rotating (60-rpm) radar with a minimal pulse width (400 nanoseconds pulse width, 5 kHz pulse repetition frequency). The diode sensor generated a reading that was 6 dB (a factor of four!) lower than the thermocouple probe. This disparity of values was to be expected, since the pulse width is short compared to the period required for a diode to perform accurate detection. In addition, other systems, such as the multi-emitter environment found at many military and telecommunication sites also require the appropriate selection of instrumentation since the RF fields around these systems are quite complex. With appropriate broadband equipment, measuring levels of RF around such systems when the RF levels are within 10 dB of the uncontrolled limits of IEEE or the public limits of the U.S. FCC is entirely feasible. Shaped probe technology, which read out in "percent of standard", are particularly useful in the multi-emitter environment. It is therefore imperative that the surveyor has comprehensive knowledge of the ambient RF environment before beginning a survey.

There are presently a number of different companies which market broadband RFR survey equipment. What, then, are some of the attributes of broadband instrumentation that the surveyor should keep in mind when selecting such equipment to perform an RFR field survey? Selection of the appropriate instrument depends on a number of factors that should be clearly understood before the instrument is obtained and include such issues as the ability to do spatial averaging, tolerance to in and out-of-band interference (particularly if magnetic fields are to be measured), and limitations of near field accuracy. Time and spatial averaging, required by the present IEEE RFR standard (IEEE 1999), is often neglected in practice. Spatial averaging offers a much more useful picture of the actual exposure situation than a single point measurement. Therefore, in performing RFR field measurements where highly localized fields exist, spatial averaging techniques should be applied. In continuous field applications, like those found around RF induction sealers, measurements may be performed at different portions of the body and averaged together manually. In the case of rotating radar where field levels are varying constantly, a time averaging module is the only way to truly measure, and average, the total exposure. There are currently modules available that can perform this

Fig. 10.7 - Roland Air Defense System Is An Example of A System Where Diode Based Detection Is Inadequate To Measure the RFR.

averaging automatically, assisting in determining time averaged exposure. There is at least one meter marketed with can perform spatial averaging over the body or workplace area and show the result automatically. This system has a "pause" feature that allows horizontal and vertical scan averages to be summed so that one can determine averages in multiple planes or for cubic areas. There is no doubt that continued advances in microprocessor technology will produce modules which will be able to perform all these functions, and because of their small size will produce minimal perturbations in the ambient RFR field. Early magnetic field probes of certain designs were not truly isotropic. A phenomena known as "spatial shadowing" existed wherein one or more of the three orthogonally mounted detector loops did not allow the same flux lines to pass through all loops. This was caused by the three loops not having a common vertex. Later designs have corrected this source of error. Nonetheless, these probes were still difficult, if not impossible, to use in multiple emitter applications because of there erroneous response to signals above their operating frequencies. Away from the controlled laboratory environment, where emitters are present throughout the frequency spectrum, large measurement errors are present when an out-of-band transmitter exists nearby a survey area. Newly designed magnetic field probes are now available that greatly enhances the accuracy and operator confidence in magnetic field measurements. Nonetheless, magnetic field measurements of low frequency emitters can be a difficult exercise for the novice surveyor and should be not be accomplished without first understanding the limitations of the instrumentation. In addition to the problems associated with magnetic field measurements is the knowledge that in many instances the surveyor will be in the near field region of the emitter, where the RFR environment is very complicated, as shown in Fig. 10.8. In order to perform near field surveys accurately, the instrument must respond to only one field with no spurious responses; not produce significant scattering; and utilize isotropic, non-directional and non-polarized probes. In conclusion, users must be aware of not only the strong points of a particular system but also its weak points. No single broadband instrument currently available will meet the needs of all users. Even though products might be specified similarly their operation in a given environment may be significantly different. It is therefore important that the radiation protection professional be aware of these differences when choosing one for a particular application. The following guidelines should be useful in selecting a probe and meter for performing a survey: field sensors employing diode detection should only be used for continuous wave (CW) signals; average power measurements on pulsed and complex modulation signals should be measured using thermocouple probes; and finally, compensated diode sensors can be utilized for multiple-signal environments, but are not recommended for pulse modulated emitter measurements.

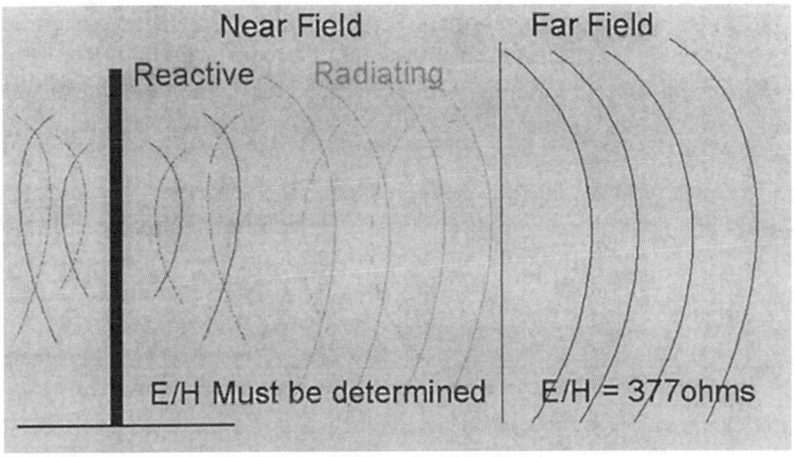

Fig. 10.8 - The RFR Environment In The Near Field Is Complicated And Requires Measurement of Both **E** and **H**

10.8 MEASUREMENT OF INDUCED AND CONTACT CURRENTS

As was previously mentioned, in addition to the measurement of **E** and **H** componenets of an RFR field, compliance with standards may also require the determination of induced and contact currents (ICC). The current IEEE standard (IEEE 1999), requires the measurement of ICCs at frequencies less than 100 MHz. The rationale for the development of ICC limits has been reviewed (Leonowich 2000). In principle, the measurement of induced currents is reasonably straightforward, but in practice can be quite difficult. Several manufacturers have produced devices to measure ICCs. The type of measurement instrument to use is still open to choice: transformer (ankle) clamps versus stand-on "bathroom scale" to measure induced currents. Examples of these types of instruments are shown in Figs. 10.9 and 10.10. Comparison of the two basic types of instruments was performed at an ICC workshop held at Brooks AFB, Texas in May 1996. The main conclusions of the workshop were that transformer/ankle probes were generally in good agreement, however, large variations were found in the stand-on probes between manufacturers and different units of the *same* manufacturer. Transformer/ankle probes also showed less variation with changes in grounding and are less susceptible to radio frequency interference (RFI). Stand-on probes usually underestimate true ankle current if not firmly in contact with ground, most likely due to the flow of displacement current around the platform probe. Because of these variations, there remains a need for rigidly standardized measurement protocols when making ICC measurements. In addition, induced current measurement of individuals standing in a uniform RFR field show markedly different results, which is a complex function of body height, girth, footwear, and tissue composition. One solution to this problem is to use a standardized phantom to make these measurements. At least one manufacturer is marketing a human equivalent antenna (HEA), which has impedance similar to a human being. The appropriateness of using an HEA as a surrogate for human exposure is still open to study.

In addition to induced current measurements, instruments to measure contact current have also been developed. These types of instruments simulate either point (single finger) or grasping (full hand) contact. An example of such an instrument is shown in Fig. 10.11.

MISSION
RESEARCH
CORPORATION
Clamp-Lite™
CLAMP-ON
CURRENT PROBE

MRC Model MG-4501B

Fig. 10.9 - Lightweight Clamp For Measuring Induced Currents (courtesy of Mission Research, Inc.)

10.9 OTHER RFR INSTRUMENTATION AND FUTURE NEEDS

The instrumentation already discussed allows the surveyor to make measurements of the exposure metrics of **E**, **H** and I which form the basis of all RFR standards. The instrumentation is analogous to ionizing radiation survey meters. Personnel dosimetry similar to TLDs for ionizing radiation does not presently exist. However, small **E** and **H** field sensors have been developed which individuals can wear to alarm at a preset field strength. These types of units have found wide applicablilty in the wireless communications industry. A personnel dosimeter for RFR at present does not exist. Diazoluminomelanin (DALM) is a unique biopolymer, originally invented by Dr. Johnathan Kiel and co-workers at the U.S. Air Force Research Laboratory, Brooks AFB, Texas.. DALM has the properties of luminescence, fluorescence, and slow fluorescence that has a broad optical absorption spectrum. When it is properly activated, it can absorb and store RFR energy, and on subsequent activation, it can release that energy via rapid luminescence followed by slow fluorescence. It has been postulated as a future personnel dosimeter for RFR (Kiel 2000).

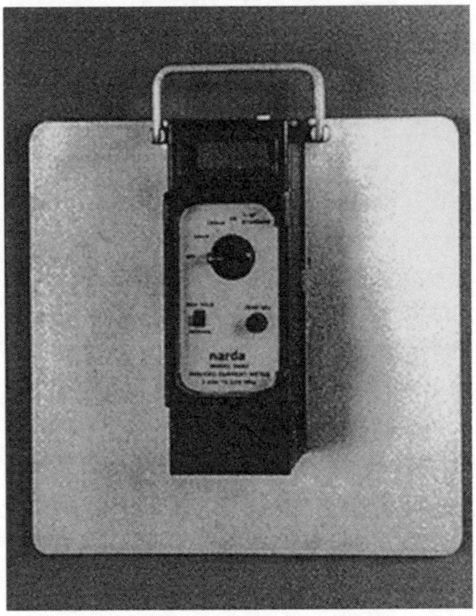

Fig. 10.10 - Stand On "Bathroom Scale" Type Meter For Measuring Induced Currents (courtesy of NARDA Microwave, Inc.)

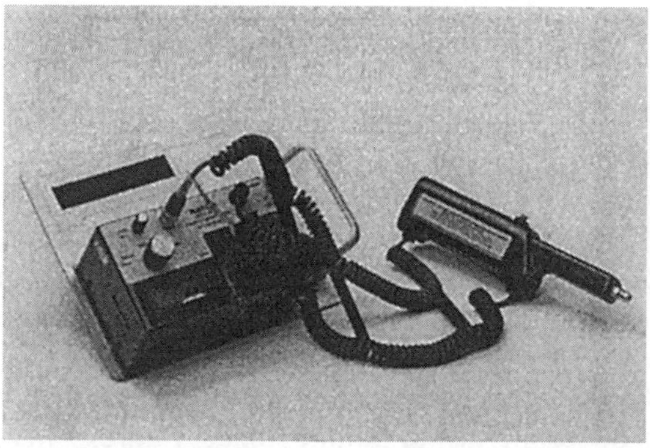

Fig. 10.11 - Contact Current Meter (courtesy of NARDA Microwave, Inc.)

Although RMS average field levels are specified in most RFR exposure situations, if peak pulse power is the desired quantity specialized diode sensors need to be employed. None of the commercial sensors reviewed in this chapter could perform this task at the present time. Future military and commercial ultra wideband (UWB) systems will undoubtedly drive the development of such peak power detectors for *general* use. Spark discharges, which can cause RF shock and burns, may occur when a person contacts a conducting object in a high electric field. The present IEEE standard (IEEE 1999) does not address protection against RF shock and burns explicitly. The U.S. Navy has used a V_{oc} of 140 volts, as measured with a high impedance "burngun," as a predictor of situations where RF shock and burns may be encountered. This type of instrument is still in prototype stage and is not commercially available.

10.10 REFERENCES

Kiel, J.L. Molecular dosimetry, In: Radio frequency radiation dosimetry and its relationship to the biological effects of electromagnetic fields, proceedings of the NATO advanced research workshop. Norwell, MA: Kluwer Academic Publishers, 227-238; 2000.

Leonowich, J.A. Measurement of radiofrequency fields. In: Non-ionizing radiation, proceedings of the 2nd international non-ionizing radiation workshop. Vancouver, UBC Press, 96-124; 1992.

Leonowich, J.A. Introduction and physics of non-ionizing radiations (NIR), In: Non-ionizing radiation: an overview of the physics and biology, proceedings of the 1997 health physics society summer school. Madison, WI: Medical Physics Publishing, 1-16; 1997.

Leonowich, J.A. Development of induced and contact current (ICC) limits in the HF and VHF regions, In: Radio frequency radiation dosimetry and its relationship to the biological effects of electromagnetic fields, proceedings of the NATO advanced research workshop. Norwell, MA: Kluwer Academic Publishers, 2000: 301-308.

Institute of Electrical and Electronics Engineers. Techniques and instrumentation for the measurement of potentially hazardous electromagnetic radiation, 3 kHz to 300 GHz, IEEE C95.3-1991. New York, NY. Institute of Electrical and Electronics Engineers; 1991.

Institute of Electrical and Electronics Engineers. IEEE standard for safety levels with respect to human exposure to radio frequency electromagnetic fields, 3 kHz to 300 GHz, IEEE C95.1-1998. New York, NY. Institute of Electrical and Electronics Engineers; 1999.

Chapter 11

RADIATION MONITORING AT NUCLEAR POWER PLANTS

David W. Miller

11.1 INTRODUCTION

11.1.1 Radiation Monitoring Systems

Radiation Monitoring Systems are designed and installed at nuclear power plants to provide important radiological information to the plant health physics staff on direct and process radiation levels in plant areas and systems. Plant-wide radiation monitoring systems are typically divided into two systems: Area Radiation Monitoring System and Process Radiation Monitoring System (including Technical Specification required Effluent Radiation Monitors).

This chapter describes the design philosophy and requirements for the process radiation monitoring systems at nuclear power plants. This chapter also describes state-owned and managed radiation monitoring systems provided to enhance public confidence in the safety of nuclear power plants operations. Finally, the use of independent, academic experts to validate utility radiation monitoring measurements is discussed in case studies provided.

Area radiation monitoring systems are designed and operated to provide measurements of direct radiation fields in plant cubicles and general access areas of the plant. Area radiation monitors are provided with local indication and remote indication in the main control room. Weekly surveys conducted by trained health physics technicians verify the radiological conditions in plant areas monitored by fixed area radiation monitors. For purposes of this chapter, the typical process monitoring systems installed at U.S. pressurized water reactors are described. Process monitoring systems at US boiling water reactors are similar in concept but have different functional requirements based on differences in plant design.

Process Radiation Monitoring Systems at nuclear power plants are designed and operated to measure, indicate and record levels of radiation and radioactivity associated with the primary coolant and various other plant process streams. Selected process radiation monitors serve as radioactive effluent monitors and accumulate data on the identities and quantities of radionuclides released from the plant. The purpose of the radioactive effluent monitors is to prevent gaseous and liquid releases from exceeding regulatory limits.

The overall purpose of area and process monitoring systems at nuclear power plants is to assist plant operators and health physicists in assuring that exposures to plant personnel and the public will be As Low As Reasonably Achievable (ALARA). To achieve this purpose, the monitoring systems are designed to activate alarms locally and in the main control room when predetermined radioactivity levels are exceeded. In some cases, automatic control actions are initiated in addition to the alarm function to isolate a process stream, terminate a release or prevent airborne radioactivity from entering plant outside air intakes.

11.1.2 Regulatory Framework

U.S. Nuclear Regulatory Commission (NRC) requires nuclear power plants to have a radiation monitoring system installed before receiving their license to operate. Radiological Effluent Technical Specifications (RETS) are the regulatory requirements that deal with effluent monitoring from the plant vent stacks and projection of dose to the public. RETS are required by federal regulations contained in 10 CFR 50.36a. The guidance for radioactive effluents is quantified in 10 CFR Part 50, Appendix I: Numerical Guides for Design Objectives and Limiting Conditions for Operation to Meet the Criterion "As Low as is Reasonably Achievable" for Radioactive Material in Light-Water-Cooled Nuclear Power Reactor Effluents.
Other federal Regulations related to radiation monitoring systems include:

- 10 CFR 20, "Standards for Protection Against Radiation."
- 10 CFR 50, "Licensing of Production and Utilization Facilities." Appendix A, "General Design Criteria for Nuclear Power Plants."
- 10 CFR 100, "Reactor Site Criteria."

The U.S. Nuclear Regulatory Commission has issued Regulatory Guides that provide guidance in the design and operation of radiation monitoring systems at nuclear power plants. These Regulatory Guides include:

- Reg. Guide 1.4, "Assumptions Used For Evaluating the Potential Radiological Consequences of a Loss-of-Coolant Accident for Pressurized Water Reactors."
- Reg. Guide 1.21, "Measuring, Evaluating and Reporting Radioactivity in Solid Wastes and Releases of Radioactive Materials in Liquid and Gaseous Effluents from Light-Water-Cooled Nuclear Power Plants."
- Reg. Guide 1.22, "Assumptions Used for Evaluating Actuation Functions."

- Reg. Guide 1.25, Assumptions Used for Evaluating the Potential Radiological Consequences of a Fuel Handling Accident in the Pressurized Water Reactors."
- Reg. Guide 1.29, "Seismic Design Classification."
- Reg. Guide 1.53, "Application of the Single Failure Criterion to Nuclear Power Protection Systems."
- Reg. Guide 1.70, "Standard Format and Contents of Safety Analysis Reports of Nuclear Power Plants, LWR Edition.
- Reg. Guide 1.97, "Instrumentation for Light-Water-Cooled Nuclear Power Plants to Assess Plant Conditions During and Following an Accident."
- Reg. Guide, 4.2, Preparation of Environmental Reports for Nuclear Power Stations," Appendix I.
- Reg. Guide 8.8, "Information Relevant to Ensuring that Occupational Radiation Exposures at Nuclear Power Stations Will Be As Low As Reasonably Achievable."

Safety Analysis Reports for each licensed plant in the U.S. contain requirements for the Area and Process Radiation Monitoring Systems in Chapter 11. The U.S. Nuclear Regulatory Commission's Standard Review Plan, Section 11.5, provides design objectives and requirements for Radiation Monitoring Systems installed at nuclear power plants. The American Nuclear Standards Institute (ANSI) and IEEE provide important technical specifications for installed radiation monitoring systems. The applicable ANSI and IEEE standards include:

- ANSI N13.1-1969, "American National Standard Guide to Sampling Airborne Radioactive Materials in Nuclear Facilities."
- ANSI N13.2-1969, "American National Standard Guide for Administrative Practices in Radiation Monitoring."
- ANSI N13.10-1974,"American National Standard Specification and Performance of On-Site Instrumentation for Continuously Monitoring Radioactivity in Effluents."
- IEEE Standard 279-1971, "Criteria for Protection Systems for Nuclear Power Generating Stations."
- IEEE Standard 323-1974, "IEEE Standard for Qualifying Class 1E Equipment for Nuclear Power Generating Station."

- IEEE Standard 338-1975,"IEEE Standard Criteria for the Periodic Testing of Nuclear Power Generating Station Class 1E Power and Protection Systems."
- IEEE Standard 344-1975, "IEEE Recommended Practices for Seismic Qualification of Class 1E Equipment for Nuclear Power Generating Stations."
- IEEE Standard 379-1975,"IEEE Trial-Use Guide for the Application of the Single-Failure Criterion to Nuclear Power Generating Station Protection Systems."
- K.G. Murphy and K.M. Campe, "Nuclear Power Plant Control Room Ventilation System Design for Meeting General Criterion 19," 13th AEC Air Cleaning Conference.

11.2 FUNCTIONAL REQUIREMENTS

11.2.1 Process Radiation Monitoring System

The Process Radiation Monitoring System at nuclear power plants is designed to continuously monitor plant process systems and plant effluents in order to warn of potential radiation hazards to plant personnel and to assist the plant operators to maintain releases to the environment within limits set forth in 10 CFR Parts 20, 50, and 100. In some cases, the process monitoring system has the additional function of warning of process or other system malfunctions.

The monitors are designed to alarm in the main control room, and in some cases locally, when radiation levels reach preset levels. The monitoring system may activate interlocks as specified for individual process system requirements. The control room is equipped with continuous readouts and recorders of radiation levels measured by the monitors. Plant health physicists are provided with trend information on past and current levels of radioactivity by system monitors.

Grab samples are taken from specific process monitors and analyzed by gamma-ray spectroscopy in the plant laboratory to identify principal gamma emitters in the process or effluent stream. Only nuclear safety-related monitors are required to remain functional in the event of loss-of-offsite power.

11.2.2 Selected Process Monitoring Systems

The functional requirements for selected process monitoring systems that illustrate various process streams being monitored are described below:

11.2.2.1 Station Vent Stack Monitor: The station vent effluent monitor continuously measures, indicates and records the level of particulate, iodine and noble gas radioactivity discharged to the environment from the stack vent. This monitor assists in fulfilling the regulatory reporting requirements contained in Reg. Guide 1.21. The monitor is capable of providing manual grab samples to support the laboratory analysis performed to comply with Reg. Guide 1.21 and plant technical specification conformance verification. Special features of the station vent stack monitor include the following:

- The particulate monitor consists of a beta scintillator and is provided with background subtraction channels for radon/thoron and also for external gamma radiation.
- The iodine monitor consists of a gamma scintillation detector with two single channel analyzers. One channel is used to monitor the 364 keV photons emitted from ^{131}I. The second channel is used to monitor a window just above the ^{131}I peak to provide a measure of background in the ^{131}I window. The purpose of the adjacent window subtraction method is to subtract the contribution of interfering noble gases which also may be present in the charcoal cartridge and are counted in the 364 keV channel. Gain stabilizing techniques are also employed by providing a small ^{241}Am source on the detector assembly to eliminate the effects of drift on the monitor.
- The noble gas monitor is provided with a duplicate monitor to subtract radiation due to the external background field.
- A by-pass sample line is provided for a tritium sampling system consisting of silica gel or equivalent absorber. Tritium release samples are removed periodically for laboratory analysis.

11.2.2.2 Fuel Handling Building Fuel Handling Accident Exhaust Monitors: The fuel handling building fuel handling accident exhaust monitors continuously measure, indicate and record the level of particulate, iodine and noble gas radioactivity in the fuel handling building exhaust in the event of a fuel handling accident. If pre-determined levels are exceeded, the system alarms in the main control room and in the fuel handling building.

11.2.2.3 Containment Building Fuel Handling Accident Monitors: The containment building fuel handling accident monitors continuously measure, indicate and record the level of gamma radiation near the fuel pool in containment. If predetermined radiation levels are exceeded, the system alarms in the main control room and in containment. Automatic control actions to isolate containment ventilation are also initiated.

11.2.2.4 Containment Atmosphere Monitors: The containment atmosphere radiation monitors continuously measure, indicate and record the particulate and noble gas radioactivity in a sample of air drawn from the containment. A charcoal sample cartridge is provided between two particulate and noble gas monitors to permit removal of the cartridge for laboratory analysis of iodine. In the event predetermined radiation levels are exceeded, the system alarms in the main control room and in containment.

11.2.2.5 Control Room Air Intake Monitors: The control room air intake monitors continuously measure, indicate and record particulate, iodine and noble gas radioactivity in the air intake louvers. In the event predetermined levels are exceeded, the monitors initiate operation of the emergency make-up system and control room alarms. Subsequent selection of the intake with the lowest measured noble gas activity will be by operator action. These monitors meet the intent of regulatory guidance concerning control room habitability requirements for 30-days post-LOCA.

11.2.3 Categories of Process Monitors

Process monitors at nuclear power plants are divided into four categories: gaseous process and effluent monitors, liquid process and effluent monitors, special process and nuclear safety related monitors. Process monitors are identified by the function they perform for specific plant buildings and systems. A typical process monitoring system for a pressurized water reactor is listed below by categories:

11.2.3.1 Gaseous Process and Effluent Monitors

Station Vent Stack Radiation Monitoring System (RMS)
Primary Auxiliary Building Exhaust RMS
Secondary Auxiliary Building Exhaust RMS
Fuel Handling Building Exhaust RMS
Radwaste Building Exhaust RMS
Containment Main Purge Exhaust RMS
Containment Atmosphere RMS
Containment Mini Purge RMS
Containment Post-LOCA Purge RMS
Gas Decay Tank Exhaust RMS
Gas Decay Tank Cubicle Exhaust RMS
Air Ejector Exhaust RMS
Equipment and Instrument Decon Room RMS

11.2.3.2 Liquid Process and Effluent Monitors

Liquid Radwaste Discharge RMS
Essential Service Water RMS

Non-essential Service Water RMS
Component Cooling Water RMS
Steam Generator Blowdown RMS
Auxiliary Steam RMS
Turbine Building Floor Drain Effluent RMS
Reactor Cavity Cleanup System Filters RMS

11.2.3.3 Special Monitors

Reactor Coolant Letdown (failed fuel) RMS
Smoke Vents RMS
Continuous Air Monitors (fixed and cart mounted)

11.2.3.4 Nuclear Safety Related Monitors

Fuel Handling Building Fuel Handling Accident RMS
Containment Building Fuel Handling Accident RMS
Control Room Air Intake RMS
External Containment RMS
ECCS Cubicle Exhaust RMS
Fuel Handling Building Fuel Handling Accident Exhaust RMS

11.2.4 Design Requirements

The design requirements for process radiation monitors are based on governing codes and standards which are applicable to process monitoring systems at nuclear power plants. The design life of the process monitoring system is 40 years with reasonable downtime for maintenance, testing, and scheduled replacement of components or systems. Safety related monitors are required to be operable for 30 days of post-LOCA environment.

11.2.4.1 Selection of Range and Lower Limit of Detectability for Non-nuclear Safety Related Monitors. The range of the process monitors for non-nuclear safety related monitors encompasses the expected operating range of the process variable being monitored. The upper limit of the range is set at least one decade above the level of radioactivity at which alarm or control action must be initiated (technical specification limit). The lower limit encompasses the desired lower limit of detectability (LLD). The term LLD is defined and corresponding state-of-the art values are given in ANSI N13.10. Generally, the LLD is based on a 95% confidence level. A 2 mrem/hr (0.02 mSv/hr) ^{137}Cs background level is assumed to be present in the monitor's general area in the plant. The detection limits are based on a 4 hour collection time for particulate and iodine channels and a continuous reading for noble gas channels and liquid channels. Background subtraction techniques are employed to meet system requirements, when needed.

The particulate radioactivity monitors have a LLD of 5 x 10^{-12} uCi/mL (0.19 Bq/m^3) for ^{137}Cs. The monitors have a sample flow rate of 4 scfm (0.113scmm) and a 4 hour sampling time. Generally, 3 inches (7.6 cm) of lead shielding is provided on the monitor skid which is sufficient shielding to achieve this LLD. A 10-minute counting time after the 4-hour collection is assumed.

The radioiodine monitors designed with a LLD of 4 x 10^{-12} uCi/mL (0.15 Bq/m^3) of ^{131}I. The monitors have a sample flowrate of 1 scfm and a 4 hour sampling time. Lead shielding 3 inches (7.6 cm) thick is generally sufficient to achieve this LLD. A 10-minute counting time after the 4-hour collection is assumed.

The radioactive noble gas monitors have a LLD of 3 x 10^{-7} uCi/mL (11.1 x 10^3 Bq/m^3) of ^{85}Kr and 5 x 10^{-7} uCi/mL (18.5 x 10^3 Bq/m^3) of ^{133}Xe. Sufficient shielding of the detector and chamber are required to achieve this LLD. A 10-minute counting time is assumed.

The radioactive liquid monitors have a LLD of 4 x 10^{-7} uCi/mL (14.8 x 10^3 Bq/m^3) of ^{137}Cs. Sufficient shielding of the detector and chamber are required to achieve this LLD. A 10-minute counting time is assumed.

11.2.4.2 Selection of Range and Lower Limit of Detectability for Nuclear Safety Related Monitors. The technical basis for selecting ranges for nuclear safety related monitors is derived from information and requirements set forth in 10 CFR 50, 10 CFR 50 and the US Nuclear Regulatory Commission Regulatory Guides. Examples of the regulatory requirement for specific monitors are provided below.

11.2.4.2.1 Fuel Handling Accident Monitors. The range for the fuel handling accident monitors is based on assumptions and parameters in Regulatory Guide 1.25. The upper range value is chosen to provide at least one decade above the doserate at which alarm and initiation of control action must be taken in time to assure compliance with exposure limits in 10 CFR 100. The lower range value is chosen to assure an on-scale reading under normal plant operating conditions.

11.2.4.2.2 Control Room Intake Monitors. The range for main control room intake monitors is based on the assumptions and parameters stipulated in Regulatory Guide 1.4, General Criterion 19: Murphy and Campe paper, and the design basis leak rate from the plant during a Loss of Coolant Accident (LOCA). The upper range values are chosen to measure concentrations which are at least one decade above the maximum concentrations expected at the control room intake louvers. The lower range value is chosen to assure on-scale readings under normal plant operating conditions. The Murphy and Campe paper provides guidance on the separation criterion to be used in the placement of the control room intake point so that operators can chose the lowest airborne concentration of outside make-up air based on wind direction and air intake location.

11.2.4.2.3 High Range Containment Monitors. The range of high range containment monitors is based on assumptions and parameters stipulated in Regulatory Guide 1.97 and TMI 2 Lessons Learned requirements (increased range of radiation monitors: section 2.1.8.b). The upper range value is provided at least one decade above the maximum doserate expected at the detector's location. This dose rate is based on the worst-case radiation level inside containment of 10^8 rads/hr (10^6 Gy/hr). A minimum of two high range containment monitors, physically separated are to be provided. The lower range is chosen to assure an on-scale reading under normal plant operating conditions.

11.2.4.2.4 ECCS Cubicle Exhaust Monitors. The range for the Emergency Core Cooling System Cubicle Exhaust Monitors (ECCS) is based on assumptions and parameters stipulated in Regulatory Guide 1.4 and on guidance provided in Section 15.6.5, Appendix B of the U.S. NRC Standard Review Plan. The upper range values are chosen to measure radionuclide concentrations which are at least one decade above the maximum concentrations expected in the discharge duct downstream of the filters. The lower range values are chosen to assure on-scale readings under normal plant operating conditions.

11.2.4.3 Design Bases for Selecting Setpoints and Control Actions. The setpoints calculated for the Non Class 1E monitors are based on the technical specification limits for the plant. For Class 1E monitors, the calculated setpoints are based on the predicted radioactivity releases for the accidents related to their function. The setpoints for these monitors are based on either technical specification for the plant or for compliance with applicable dose limits to station personnel and the public. For both types of monitors, setpoints are set high enough to avoid spurious alarms. Also, the calculated setpoints determined prior to plant commercial operation are adjusted based on operating experience and changes in the regulatory guidance documents.

11.2.4.4 Energy Response of Channels The energy response curves must be supplied by the instrument vendor for each type of beta and gamma detector used in the process radiation monitoring system.

11.2.4.5 Channel Accuracy The accuracy of effluent monitors and process monitors having an automatic control function is based on the best available technology at the time of licensing of the plant. Other process monitors are provided with channel accuracy within $\pm$ 20% of the reading over the upper 80% of their dynamic range. The required energy range for beta detectors is 0.07 to 3 MeV beta. The required range for gamma detectors is 0.08 to 3 MeV gamma.

11.2.4.6 Precision The reproducibility of any channel of any process monitor for a given measurement, over its stated range, must be within $\pm$ 10% at the 95.4% confidence level.

11.2.4.7 Response Time The response time of the process monitor should be not less than that time required to maintain the design basis background or detection limit within the required accuracy. Also, it should be not more than the time required for control actions, where appropriate, or reasonable alarm response times based on required operator action.

Nuclear safety-related process monitors should have response times that are short with respect to the overall plant system response time. The monitor response time should be on the order of one second unless otherwise calculated per the process monitor specification.

11.2.4.8 Microprocessor Units In plant-wide radiation monitoring systems, local microprocessors are provided to process the signals generated by the detectors and to operate the local alarms. One microprocessor is provided per process radiation monitor (for one, two or three channel monitors). The microprocessor units transmit all radiation data, alarm information and monitor status to the centralized computer in the main control room. Controls available at the local microprocessor include:

- Power on/off switch
- Check source insertion
- Means for alarm setpoint entry
- Means for entry of numerical constants
- Provision for switching the monitor status to "test and calibration"

Portable readout capability is provided for the local microprocessors. Local microprocessors must be capable of stand alone operation (independent of the centralized equipment).

11.2.4.9 Centralized Equipment The process radiation monitoring system utilizes and is integrated with centralized digital equipment located in the main control room. The centralized equipment is shared with the area radiation monitoring system and includes:

- Data processing unit (computer)
- Cathode ray tube(CRT) display unit
- Equipment to initiate control room alarms
- Printer/typer
- Long term storage unit

The processing unit receives and processes all signals representing the radiation monitoring parameter of interest, alarm information and operating status of each process monitor. The processing unit also transmits appropriate information to the CRT, the printer/typer and the long-term storage unit. In the event predetermined levels are exceeded, alarms are activated in the main control room by the processing unit. Readout information that is transmitted and displayed on the CRT include the following:

- Identification of the monitor
- Radioactivity level
- Effluent flow
- Power on/off status
- Alarm status (high or low)
- Alarm setpoints
- Check source response
- High voltage level
- Time history graphic display

Control functions which are available from the central location include:

- Initiation of check source insertion
- Initiation of reset action

It should be possible to manually initiate all control actions performed by the process radiation monitoring system. The above information can be displayed in four modes:

- Overall status grid
- Trend display
- Group Display
- Specification Display

The printout of the above information must always be available on demand. Data transmitted to the long-term storage unit must be stored on a suitable mass storage media for retrieval on demand for the life of the plant. Safety related monitors are also provided with independent and redundant safety related readouts in the main control room.

11.2.5 Equipment Design Considerations

Process Monitoring Systems at nuclear power plants must meet specific design requirements due to the in-plant environment and normal, abnormal and accident design conditions. Equipment design considerations for process monitors at nuclear power plants include the following requirements:

11.2.5.1 Particulate Filters. Particulate filters must have an efficiency of 98% or better for 0.3 micron DOP (di-octyl-phtalate) particles.

11.2.5.2 Iodine Collectors. Iodine collectors consist of activated, impregnated charcoal cartridges in metal canisters. Prefilters are installed upstream of the cartridges to remove particles. The iodine filters have a design basis iodine removal efficiency of at least 95%. To avoid noble gas interference in the iodine channel, silver zeolite adsorption canisters have been placed in gaseous effluent monitors since the 1979 Three

Mile Island accident. The silver zeolite adsorption canister retains the radioiodine from the monitored air and does not retain significant amounts of noble gases (Cline 1980). Tests have shown that the use of nitrogen gas purging cycles allow the radioactive noble gases to be purged from the iodine chamber and the detection of lower levels of radioiodine accumulated on the canister is improved (Tseng, 1986).

11.2.5.3 Sampling Mode. Liquid radiation monitors generally utilize off-line detector assemblies to facilitate maintenance, calibration and to minimize the effects of background. Gaseous radiation monitors are also generally off-line detector assemblies with sample lines designed to comply with ANSI N13.10 guidance. Inlet and outlet connections for sampling system components must allow for easy disconnection and connection of maintenance. The contemporary sample streams are designed to return the sample to the stream sampled. When this is not possible, the gaseous streams are routed to the appropriate ventilation duct upstream of the final effluent monitor. The liquid streams are be routed to a radwaste tank.

11.2.5.4 Supplementary Shielding. Process monitors need to be located in plant areas where the design basis radiation levels are not more than 2.0 mrem/hr (0.02 mSv/hr). If detectors are required to operate in a higher ambient background radiation level, thicker detector shielding must be provided to prevent interference with the ability to detect low concentrations (LLD).

11.2.5.5 Periodic Testing. The process monitoring channels should be capable of being checked, tested and calibrated periodically to verify proper operation. Suitable check sources, test signals and calibration sources should be provided and properly stored for periodic calibration tests. Trip circuits should be capable of convenient operational verification by means of test signals or through the use of portable gamma or beta sources, as appropriate. Trips which could jeopardize the plant operations should not be tested by a single test which could include actuation of the trip.

11.2.5.6 Alarms, Annunication and Saturation. The process radiation monitoring indication and recording equipment are located in the main control room and initiate alarms when predetermined radiation levels are exceeded. Trip circuits remain in the tripped mode until the cause ceases and the alarm has been manually reset. Radiation detectors are designed so that they will not saturate when exposed to activity levels up to 100 times full scale indication. Process monitors should remain upscale even when saturated and should be fully operational following reset action.

11.2.6 Training of Monitoring System Operators

Training of health physics, control and instrumentation, and plant operators on the proper response to radiation monitor alarms is key to safe and efficient operations of the nuclear power plant. The radiation monitoring systems at the nuclear power plant, if properly calibrated and maintained, play an essential role in the successful operation of the plant.

INPO accredited training programs at U.S. nuclear power plants provide the initial comprehensive training on the radiation monitoring system installed at nuclear power plants. Periodic retraining on the radiation monitoring system is important to assure that the plant staff maintains a high level of readiness to respond to any abnormal plant condition. Careful review and self-assessment of trend data from the radiation monitoring systems can provide valuable information to the Radiation Protection Manager and the plant operator on changes in the plant's radiological environmental.

11.3 STATE RADIATION MONITORING PROGRAMS FOR NUCLEAR POWER PLANTS

11.3.1 Illinois Radiation Monitoring Program Overview

Eleven nuclear power plants in Illinois have been equipped with a remote radiation monitoring system owned and operated by the state. The system was designed and installed on each Illinois nuclear power plant by the Governor and state legislature as a direct result of the Three Mile Island Accident. The Illinois Department of Nuclear Safety (IDNS) was established in 1980 to develop and operate the remote radiation monitoring system. A key responsibility of IDNS is to recommend protective actions to the Governor and his staff in the unlikely event of a nuclear accident at one of the 11 Illinois nuclear power plant units.

Each nuclear utility in the state is required by state law to pay an annual fee to IDNS for the development and maintenance of the remote radiation monitoring system. The state's remote radiation monitoring system is fully owned and managed by state health physicists including annual calibration of the instrumentation.

The remote monitoring system includes the following:

11.3.2 Gaseous Effluent Monitoring System (GEMS*)*

consisting of six sub-systems:

- Isokinetic probe, sample line, and input flow control
- Particulate filter and detector channel
- Iodine cartridge and automatic cartridge change-out /counting system
- Noble gas counting chamber (6 liters) and detector channel
- Allen-Bradley controller

- Computer/multichannel analyzer

The GEMS instruments are housed in specially constructed buildings near the plant gaseous effluent release point (vent stack). An isokinetic probe and velocity sensors are installed in the vent stack and routed in heat-traced sample lines to the GEMS building. GEMS uses liquid nitrogen cooled, high-purity Germanium detectors supplied by Canberra. The 3-channel stack monitoring system was supplied by SAIC with extended noble ranges including low, mid, and high ranges (maximum of 10^5 uCi/mL (109 MBq/m3) of ^{133}Xe).

11.3.3 Gamma Detection Network (GDN)

Reuter-Stokes pressurized ion chambers are installed in a circle at a distance of one to three miles (1.6 to 4.8 km) from the center of each plant site. Generally, 16 environmental micro-R meters are installed at each site including site boundary, two-mile (3.2 km) radius, and in local communities within the 10-mile (16 km) Exclusion Area Boundary of the site, where possible. The IDNS system polls each pressurized ion chamber every eight minutes (integrated reading) and receives the radiation level measurements in Springfield, the state's capital city.

11.3.4 State System Results

The IDNS remote monitoring system was operational for several years prior to the start-up of some of the Illinois nuclear power plants. No significant differences in environmental radiation levels have been measured by the GDN between pre-operational and operational periods. However, slight increases in the pressurized ion chambers have been observed when high-pressure weather fronts travel across the state. Also, radioactive waste shipments passing by a pressurized ion chamber will show a brief increase in the ambient radiation levels. The observation was most frequently measured around the Dresden nuclear power plant due to frequent radioactive waste shipments to the General Electric (GE) spent fuel storage facility adjoining the site.

The IDNS remote monitoring system installed in the state in the U.S. with the most nuclear power plants has provided the Governor of Illinois with radiological and plant status data to assist in the early determination of significant abnormal power plant conditions and assessment of radiological risk to the general public.

11.4 CONFIRMATORY MEASURES FOR PLANT EFFLUENTS

11.4.1 Plant Effluent Monitoring System, RETS and REMP

The plant health physicists use stack and liquid discharge radiation monitoring systems to project the potential doses to the public. The US NRC requires documentation of the estimated doses to the public due to day-to-day operations of the plant. These doses are calculated based on projections from the discharge points effluent monitoring systems. Radiological Effluent Technical Specifications (RETS) deals with the projected value.

US NRC requires nuclear power plants to have a radiation monitoring system installed before receiving their license to operate. RETS are the regulatory requirements that deal with effluent monitoring from the stacks and projecting dose to the public. These specifications are standard with slight variations according to plant design. RETS are required by federal regulations contained in 10 CFR 50.36a. The guidance for radioactive effluents is quantified in 10 CFR Part 50, Appendix I: Numerical Guides for Design Objectives and Limiting Conditions for Operation to Meet the Criterion "As Low as is Reasonably Achievable" for Radioactive Material in Light-Water-Cooled Nuclear Power Reactor Effluents.

RETS requires the operator to produce a plant-specific Offsite Dose Calculation Manual (ODCM), which includes the methodology and detailed procedures for monitoring environmental radiation. The guide to preparing the ODCM is Regulatory Guide 1.21: Measuring, Evaluating, and Reporting Radioactivity in Solid Wastes and Releases of Radioactive Materials in Liquid and Gaseous Effluents from Light-Water-Cooled Nuclear Power Plants.

Environmental monitoring corroborates the dose projections made from gaseous effluent monitors by taking air, TLD, aquatic and terrestrial environmental samples and making radionuclide measurements of these samples. This program is called the Radiological Environmental Monitoring Program (REMP). The federal requirement is described in the Code of Federal Regulations 10 CFR 50, Appendix A. Federal guidance in the development of the REMP program is contained in Regulatory Guide 4.1: "Programs for Monitoring Radioactivity in the Environs of Nuclear Power Plants."

During an emergency condition at a U.S. plant, REMP changes to EREMP and serves as the Emergency Radiological Environmental Monitoring Program. Under EREMP conditions, environmental samples would be split between the utility and the regulatory agency. For example, particulate filters from air sampling stations and composite water samples from the liquid discharge canal from a nuclear power plant could be split between the utility and regulatory health physicists. (The composite water sampler continuously collects approximately 10 ml of water per hour into a sample bottle which is collected monthly for laboratory analysis.) NUREG-0473 also provides specific regulatory guidance on the administration of the ODCM.

11.4.2 Radiation Monitors Accessible to the Public

Public credibility of the accuracy of ambient radiation measurement taken near the plant is important for good relations with the local community. Some nuclear utilities have funded the placement and maintenance of radiation monitors in public areas near the plant. The location of the monitors vary. Some utilities have decided to place monitors on public buildings, such as schools or city hall. Current dose measurements are compared with pre-operational levels in trend graphs available to the public.

11.4.3 Case Study: City Hall Monitor

Pennsylvania Power and Light Company provided a micro-R meter in the main corridor of the city hall for the town downwind of the nuclear power plant. The town had a population of 12,274 in 1980. The total population in the counties included in the 50-mile radius of the plant was approximately 325,000 in 1980. An independent contractor calibrates the micro-R meter. The micro-R meter readings are available to be read in the main corridor or on the outside wall of the city hall by the public. The outside display allowed 24-hour access of the real-time readings for the public.

11.4.4 Case Study: Local College Independent Monitoring Program

Pennsylvania Power & Light Company also provided annual research funds for radiation monitoring instrumentation at two local colleges. The colleges were located 20 miles upstream of the Susquehanna River (Wilkes College) and 12 miles downstream of the Susquehanna River (Bloomsburg State College). The monitoring equipment included NaI counting systems, micro-R survey meters, rain and snow collectors and meteorological sensors for wind speed and wind direction.

The environmental monitoring instrumentation was intended for independent use by the physics or science professors at the colleges to monitor ambient radiation levels. The college professors provided annual reports to the local newspapers that described the radiation measurements taken at the two colleges prior to plant operations and during the life of the plant. The utility scientists were provided copies of the independent annual reports prior to the release of the reports to the local newspapers. The local communities' response to this program was favorable. The college professors were pleased that they could provide a service to the local community and, at the same time, gain radiation-monitoring instrumentation and learning opportunities for their students.

The professors from the local colleges were invited to make presentations on ambient radiation measurements at the annual physicians' seminars on radiological health also sponsored by the nuclear utility.

11.4.5 Ginna Case Study

Having independent, academic experts involved in assessing utility radiological monitoring programs helps to give the public confidence in the plant operator's concern with the public's health and safety. Other utilities' experiences show that having independent experts available can be helpful in handling questions from the media when a release of radioactivity occurs. For example, when there was a small airborne release of radioactive iodine at the Ginna nuclear power plant in the 1980's, two local professors responded to answer questions from the press about the release. These two professors had been selected two years earlier by the local news media at the specific request of the Ginna plant management. They did an excellent job in handling the technical questions from the local news media. The selection process and use of these local academic experts was mentioned in Forbes magazine as a good industrial practice.

11.4.6 Availability of US NRC Effluent Reports

All the effluent radiation monitoring reports are publicly available through the U.S. NRC's public document rooms at headquarters and often at public libraries near the plant sites. Although few members of the public take the time to read these technical documents, just letting the public know that they are available helps to build trust. The low levels of airborne emissions from U.S. nuclear power plants is helping the American public take a much more positive view of nuclear power plants.

11.5 CONCLUSIONS

A successful radiation safety program at nuclear power plants must have the proper equipment and a trained staff to support safe plant operations. The radiation monitoring systems at nuclear power plants play the important role of continuously measuring, indicating and recording the in-plant radiological conditions and releases to the environment for the Radiation Protection Manager and Plant Manager.

State-owned radiation monitoring systems have been shown to enhance public confidence in the safety of plant operations and public health. With over three decades of successful U.S. nuclear power plant operations completed in the last century, the installed radiation monitoring systems have proven their value in supporting electrical production from nuclear power plants.

11.6 REFERENCES

Cline, E. Retention of Noble Gases by Silver Zeolite Iodine Samplers. Health Phys, 40; 70, 1996.

Mattson R.J. TMI-2 Lessons Learned. Washington DC; US NRC; NUREG –0578; 1997.

Miller, D.W.; Nestel, W. A.; Davidson, G.R.; Coley, R.F. Design Alternatives for Plant-Wide Digital Radiation Monitoring Systems. IEEE Transactions on Nuclear Science. 1979.

Miller, D.W.; Nestel, W.A.; Davidson, G.R.; Coley, R.F. A PWR Containment Air Monitoring System (CAMS) for Normal and Accident Conditions. IEEE Transactions on Nuclear Science. Vol. NS-27, No. 1; February 1980.

Miller, D.W.; Nestel, W.A.; Lehmann, T.H. Post-Accident Liquid and Containment Atmosphere Sampling Systems. Las Vegas, Nevada. American Nuclear Society. 1980 Annual Meeting; 1980.

Miller, D.W.; Harris, J.T. North American Gaseous and Liquid Effluent Databases for UNSCEAR Reports and Utility Benchmarking. Falmouth, MA. Tenth Annual RETS/RE<P Workshop; 2000.

Parker, M,; Niziolek, F. Power Reactor Accident Monitoring and Risk Analysis. San Francisco, California. Proceeding of the Sixth Topical Meeting on Emergency Preparedness and Response. Lawrence Livermore National Laboratory. 1997.

Tseng, T.T.; Jester, W.A., Baratta, A.J., ; McMaster, I. B.; Miller, D. W. The Development and Testing of a Prototype On-line Radioiodine Monitor for Nuclear Power Stations. Health Phys. Vol. 50:1986.

Chapter 12

HEALTH PHYSICS INSTRUMENTATION AT MEDICAL CENTERS

David J. Derenzo

12.1 INTRODUCTION

A wide variety of radiation sources are used at the typical medical center. Clinical uses include diagnostic radiography, nuclear medicine, radiation therapy, blood irradiation, dental radiography, and clinical laboratory testing. Medical centers with positron imaging facilities may have an on-site cyclotron and a radiopharmaceutical manufacturing operation. Biomedical research programs frequently include the uses of radioactive chemicals, sealed sources, irradiators, and analytical x-ray equipment. This can be a challenging environment for radiation protection personnel. In this chapter, radiation instrumentation needs will be discussed in relation to personnel safety, environmental safety, and regulatory compliance.

12.2 THE REGULATORY ENVIRONMENT

The Atomic Energy Act (AEA) as amended in 1954 required the licensing and regulation of radioactive material used in the public sector. Because Congress considered it inappropriate to regulate radiation sources that were in existence before the advent of atomic energy, the AEA did not include the mandate to regulate naturally occurring and accelerator produced radioactive material (NARM) or radiation producing equipment (Blatz 1964). This has resulted in federal regulation of reactor produced radionuclides by the Nuclear Regulatory Commission (NRC) and state or local regulation of NARM and radiation producing machines. See Table 12.1 for a list of reactor and accelerator produced radionuclides. Proposals have recently been made to extend the NRC's authority in the AEA to include NARM (NRC 2000).

There are currently 35 States that have elected to assume most of the NRC's regulatory functions. These Agreement States are required to maintain regulations that are compatible with the NRC's. Other government agencies that regulate some aspect of radiation use at medical institutions include the Environmental Protection Agency (EPA), the Occupational Safety and Health Administration (OSHA), the Food and Drug Administration (FDA), and the Department of Transportation (DOT). Lastly, medical

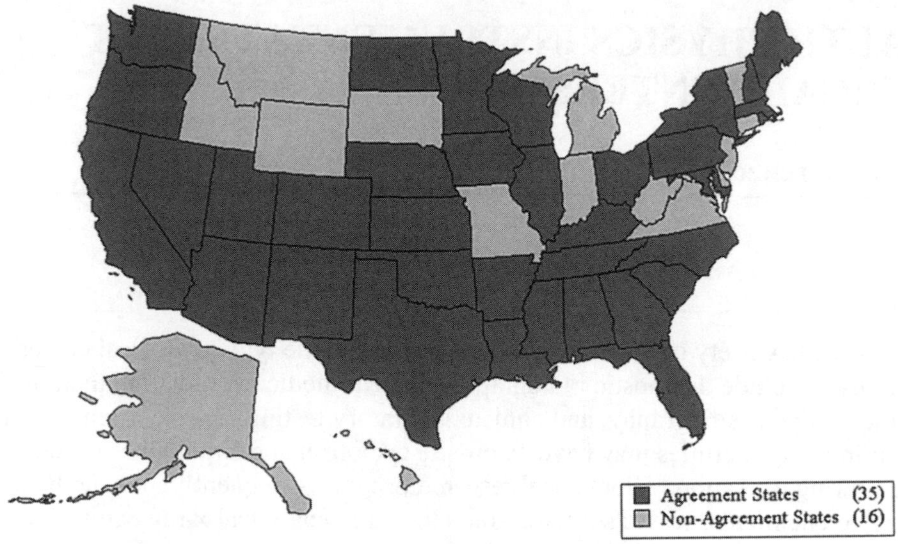

Fig. 12.1 - NRC Agreement and Non-Agreement States.

centers must attain and maintain accreditation from organizations such as the Joint Commission on Accreditation of Health Care Organizations (JCAHO) and the College of American of Pathologists (CAP), which establish requirements that are intended to ensure patients receive high quality medical care.

Table 12.1. List of reactor and accelerator produced radionuclides

NRC regulated radionuclides	**State or locally regulated radionuclides**	
Reactor produced	**Accelerator produced**	**Naturally occurring**
^{3}H, ^{14}C, ^{32}P, ^{60}Co, ^{90}Sr/^{90}Y, ^{99}Mo/^{99m}Tc, ^{125}I, ^{131}I, ^{133}Xe, ^{137}Cs, ^{153}Sm, ^{192}Ir	^{11}C, ^{15}O, ^{18}F, ^{22}Na[a], ^{51}Cr[a], ^{57}Co, ^{67}Ga, ^{81}Rb, ^{89}Sr, ^{111}In, ^{123}I, ^{201}Tl	^{226}Ra, ^{210}Pb

[a] May also be reactor produced

12.3 PERSONNEL MONITORING

12.3.1 Introduction

Personnel monitoring is one of the most important aspects of a medical center radiation protection program. This section will outline monitoring requirements and will consider the types of instrumentation needed to measure external and internal doses.

12.3.2 Radiation Dose Limits

Table 12.2 summarizes the radiation dose limits that are established in the NRC regulations. Licensees are responsible for ensuring these limits are not exceeded and that radiation exposures are maintained as low as reasonably achievable (ALARA). These limits also apply in the Agreement States and states that regulate the use of x rays.

Table 12.2 Summary of NRC radiation dose limits *

Exposure category		Annual limit
	Total Effective Dose Equivalent (TEDE)	0.05 Sv (5 rem)
	-or-	
Occupationally Exposed Adults	Total Organ Dose Equivalent (TODE)	0.5 Sv (50 rem)
(10 CFR 20.1201)	Lens of Eye (LDE)	0.15 Sv (15 rem)
	Skin of Whole Body (SDE,WB)	0.5 Sv (50 rem)
	Extremities (SDE,ME)	0.5 Sv (50 rem)
Occupationally Exposed Minors (10 CFR 20.1207)		10% of adult limits
Embryo/Fetus of a Declared Pregnant Woman (10 CFR 20.1208)		0.005 Sv (0.5 rem)
Members of the Public(10 CFR 20.1301)	Annual Limit	0.001 Sv (0.1 rem)
	In any one Hour	0.02 mSv (2 mrem)

* Excludes doses received from medical procedures, background, and non-hospitalized patients containing radioactive material.

12.3.3 Measurement of External Doses

Groups of personnel who are likely to receive the highest external exposures include cardiac catheterization physicians, physicians performing angiography, cyclotron operators, and personnel who prepare radiopharmaceuticals. Smaller external exposures should be received by radiology and nuclear medicine technologists, radiation therapy physicists and physicians who prepare and apply brachytherapy sources, nurses caring for brachytherapy and radiopharmaceutical therapy patients, analytical x-ray equipment operators, and radiation safety personnel. Some external exposure may occasionally be received by biomedical research personnel who use relatively small quantities of radioactive materials.

Medical licensees are required to monitor occupational exposure of:

- Adults likely to receive in one year a dose from external sources in excess of 10 percent of the limits established in 10 CFR §20.1201(a),
- Minors likely to receive, in 1 year, from radiation sources external to the body, a deep dose equivalent in excess of 1 mSv (0.1 rem), a lens dose equivalent in excess of 1.5 mSv (0.15 rem), or a shallow dose equivalent to the skin or to the extremities in excess of 5 mSv (0.5 rem),
- Declared pregnant women likely to receive during the entire pregnancy, from radiation sources external to the body, a deep dose equivalent in excess of 1 mSv (0.1 rem); and
- Individuals entering a high or very high radiation area.

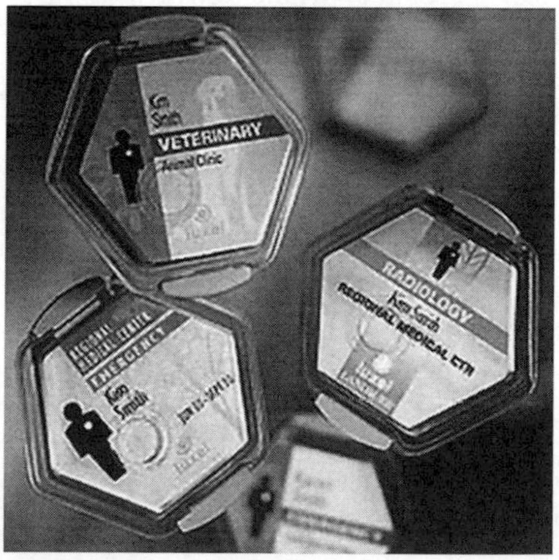

Fig. 12.2 - Landauer Luxel dosimeters (courtesy Landauer, Inc.).

Whole body and ring dosimeters are usually used for the purposes listed above. Most medical institutions rely on commercial dosimetry services, which must be NVLAP approved, for external radiation monitoring. Film badges, thermoluminescent dosimeters (TLD), and optically stimulated aluminum oxide dosimeters are available for this purpose. They can be configured to monitor whole body and extremity exposures.

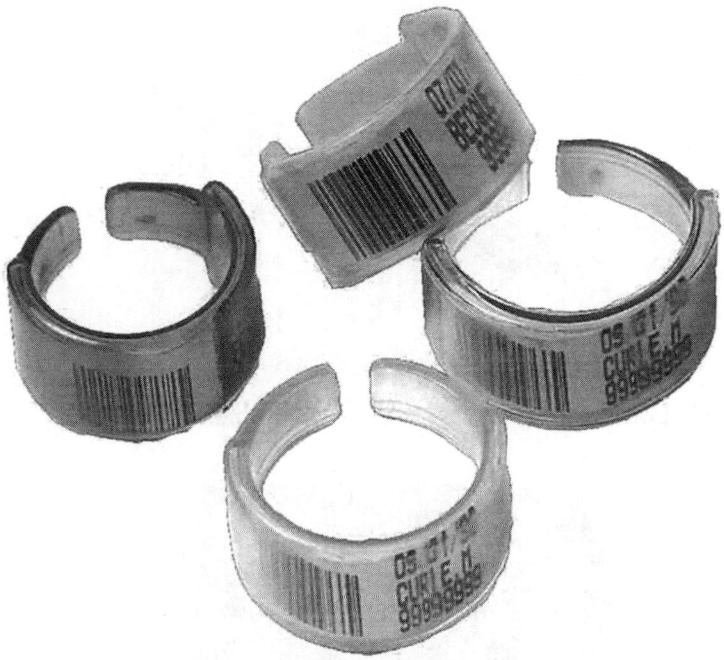

Fig. 12.3 - ICN TLD ring dosimeters (courtesy ICN Dosimetry Services.

Medical personnel who are in the procedure room during fluoroscopy examinations are required to wear lead aprons. This is a difficult monitoring situation because the radiation fields causing the exposure are not uniform, static, or of the same radiation quality. If only one dosimeter is used, there may be a regulatory requirement to wear it on the collar outside of the lead apron. It has been shown that wearing the dosimeter in this manner will overestimate the effective dose equivalent, H_E, by a factor of 8 to 23 (Rosenstein 1994). The NCRP (NCRP 1978) recommends that two dosimeters be worn, one under and one over the apron. If this method is used it is very important to ensure the dosimeters are not reversed from one use to the next. It has been shown that the dose equivalents measured by a dosimeter under the apron at the waist level, H_1, and the dosimeter worn at the collar, H_2, can be used to estimate the dose equivalent, H_E, by using what is sometimes referred to as the Webster equation (Rosenstein 1994):

$$\text{estimate of } H_E = 1.5\, H_1 + 0.04\, H_2. \qquad (12.1)$$

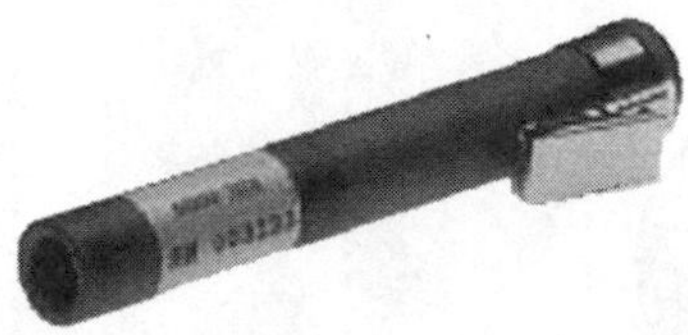

Fig.12.4 - Pocket dosimeter (courtesy Ludlum Instruments).

Fig. 12.5 - Electronic personal dosimeter (courtesy Siemens).

Pocket dosimeters are small ion chambers that discharge in direct proportion to the dose received from penetrating ionizing radiations. The can be read instantly but are relatively fragile. If the quartz fiber, which indicates the accumulated dose, goes off-scale during use, the cause may be a real exposure, a breakage of the fiber, or excessive charge leakage. Calibration and charge leakage testing requirements are burdensome.

For these reasons pocket dosimeters are seldom used as a primary monitoring device, but are sometimes used as a secondary device. Electronic dosimeters, which are relatively expensive, should be considered if a reliable direct reading dosimeter is needed.

12.3.4 Monitoring of Internal Radiation Exposures

In 10 CFR 20.1502(b) the NRC requires licensees to monitor the occupational intake of radioactive material by:

- Adults who are likely to receive an annual intake in excess of 10 percent of any applicable Annual Limit on Intake (ALI);
- Minors likely to receive an annual committed effective dose equivalent in excess of 0.1 rem (1 mSv); and
- Declared pregnant women likely to receive, during the entire pregnancy, a committed effective dose equivalent in excess of 1 mSv (0.1 rem).

Fig. 12.6 – Measurement of thyroid burden (photo by author).

The most likely medical procedures that could cause significant intakes in workers involve treatment of the thyroid gland with therapeutic quantities of ^{131}I. It is not unusual for a patient being treated for thyroid carcinoma to receive an oral dosage of as much as 7.4 GBq (200 mCi). This situation is directly addressed in the NRC regulations, which require the thyroid burden of personnel who helped prepared or administered a therapeutic ^{131}I dosage to be measured within three days (10 CFR 35.315(a)(8)). This procedure can be performed in-vivo using a single- or multi-channel analyzer equipped with a NaI(Tl) crystal. Using the interval between the intake and the activity in the thyroid, the amount of inhaled ^{131}I can be calculated (Lessard 1990). Measurement of the intake can then be compared to the ALI and the thyroid dose can be estimated. Fig. 12.6 shows a person having his thyroid burden measured. The monitoring system utilizes a 3" x 3" NaI(Tl) crystal, a shielded collimator, and a single channel analyzer. A computer program initiates counting, inputs the counting data, calculates the thyroid burden, computes the dose to the thyroid, and prints a dosimetry report. The object that is attached above the detector assembly is a thyroid phantom that contains a 0.37 kBq (0.01 μCi) ^{129}I source used for daily quality control testing. When measuring thyroid burdens it is important to maintain a low background and ensure that monitored personnel and their clothing are not contaminated.

Biomedical research usually involves the use of relatively small quantities of radioactive material. The most common procedures that could require performance of bioassays are iodinations of proteins and other biochemicals with ^{125}I sodium iodide. Thyroid counting and intake calculations for this radionuclide are essentially identical to those for ^{131}I, however the counting threshold and window must be adjusted for the lower photon energies. In-vitro bioassay procedures are performed for most other radionuclides by collecting and analyzing the radioactivity in urine, feces, blood, or breath. Guidance regarding bioassay programs can be found in NCRP Report 87 (NCRP 1987) and in NRC Regulatory Guides 8.9 (USNRC 1979) and 8.20 (USNRC 1993).

12.4 RADIATION SURVEY METERS

12.4.1 Introduction

Radiation surveys are used to determine the need for radiation protection measures and to confirm the effectiveness of the radiation protection measures that are in place. It is not surprising that the NRC has many prescriptive regulations regarding surveys, many of which are summarized in Table 12.3.

Table 12.3. Radiation survey requirements applicable to medical licensees.

Regulation	Requirement
10 CFR 20.1302(a)	Conduct surveys in unrestricted areas
10 CFR 20.1501(a)	General survey requirements to demonstrate compliance
10 CFR 20.1502(b)	Monitor internal exposure of occupationally exposed personnel likely to receive ≥ 10% of any applicable limit, minors and declared pregnancies
10 CFR 20.1906(b)	Monitor labeled and damaged radionuclide packages within 3 hours of receipt for surface contamination and external radiation level at surface and 1 meter
10 CFR 35.51(a) & 35.49(a)	Calibrate survey meters before first use, annually thereafter and following repairs
10 CFR 35.53(a)	Assay each dosage of a photon-emitting radionuclide prior to medical use
10 CFR 35.53(b)	Assay alpha and beta emitting radiopharmaceuticals that are not obtained as unit doses
10 CFR 35.59(a)	Leak test sealed sources every 6 months, every 3 months if designed to emit alpha particles
10 CFR 35.90	Test hoods used to store radioactive gases (not an explicit requirement)
10 CFR 35.70(a)	End of day surveys in areas where radiopharmaceuticals were prepared for used or administered (patient rooms are excepted)
10 CFR 35.70(b) & (e)	Weekly area and contamination surveys in areas where radioactive wastes are stored and where radiopharmaceuticals are used, administered, or stored
10 CFR 35.75(a)	Survey of patients being treated with therapeutic dosages of radiopharmaceuticals or permanent implants prior to release from hospitalization
10 CFR 35.205(b)	Test negative pressure in areas in which radioactive gases are administered (not an explicit requirement)
10 CFR 35.205(e)	Check proper operation of reusable radioactive gas or aerosol collection systems monthly
10 CFR 35.315(a)(4)	Perform area surveys in and around rooms of radiopharmaceutical therapy patients
10 CFR 35.315(a)(5)	Monitor items being removed from rooms of radiopharmaceutical therapy patients
10 CFR 35.315(a)(7)	Perform contamination surveys in radiopharmaceutical therapy patient rooms prior to reuse
10 CFR 35.315(a)(8)	Measure thyroid burden of individuals who helped prepare or administer I-131 therapy dosages
10 CFR 35.404(a)	Survey patient and the patient's room immediately after removal of temporary implants
10 CFR 35.406(c) & 35.415(a)(4)	Immediately survey each brachytherapy patient, area around the patient, and the contiguous restricted and unrestricted areas
10 CFR 35.615(d)	Install a permanent monitor at the entrances to teletherapy room
10 CFR 35.641(a)	Survey teletherapy facilities after source changes and facility modifications
10 CFR 71.5(a)	Survey radioactive material packages that are being sent (by reference)
License Condition	Survey HDR facilities after source changes and facility modifications

12.4.2 Requirements to Possess Portable Survey Instruments

Part 35 of 10 CFR contains specific requirements for possession of portable radiation detection survey meters by medical licensees. A licensee that conducts uptake, dilution or excretion studies must possess a survey instrument capable of detecting dose rates over the range of 0.001 mSv h^{-1} (0.1 mrem h^{-1}) to 1 mSv h^{-1} (100 mrem h^{-1}). If the licensee is authorized to use radioactive material for imaging and localization studies, radiopharmaceutical therapies or implant therapies, a meter as described above and a meter capable of detecting dose rates over the range of 0.01 mSv h^{-1} (1 mrem h^{-1}) to 10 mSv h^{-1} (1,000 mrem h^{-1}) must also be possessed. Instruments that satisfy these requirements are discussed in the following sections.

12.4.3 Radiation Level Surveys Meters

12.4.3.1 Geiger Counters. Geiger-Müller (GM or Geiger) counters, such as the one as shown in Fig. 12.7, consist of a GM detector connected to a rate meter with a coaxial cable. A sensitivity of $\leq 1.23 \times 10^{-8}$ C kg^{-1} h^{-1} (0.05 mR h^{-1}) with a maximum reading of about 5.15×10^{-5} C kg^{-1} h^{-1} (200 mR h^{-1}) is typical. Some Geiger counters also have a small GM tube mounted inside the rate meter case that extends the range by a factor of 10. Energy compensated thin wall detectors have an typical energy response of $\pm$15% between 50 keV and 1.25 MeV. Since any ionizing event that occurs within the Geiger tube causes it to fully discharge, the pulse is not proportional to the energy of the radiation being measured. There is also a fairly long recovery time between pulses (on the order of 100μs) during which the detector and rate meter cannot respond, thereby resulting in an undercount when the meter indicates thousands of counts per minute. Typical uses include surveys of nuclear medicine facilities, source storage rooms, and waste storage areas. Geiger counters are not suitable for use in pulsed photon fields typical of linear accelerators and x-ray equipment. In this situation they will usually indicate the pulse rate of the radiation source rather than an accurate indication of the exposure rate (Golnick 1983). In addition, high radiation fields can saturate the Geiger tube rendering the meter almost totally unresponsive to the radiation field, i.e., it will indicate an exposure rate of about zero. The exposure rate at which this occurs depends on the design of the detector and the rate meter circuitry. Desirable construction characteristics include:

- use of standard batteries;
- standard cable connectors;
- separate calibration potentiometers for each scale;
- a speaker that can be switched off;
- a selector switch for slow or fast response time;

- an overload indicator; and
- a rugged design.

Fig. 12. 7 – Ludlum Model 3 (photo by author).

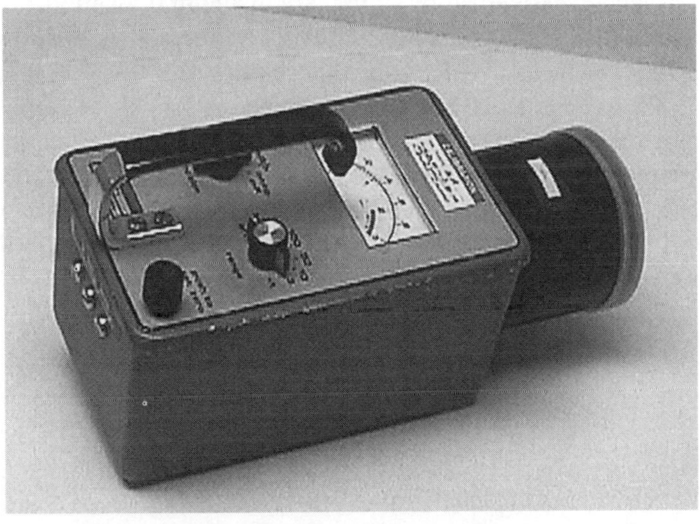

Fig. 12.8 – Victoreen 471 Ion Chamber (photo by author).

12.4.3.2 Ion Chamber Survey Meters. Ion chamber survey meters, such as the one shown in Fig. 12.8, are very versatile instruments. Low range models can measure dose rates of ≤ 0.001 mGy h^{-1} (0.1 mrad h^{-1}). High range models can measure dose rates of ≥ 0.5 Gy h^{-1} (50 rad h^{-1}). When an ionizing event occurs in the ionization chamber, the current that is created is directly proportional to the energy deposited by the event. Ion chamber response is therefore proportional to absorbed dose. A well designed ion chamber has a fairly flat energy response curve from below 10 keV to 2 MeV or higher. Ion chambers do not readily saturate in a high radiation field and are suitable for making measurements in pulsed radiation fields. They are typically used to perform external radiation level surveys in nuclear medicine facilities and brachytherapy source storage rooms; rooms of patients undergoing brachytherapy or radiopharmaceutical therapy; surveys of incoming and out going radioactive material shipments; and surveys of accelerator, cyclotron, and teletherapy facilities. Some ion chambers may need to be returned to the manufacturer for calibration, so inquire before purchasing if this is an issue. Desirable characteristics of portable ion chambers include:

- use of standard batteries;
- a sliding window or removable cap that exposes a thin aluminized mylar window that allows penetration of beta particles and low energy x-rays into the chamber;
- readily accessible calibration potentiometers; and
- a rugged design.

Air filled ion chambers are not usually sealed and some of the electrical components are sensitive to moisture. If you are in a humid location, a better choice might be a meter with a sealed chamber. The tissue equivalent pressurized ion chamber shown in Fig. 12.9 can measure background dose rates of ≤ 0.1 μGy h^{-1} (10 μrad h^{-1}) and can measure fields as large as 10 mGy h^{-1} (1000 mad h^{-1}). This meter is ideal for making environmental level measurements around linacs, irradiators, teletherapy installations, and brachytherapy source storage rooms. Since it has an integrate mode, it can make measurements in the areas surrounding diagnostic x-ray rooms where the beam is only on for a fraction of a second. With these measurements and a work load estimate, doses to surrounding personnel can be projected. This is an excellent way to verify the adequacy of facility shielding.

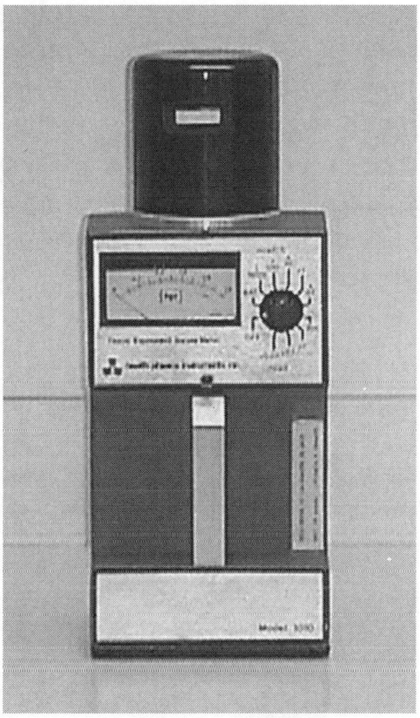

Fig. 12. 9 – Health Physics Instruments Model 1010 pressurized ion chamber (photo by author).

Portable meters with internal NaI(Tl) detectors are also available for low-level gamma surveys. These meters have a range of detection of < 0.01 μGy h^{-1} (1 μR h^{-1}) to about 0.5 μGy h^{-1} (500 μR h^{-1}), however they are energy dependent, i.e., the response of the instrument is dependent on the energy of the detected radiation. Unless the calibration source and the radiation being measured have similar energies, measurements can be significantly in error. Portable rate meters and single channel analyzers utilizing external NaI(Tl) detectors are also available. These instruments are very sensitive, therefore they must be used in low background areas. Both variations are ideal for locating lost sealed sources, verifying that decayed waste is no longer radioactive, and monitoring medical wastes prior to disposal.

12.4.3.3 Radiation Area Monitors. The use of alarming radiation area monitors and entrance interlocks are required for cobalt teletherapy, linac, high dose rate afterloader, and most non-self contained irradiator facilities. They can also be useful in radio-pharmacies, instrument calibration facilities, and other facilities utilizing potentially hazardous radiation sources. Area monitors warn operators of the existence of hazardous radiation fields when they are not anticipated, e.g., during interlock malfunctions or power failures. Area monitors are usually installed within the room near the entrance and a remote indicator is frequently mounted outside the entrance in a prominent location. The monitor and remote indicator should be equipped with visual and audible alarms to warn personnel who are entering the room of any high radiation level that might be present. Area monitors should be provided with emergency backup power and need to be tested for proper operation frequently. In the event the monitor malfunctions or is gone for calibration or repair, persons entering the facility are usually required to enter carrying a properly operating survey meter.

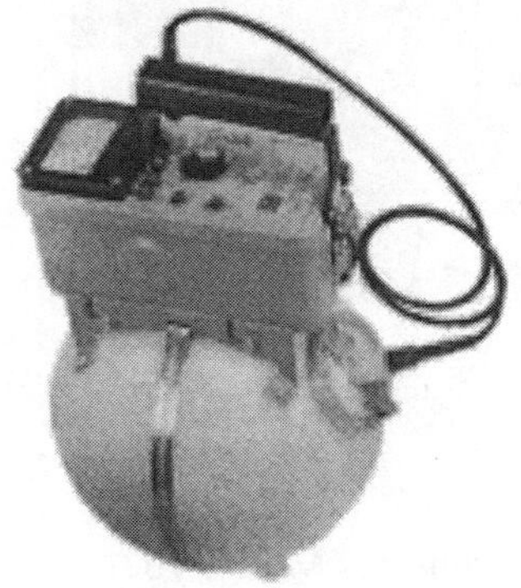

Fig. 12.10 - Neutron survey meter (courtesy Ledlum Instruments).

12.4.3.4 Neutron Survey Meters. Neutron sources are rarely used at medical institutions, however, linacs operating at energies above 12 MeV are common. These sources have the potential to produce significant quantities of neutrons. At the time such facilities are put into service, and whenever modifications are made that could effect the shielding characteristics of the facility, a neutron survey needs to be conducted. The instrument of choice is a survey meter like the one shown in Fig. 12.10. This instrument is costly and if its use will not be frequent, a loaner or rental unit should be sought. If a meter cannot be obtained, neutron dosimeter badges can be placed in key locations. It is also recommended that neutron badges be provided to potentially exposed personnel when a new accelerator facility is being put into service. After reviewing the results of neutron monitoring for a few months, a decision can be made regarding the need for continued monitoring.

12.4.4 Radioactive Material Contamination Surveys

Contamination surveys are performed to determine the quantity of fixed or removable radioactive material surface contamination in a given location. Some contamination limits are established in the NRC regulations and others are established during the licensing process. For instance, 10 CFR 35.315(a)(7) prohibits the reassignment of rooms used to house radiopharmaceutical therapy patients until the removable contamination is less than 3.33 Bq (200 dpm) per 100 square centimeters. Portable Geiger counters and ion chambers cannot detect this low level of contamination. Instead, wipes must be taken and analyzed in appropriate laboratory instrumentation. Limits for floors, hoods, glove boxes, clothing, and other work surfaces can be established during licensing. Table 12.3 lists the surface contamination limits that are suggested in Regulatory Guide 8.23 (USNRC 1981).

Table 12.4 Recommended action levels for removable surface contamination in medical institutions, adapted from Table 2 of NRC Regulatory Guide 8.23.

	Type of radioactive material [a]		
Type of Surface	Alpha Emitters Bq dpm per 100 cm^2	Beta or X-ray Emitters Bq dpm per 100 cm^2	Low Risk Beta or X-ray Emitters [b] Bq dpm per 100 cm^2
Unrestricted areas	0.367	3.67	36.67
Restricted areas	3.67	36.67	366.67
Personal clothing worn outside restricted areas	0.367	3.67	36.67
Protective clothing worn only in restricted areas	3.67	36.67	366.67
Skin	3.67	3.67	36.67

[a] Averaging over 100 cm^2 is acceptable for floors, walls, ceiling, and skin. Averaging over 300 cm^2 is acceptable over the whole area of the hands.

[b] Low risk radionuclides include ^{14}C, ^{3}H, ^{35}S, ^{99m}Tc, and others whose beta energies are less than 0.2 MeV maximum, whose gamma- or x-ray emission is less than 6.973×10^{-7} C kg^{-1} h^{-1} GBq^{-1} (0.1 R h^{-1} Ci^{-1}) at 1 meter, and whose permissible concentration in air is greater than 3.7×10^{-5} kBq ml^{-1} (10^{-6} µCi ml^{-1}).

Radioactive contamination can be tightly or loosely bound to a surface. When a contaminant chemically bonds to or is absorbed by a surface it can be very difficult to remove. On the other hand, contamination can be loosely bonded to a surface making it readily removable with little or moderate effort. Most contamination is somewhere between these extremes. The most significant contamination problem encountered in medical institutions is usually in rooms of ^{131}I therapy patients, particularly when the patient is unable to follow radiation safety instructions. While most of this contamination is readily removed, a significant effort is usually needed to satisfy the release limit of 3.33 Bq (200 dpm) per 100 cm^2.

12.4.4.1 Direct Surface Surveys. Direct monitoring will detect the total amount of contamination on a surface, i.e., both the fixed plus the removable quantities. Meters equipped with NaI(Tl) detectors, discussed above, can be used in low background areas to detect contamination from gamma emitters that do not have a beta component such as ^{125}I. Fig. 12.11 depicts a pancake detector with a 1.7 ± 0.3 mg cm^{-2} thick mica window that has an effective surface area of 12 cm^2. This type of detector can detect low level contamination in low background areas such as in patient administration areas. It is ideal for monitoring hands, shoes, and lab coats. Large areas can be monitored in a short amount of time and hot spots are readily identifiable. Typical counting efficiencies are 5% for ^{14}C, 22% for $^{90}Sr/^{90}Y$, and 32% for ^{32}P, making this detector ideal for contamination surveys in biomedical research laboratories.

Fig. 12.11 - Pancake detector (courtesy Ludlum Instruments).

12.4.4.2 Wipe Testing. Wipe testing, also known as smear testing or swipe testing, is a common technique used to monitors surfaces for removable contamination. Using a moderate amount of pressure, a small piece of filter paper or a cotton swab is wiped across about 100 cm^2 of the surface being tested. The wipe sample is then counted using appropriate instrumentation. In the field, wipe samples taken in high background areas can be removed to a low background area and counted with a portable Geiger counter equipped with a pancake or thin end-window detector. A more sensitive method of analyzing wipe samples is to count them in a liquid scintillation counter such as the one shown in Fig. 12.12, a gas-flow proportional counter, an auto-gamma counter, a single- or multi-channel analyzer, or some other low-level laboratory counting system. Wipe test results are usually reported in units of Bq or dpm per 100 cm^2.

In a study to determine the best material for taking wipe tests of beta emitting radionuclides, glass fiber filter disks dampened with distilled water or 70% ethanol were found to yield the highest contamination collection efficiencies from tile, glass, and lead foil (Klein 1992). In a study of wipe tests sample analysis using a liquid scintillation counter, a solution of counting fluid and 2% water was found to have the highest recovery efficiency (Kobayashi 1992). For additional information see the article entitled *Swipe Assays* on the Packard web site at the following URL: http://packardinst.com/.

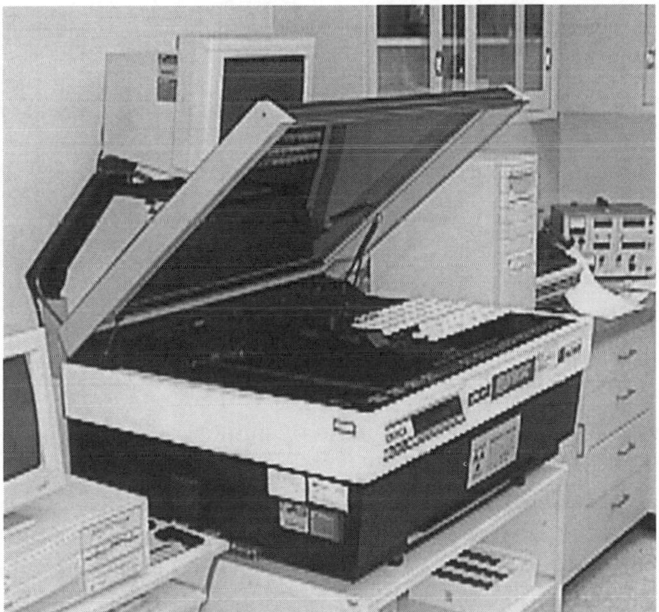

Fig. 12.12 – Packard Model 2000CA liquid scintillation counter (photo by author).

Fig. 12.13 is a picture of a counting system used to analyze wipes test samples taken from incoming and outgoing radioactive material shipments. It consists of an alarm rate meter, a shielded GM pancake probe, and a sample holder. This system would also be good for analysis of wipe test samples taken from laboratories using alpha, beta, or beta-gamma emitting radionuclides. This system cannot detect the very low energy beta particles emitted by tritium (^3H) or the low energy x-rays and electrons emitted by ^{55}Fe. Tritium is best detected using a liquid scintillation counter or a windowless gas flow proportional or GM counting system. A liquid scintillation counter or a NaI(Tl) detection system designed to detect very low photon energies is needed to analyze ^{55}Fe samples.

12.4.4.3 Sealed Source Leak Testing. Requirements for testing sealed radioactive material sources for leakage are established in 10 CFR 35.59. The most common method used to perform leak tests is by analyzing a wipe test sample from the surface of the source, the inside of the storage container, or a surface where radioactive contamination would be expected to accumulate.

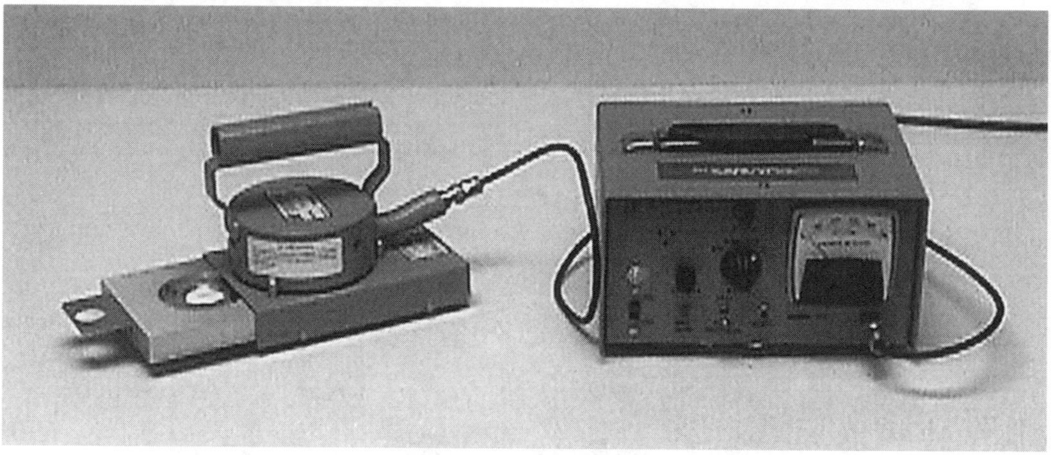

Fig. 12.13 – Wipe test counting system (photo by author).

12.4.4.4 Determining the Minimum Detectable Activity of a Counting System. Counting systems used to analyze leak test samples must be able to detect 185 Bq (0.005 μCi) of the radionuclide in the source being tested. Also, licensees need to be able to detect 33.3 kBq (2,000 dpm) on wipe tests in radiopharmaceutical preparation, administration, and storage areas (10 CFR 35.70(f)). When determining the minimum detectable activity of a counting system for regulatory purposes, use of the limit of detection, L_d, has been recommended (Lochamy 1976). This value is calculated as follows:

$$L_d = \frac{k^2}{T_t} + 2k\sigma_b\left(1+\frac{T_b}{T_t}\right)^{\frac{1}{2}} \tag{12.2}$$

where: T_t = sample counting time
T_b = background counting time
k = the one sided confidence level (1.65 for 95% confidence level)
σ_b = background standard deviation

12.4.5 Survey Meter Calibrations and Operational Checks

An important aspect of every medical center radiation safety program is the calibration and operational testing of portable survey meters. The applicable requirements can be found in 10 CFR 35.51, which state in part:

- Survey meters must be calibrated before first use, annually, and after repairs;
- All scales with readings up to 10 mSv h^{-1} (1000 mrem h^{-1}) must be calibrated with a radiation source;
- Two separated readings on each scale must be calibrated; and
- A licensee shall check each survey instrument for proper operation with the dedicated check source each day of use.

These requirements can be satisfied by developing an instrument calibration program within the institution or by employing the use of an instrument calibration service that is approved by the regulatory body that issued the license to the medical institution. In either case the person performing the calibration should thoroughly review the instrument manual, in particular all instructions regarding calibration.

12.4.5.1 Calibration of Rate Meters. As discussed above, pancake GM detectors, thin end window GM detectors and NaI(Tl) crystals attached to portable rate meters are frequently used to monitor surfaces or analyze wipe test samples. Calibration of these meters is performed in two steps: calibration of the rate meter followed by determination of the system's counting efficiencies.

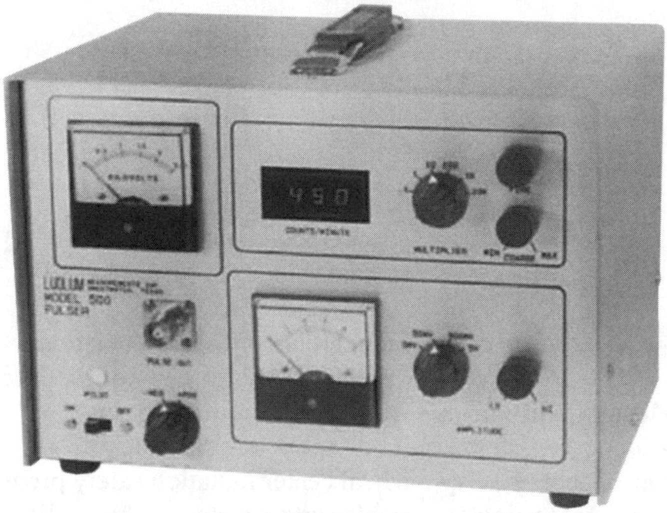

Fig. 12.14 – Ludlum Model 500 Pulser (courtesy Ludlum Instruments).

Rate meter calibration is usually performed with a pulser such as the one shown in Fig. 12.14. The pulser is attached to the rate meter with an appropriate cable and a known pulse rate is provided to the ratemeter's input. A good pulser will have the ability to select the pulse shape, timing characteristics, and electrical polarity. Each scale of the rate meter is adjusted to read the correct pulse rate. The pulser may also have a built-in voltmeter, which is designed to measure the high voltage being provided to the detector. The high voltage should be tested to ensure that it is stable and set correctly for the detector being used. Do not use common volt meters to measure the voltage of portable survey meters because they place an unacceptably high load on the high voltage circuit of the rate meter. This will result in significant voltage drop due to the excessive current drawn by the volt meter and could damage the high voltage circuit. Use only volt meters with an impedance of at least 2,500 megohms. The ideal high voltage measurement instrument is an electrostatic volt meter; however they are costly.

The second step of the calibration is performed by counting calibration standards to determine the counting efficiencies of the system. The standards should be counted in a geometry that is very similar to the geometry that will be used for the samples. Sources of various energies can be counted and an efficiency curve in units of counts per disintegration can be made. Using this curve, counting data can be converted into the activities found on wipe test samples. For instance, if the efficiency of a pancake detector for a particular beta particle energy is ϵ, the background count rate is R_b and the sample count rate is R_t, the activity, Q, on a wipe test sample is calculated as follows:

$$Q = \frac{R_t - R_b}{\in} \tag{12.3}$$

12.4.5.2 Calibration of Exposure and Dose Rate Meters. The primary instruments used to make exposure and dose rate measurements at medical institutions are Geiger counters and ion chambers. Calibration of these instruments requires the use of a radiation source of known activity or radiation output. Sealed ^{137}Cs sources are frequently selected for this purpose because they are readily available, have a half-life of 30 years, and emit 662 keV gamma rays . Only sources that are traceable to NIST standards should be used for these calibrations.

The exposure rate, X, emitted from a ^{137}Cs point source of activity A can be calculated using the specific gamma ray constant, Γ, for the calibration source (Cember 1996):

$$\dot{X} = \frac{\Gamma \times A}{d^2} \tag{12.4}$$

For instance, the exposure rate emitted from a 37 GBq (1 Ci) ^{137}Cs source at a distance of 100 cm is given by:

$$\dot{X} = \frac{\Gamma_{^{137}Cs} \times A}{d^2} = \frac{2.262\times10^{-2} C\,kg^{-1}\,h^{-1}\,cm^2\,GBq^{-1}\times 37\,GBq}{(100\;cm)^2} = 8.369\times10^{-5} C\,kg^{-1}\,h^{-1}$$

$$or$$

$$\dot{X} = \frac{3.244\,R\,cm^2\,mCi^{-1}\,h^{-1}\times 1000\,mCi}{(100\,cm)^2} = 0.3244\,R\,h^{-1} \tag{12.5}$$

This equation assumes that the source and the detector are single points in space, which can never be true. If a small detector is in close proximity to an extended source, the radiation contribution from the closest and farthest points of the source are very different due to the inverse square law. This violates the assumptions made in Eqn. 12.4. If a large detector is in close proximity to a small source, there will be a significant exposure rate gradient across the detector due to the inverse square law. Once again this violates the assumptions inherent in the equation. Proper calibration of a detector requires that it be in a uniform radiation field. This occurs when the distance between the effective centers of the source and detector is at least 7 times the maximum dimension of the detector or the source, whichever is larger (ANSI 1978). If the source must be placed closer, a correction factor should be used. For instance, consider an ion chamber with a dimension of 10 x 4 cm that is being calibrated using a sealed source with an active length of 3 cm. Since the chamber is larger than the source, the center of the chamber should be placed no closer than 7 x 10cm = 70 cm from the radiation source during calibration. At that distance the front of the detector will be in a radiation field that is about a 28% greater than the field at the back.

Fig. 12.15 - Survey Meter Calibrator (photo by author).

Even though calculating the exposure rate from a source will give a close approximation of its radiation output and this may satisfy regulatory requirements, it does not take into account scattering of the gamma rays from surrounding objects, the attenuation of gamma rays by the interposing air, or the backscatter of gamma rays within the source shield. For this reason, a more accurate way to determine the output at a given location is to measure it with a transfer instrument, such as a condenser R-meter or a precision ion chamber, that is calibrated for that radionuclide by NIST or an accredited laboratory. If the source is permanently installed in a facility, this procedure only needs to be performed once along the length of the useable beam path. Fig. 12.15 shows a calibration device that contains six separate ^{137}Cs sources of varying sizes. A cart rides along the beam center line on an inverted angle iron track. The center of the detector is positioned in the beam vertically and horizontally with instrument specific platforms and positioning devices. The exact distance between the center of the detector and the source is taken from a tape measure permanently mounted to the floor. A video monitoring system is used to view the instrument's scale during calibration. Safety features include lighted warning signs, an interlocked entrance gate, and a pressure sensitive pad that will sound an alarm if stepped on while the higher activity sources are exposed. After the range was calibrated, a computer program was written that calculates the best sources and distances to use when calibrating survey meters.

Because most air filled ion chambers are not sealed, the amount of air within them will vary with temperature and barometric pressure. For this reason, an accurate thermometer and barometer are needed to calculate correction factors for use during calibrations. Humidity is also a concern and should be regulated if possible. Unsealed ion chambers also may have an internal container of desiccant to remove moisture from the air that enters the chamber and to keep moisture away from sensitive electronic components. This desiccant needs periodic replacement or drying in a conventional or microwave oven.

12.5 OTHER INSTRUMENTATION NEEDS

12.5.1 X-Ray Equipment Inspections

Medical and non-medical radiation producing machines require periodic surveys and inspections to verify their safe operation. Surveys of non-medical equipment such as cabinet x-ray machines, and x-ray diffraction and spectroscopy equipment can be conducted with the equipment described above in section 12.4.3. Several additional instruments may be needed by health physicists who inspect diagnostic medical x-ray equipment. Condenser R-meters or in-beam precision electrometer ion chambers (sometimes called dose meters) are used to determine half-value layers and to measure patient doses for standard procedures. In addition, this type of meter can usually measure the x-ray beam duration. Noninvasive kVp meters are sometimes useful when there is a need to verify the maximum kVp of a beam. Miscellaneous equipment that might also be needed include aluminum and copper filters, test stands, phantoms, and radio-luminescent strips.

12.5.2 Assay of Doses Prior to Administration

With a few minor exceptions, all doses of radiopharmaceuticals must be assayed prior to administration (10 CFR 35.50). The instrument used to perform this task is called a dose calibrator, which is usually a well type ionization chamber. Internal potentiometers or potentiometer settings are provided for each radionuclide to be measured so that the instrument will read the correctly in units of activity for that radionuclide. Daily constancy checks with a long-lived source must be performed on each setting that will be used. In addition quarterly linearity, annual accuracy, and initial geometry testing must be performed.

12.5.3 Medical Use of Radioactive Aerosols and Gases

Almost every nuclear medicine department uses radioactive gases or aerosols. The most prominent use is ^{133}Xe gas for lung ventilation and clearance studies. Inhalation of aerosolized ^{99m}Tc is also encountered. To practice these procedures safely, the NRC requires that the gas exhaled from the patient be vented directly to the atmosphere using a ventilation system or that it be collected and decayed in a xenon trap (10 CFR 35.205(a)). If a trap is used, it must be tested for proper operation at least monthly. These tests can be conducted by collecting the exhaust from the unit in a large plastic bag and counting it in a low background area with a gamma camera, Geiger counter or other appropriate survey meter.

There is also a requirement that radioactive gases are to be administered in rooms that are at negative pressure with respect to the surrounding areas (10 CFR 35.205(b)). This can be verified by determining the direction of air flow at each room entrance using a velometer or smoke tube. Since air pressure differences are sometimes slight, the door may need to be partially closed to make this measurement. It is important to check this in different seasons of the year because of seasonal variations in building heating and cooling systems.

Velometers are also used to measure the face velocity of fume hoods used to prepare and store radioactive gases and volatile radioactive compounds such as therapeutic quantities of Na^{131}I. Face velocities should be measured at least once each year to ensure they meet established standards.

12.5.4 Radioactive Waste

12.5.4.1 Incineration of Radioactive Waste. Some medical institutions are licensed to incinerate low-level radioactive waste. Effluent released from the incinerator exhaust stack must meet concentration limits for releases to unrestricted areas. The method to determine compliance with this requirement is usually determined during licensing. One of the options is to monitor the effluent using stack monitoring techniques. Iso-kinetic sampling of the stack using a pitot tube and filter can be performed and the samples can be analyzed using a multi-channel analyzer, liquid scintillation counter, auto-gamma counter, or other appropriate instrumentation. Alternatively, entire monitoring systems can be purchased from commercial manufacturers.

12.5.4.2 Sewage Disposal of Radioactive Liquids. Sewage effluent from hospitals can contain significant quantities of radioactive materials. These releases result from the radioactive material present in patient excreta, which is exempt from NRC regulatory control (20.2003(b)). In particular, patients being treated with therapeutic quantities of ^{131}I will excrete a large fraction of the material in their urine within the first 72 hours following administration of the dosage. A smaller quantity of activity is excreted from hospitalized patients containing diagnostic quantities of ^{99m}Tc, ^{123}I, ^{111}In, ^{67}Ga, and others. A recent pilot study of the concentration of radioactive material in sludge and ash at nine publicly owned treatment works (POTWs) was conducted jointly by the EPA and NRC (USEPA 1999). Twenty two of thirty dried sewage sludge samples tested positive for ^{131}I with concentrations up to 2.58 Bq g^{-1} (69.8 pCi g^{-1}) dry weight. The stated objective of the study was to "estimate the extent to which radioactive contamination comes from either NRC/State licensees or naturally-occurring radioactivity . . and to support potential rulemaking decisions." The full study will include 300 POTWs. The results of this study may play a role in future instrumentation needs of medical institutions. If rulemaking establishes the need to monitor sewage for radioactivity, in-line sewage monitors or grab sampling techniques may need to be implemented. These methods will require counting of samples or interpretation of detector readings as well as instrument calibrations.

12.6 CONCLUSION

A wide variety of potentially hazardous radionuclide and machine produced radiation sources are found in a typical medical center. With adherence to the established radiation safety regulations and practice, and with the proper use of appropriate health physics instruments, medical administration of radionuclides and radiation can be done safely for the benefit of patients and biomedical research can be conducted safely for the benefit of society.

12.7 REFERENCES

American National Standards Institute. Radiation protection instrumentation test and calibration. New York: The Institute of Electrical and Electronics Engineers, Inc.; ANSI N323-1978; 1978.

Blatz, H. Introduction to radiological health. New York: McGraw-Hill Book Company; 225; 1964.

Cember, H.; Introduction to health physics. New York: McGraw-Hill Book Company; 420; 1996

Golnick, D.A.; Basic radiation protection technology. Temple City, CA: Pacific Radiation Press; 334; 1983.

Klein, R.C.; Linins, I; Gershey E.L. Detecting removable surface contamination. Health Phys. 186-189; Feb 1992.

Kobayashi, Y. Notes on swipe testing (unpublished article). Wellesley, MA: KO-BY Associates; 1992.

Lessard; E.T.; Yihua, X.; Skrable, K.W.; Chabot, G.E.; French, C.S.; Labone, T.R.; Johnson, J.R.; Fisher, D.R.; Belanger, R.; Lipsztein, J.L. Interpretation of bioassay measurements, prepared for the U.S. Nuclear Regulatory Commission. New York; Brookhaven National Laboratory; 1990.

Lochamy, J. The minimum detectable activity concept. In: Proceedings of NBS symposium on measurements for the safe use of radiation, March 1-4, 1976. National Bureau of Standards, Report No. NBS-SP456; 169-172; 1976.

National Council on Radiation Protection and Measurements. Limitation of exposure to ionizing radiation. Washington , D.C.: NCRP Publication 57; 1978.

National Council on Radiation Protection and Measurements. Use of bioassay procedures for assessment of internal radionuclide deposition. Bethesda, MD: NCRP Publication 87; 1987.

Rosenstein, M.; Webster, E.W. Effective dose to personnel wearing protective aprons during fluoroscopy and interventional radiology. Health Phys. 88-89; Jul 1994.

U.S. Environmental Protection Agency and U.S. Nuclear Regulatory Commission. Joint NRC/EPA sewage sludge radiological survey: survey design and test site results. EPA 832-R-99-900; Mar 1999.

U.S. Nuclear Regulatory Commission. Regulatory Guide 8.20; Applications of bioassay for I-125 and I-131; Revision 1: Sept 1979.

U.S. Nuclear Regulatory Commission. Regulatory Guide 8.23; Radiation surveys at medical institutions; Revision 1: Jan 1981.

U.S. Nuclear Regulatory Commission. Regulatory Guide 8.9; Acceptable concepts, models, equations, and assumptions for a bioassay program; Revision 1: Jul 1993.

U.S. Nuclear Regulatory Commission. NRC staff requirements memorandum - CMEXM-00-002. Dec 5, 2000.

Chapter 13

ENVIRONMENTAL INSTRUMENTATION

Joseph L. Alvarez

13.1 INTRODUCTION

Environmental radiation measurements in general deal with the natural radiation background and with man's additional contribution to environmental radiation. This paper is directed at anthropogenic activities. Current radiation protection (regulation) is based on risk, but defined on the basis of dose. Risk is not a directly measurable quantity, and generally, neither is dose. To make matters worse, all definitions of dose and risk are not equal. The problem with measuring risk and dose at near natural background levels is that the definitions based on risk involve very small risks making the consequent measurement of dose very tedious. Even the usually simple measurement of external dose requires calculation by a model to be precise. In general, environmental radiation measurements are performed as input to models or to confirm that radionuclide concentrations are below modeled concentrations. Nevertheless, there are other reasons for measuring the natural background, including radon and cosmic radiation. These will not be the topic of this discussion.

Three major guidelines on environmental measurements are used as a basis for meeting regulatory requirements for environmental clean-up levels:

1. NRC and EPA (NRC 1997a, EPA 1994, EPA 1997) specify final clean-up criteria in terms of an annual total external and internal effective dose equivalent (TEDE) limit.
2. RESRAD (Yu 1993) is a computer program for quantitatively relating the TEDE to environmental measurements.
3. MARSSIM (NRC 1997b) deals with measurements techniques for providing input data to RESRAD.

These guidelines practically ensure that all environmental measurements are radionuclide concentrations rather than dose rate.

Dose reconstruction and clean up regulations require consideration of dose from all pathways. The external dose from radionuclides may be a minor, yet, significant part of the total dose. It may be important to measure the external dose correctly as effective dose, especially when limits are approached or legal considerations are necessary. The conversion of dose rate in air in Gy h^{-1} or rad h^{-1} to effective dose rate in tissue in Sv h^{-1} or rem h^{-1} may not be obvious for a given measurement device, especially when the device is restricted to measuring ions or counting events.

The ICRP 60 (ICRP 1991) currently recommends three types of quantities that are necessary for correct measurement of external dose. These are protection quantities, operational quantities, and physical quantities. These quantities and their use in measuring external dose will be discussed based on the ICRU 57 (ICRU 1998). Most regulations in the US are based on ICRP 26/30 (ICRP 1977, ICRP 1988) rather than ICRP 60. The differences between these quantities are usually small and negligible, but for some energies and geometries the difference can be as high as 68% (Zankl 1992). This discussion will use the ICRP 60/ICRU 57 definitions and quantities.

The radiation protection quantity currently recommended by the ICRP is the effective dose, E. Two other protection quantities recommended are the mean absorbed dose in an organ or tissue, D_T and the equivalent dose in an organ or tissue, H_T.

Effective dose is the summation of the equivalent doses in tissue or organs, each multiplied by the appropriate tissue-weighting factor. It is given by the expression

$$E = \sum_T w_T \bullet H_T \qquad (13.1)$$

where H_T is the equivalent dose in tissue or organ, T, and w_T is the tissue-weighting factor for tissue, T. The unit of effective dose is joule per kilogram (J kg^{-1}) and its special name is sievert (Sv).

The equivalent dose is the product of the absorbed dose in an organ or tissue multiplied by the relevant radiation-weighting factor

$$H_T = w_R \bullet D_T$$

If the tissue or organ is irradiated by more than one radiation, then the equivalent dose is the sum of the equivalent doses from each radiation.

$$H_T = \sum_R w_R \bullet D_{T,R}$$

where $D_{T,R}$ is the absorbed dose averaged over the tissue or organ, T, due to radiation R, and w_R is the radiation-weighting factor for radiation, R. The unit of equivalent dose is joule per kilogram (J kg^{-1}) and its special name is sievert (Sv).

The mean absorbed dose, D_T, in a specified tissue or organ of the human body, T, given by

$$D_T = (1/m_T)\int_{mT} D\,\mathrm{d}m \qquad (13.2)$$

$$D_T = \varepsilon_T / m_T \qquad (13.3)$$

where m_T is the mass of tissue or organ, D is the absorbed dose in the mass element dm, and ε_T is the total energy imparted in the tissue or organ. The unit of absorbed dose is joule per kilogram (J kg^{-1}) and its special name is gray (Gy).

A set of operational quantities is defined that allows for calibration of instruments for measurements to show compliance with the system of protection quantities. These measurable quantities are the ambient dose equivalent, $H^*(d)$, for specifying a field of highly penetrating radiation the directional dose equivalent, $H(d,\Omega)$ for specifying a field of weakly penetrating radiation, and the personal dose equivalent, $H_p(d)$, which is the dose to a person in a radiation field. The first two quantities are defined for a parallel beam of photons at a depth d in a 30 cm diameter tissue equivalent sphere, while $H_p(d)$ is defined for a parallel beam of photons on a 30 cm x 30 cm x 15 cm tissue equivalent slab. The depth, d, for deep or whole body dose is 10 mm. These equivalent doses are in units joule per kilogram (J kg^{-1}) and the special name is sievert (Sv).

The most commonly measured radiation physical quantity is the measurement of air kerma, K, free-in-air. Nevertheless, the measurement is indirect and rarely does one claim to be measuring kerma.

Kerma is the quotient of dE_{tr} by dm, where dE_{tr} is the sum of the initial kinetic energies of all the charged ionizing particles liberated by uncharged ionizing particles in a volume element of mass dm,

$$K = \mathrm{d}E_{tr}/\mathrm{d}m. \quad (13.4)$$

The unit of kerma is joule per kilogram (J kg^{-1}) and its special name is gray (Gy).

The standard device for measuring air kerma, free-in-air, is the free-air ionization chamber. This device and calorimeters are basic instruments used to develop and characterize calibration systems. A third physical quantity, the fluence, is used to completely develop calibration systems, but fluence is most important in calculating the expected response of devices designed to measure the operational quantities.

Fluence is the number of particles incident on a sphere of unit cross-sectional area. The fluence, Φ, may be defined as the number of particles, dN, divided by the area, da , which they irradiate.

$$\Phi = \mathrm{d}N/\mathrm{d}a. \quad (13.5)$$

The three physical quantities (mean absorbed dose, kerma, and fluence) form the basis of radiation measurement, while calculation, based on empirical evidence of risk, forms the basis of radiation safety criteria. Measurement systems for the operational quantities rely on calculation and calibration. The use of these two sets of quantities often leads to confusion as to what is being measured at any given time. There is a conviction among some that the only unit of measurement is the Gy. While this is technically true, calibration allows measurement in Sv. Unfortunately, neither unit may be correctly applied unless the device used for measurement is used only under the calibrating conditions.

13.2 MEASUREMENT OBJECTIVES

The conduction of environmental measurements requires a clear objective and a strategy to meet that objective. The strategy includes the selection of instruments and the employment methods for the instruments. The development of the strategy for either measuring dose rate or concentration requires knowledge of the measurement characteristics of the instruments.

The needs of the measurement usually determine the objective. These needs may be nuclide specific as determining soil concentration and depth of concentration. The usual reason for nuclide specific measurements is that a nuclide specific clean up requirement has been established or to calculate a dose rate in order to establish a clean up requirement.

The objective may depend upon the progress of an operation. The operation may be in the investigation and planning stage or at any stage up to clean closure. Simple dose rate measurements may suffice for locating areas of contamination, while measurements of soil concentrations may be necessary for clean closure. If the contaminants are few and the distribution well known, dose rate measurements may be sufficient for concentration measurements.

Concentration measurements may be made using dose- or count-rate instruments based on modeling and/or calibration. The precision and accuracy requirements will determine how closely the measuring geometry must be established and maintained. Measurement geometry may include detector and detector array, height above the ground, detector shielding, and measurement location.

There are essentially three basic objectives:

1) Mapping relative levels of contamination or dose rate. This objective will not be addressed explicitly. The specific instrumentation needs and problems will be addressed under the other two objectives.
2) Measuring external dose rate as a component of total dose.
3) Measuring concentration for calculation of total dose.

13.3 MEASURING EXTERNAL DOSE RATE AS A COMPONENT OF TOTAL DOSE

The measurement of external dose rate is dependent upon the photon energy (alpha and beta are usually not external concerns), irradiation geometry, and detector response. Most detectors are calibrated to $H_p(10)$ using a nearly parallel beam of a single energy. The calibration is usually not the best for environmental measurements. The environmental dose rate involves several energies and irradiation other than a parallel beam.

Effective dose can be modeled in the human body based on measurements in air and use of appropriate human equivalent models or phantoms. The computational phantom for effective dose and the various modeling geometries are shown in Fig. 13.1.

The energy dependence of effective dose for the various standard geometries is shown in Fig. 13.2. The figure shows there are major geometric dependencies in the effective dose. A simple measurement of photons in air does not readily lead to effective dose in tissue.

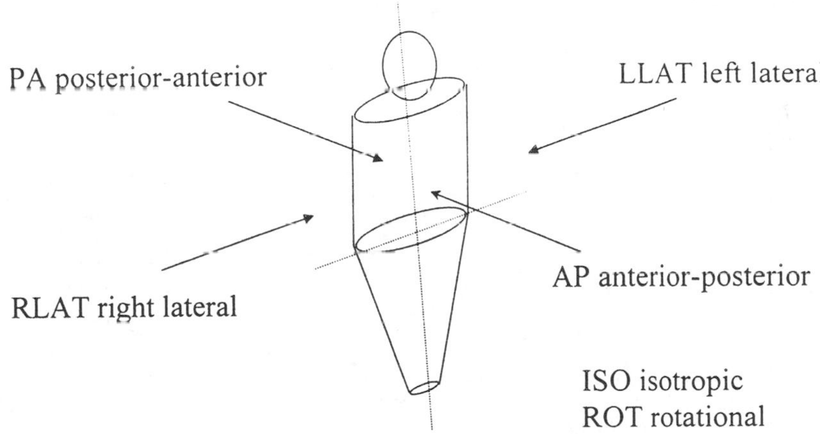

Fig. 13.1. The computational phantom for effective dose and the standard geometries for modeling effective dose

The energy dependence of effective dose for the AP geometry is shown in Fig. 13.3. The figure shows the relationship of air kerma to effective dose for a human phantom in a uniform parallel field in anterior incidence and $H_p(10)$ for a slab phantom at perpendicular incidence. Dose is less than air kerma for energies less than 47 keV for AP effective dose. Dose is greater than kerma from 40 to 1000 keV with a maximum at 80 keV. Dose is slightly less than kerma for energies higher than 1000 keV. The $H_p(10)$

dose is greater than AP effective dose at all energies. $H_p(10)$ is the usual calibration curve. An ion chamber may be easily calibrated for energies 300 keV and above with little concern for energy dependence (within 10%). Energies less than 300 keV may prove problematic.

13.3.1 Commercial Instruments

The obvious type of instrument for performing measurements of effective dose is an air ionization chamber since coefficients are available for converting air kerma in air to dose in tissue and effective dose. Many air ionization devices are commercially available that are calibrated in Sv h^{-1} or rem h^{-1}. The response of these instruments above 300 keV should be easily calibrated to read in effective dose in the AP direction, personal dose $H_P(10)$, Fig. 13.3. Several practical geometries (cloud, uniform distribution on the ground plane, and uniform distribution in soil) are shown in Fig. 13.4. Fig. 13.4 shows that other environmental geometries may present problems for an instrument calibrated to $H_p(10)$. A compromise calibration may be rotational irradiation since many exposure situations may be effectively rotational (Fig. 13.2).

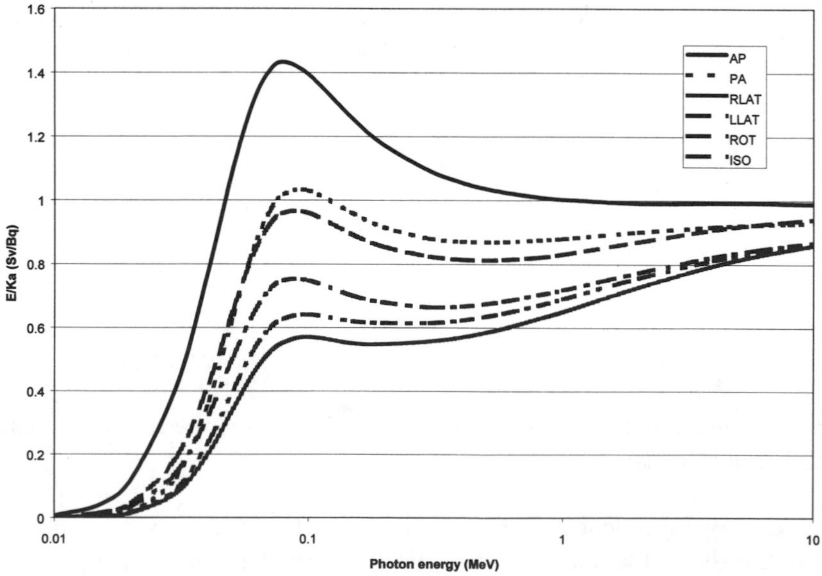

Fig. 13.2 The geometry dependence of effective dose for the phantom and geometries of Fig. 13.1 relative to air kerma.

Commercial instruments for measuring environmental exposures may be constructed from materials other than air. These other materials may have sufficiently different characteristics to cause problems for calibration to effective dose. The mass attenuation coefficients for several materials are compared to muscle for various energies in Fig. 13.5.

The proper calibration is most likely to personal dose equivalent, since this is the most practical measurement definition for safety assessment and for regulatory

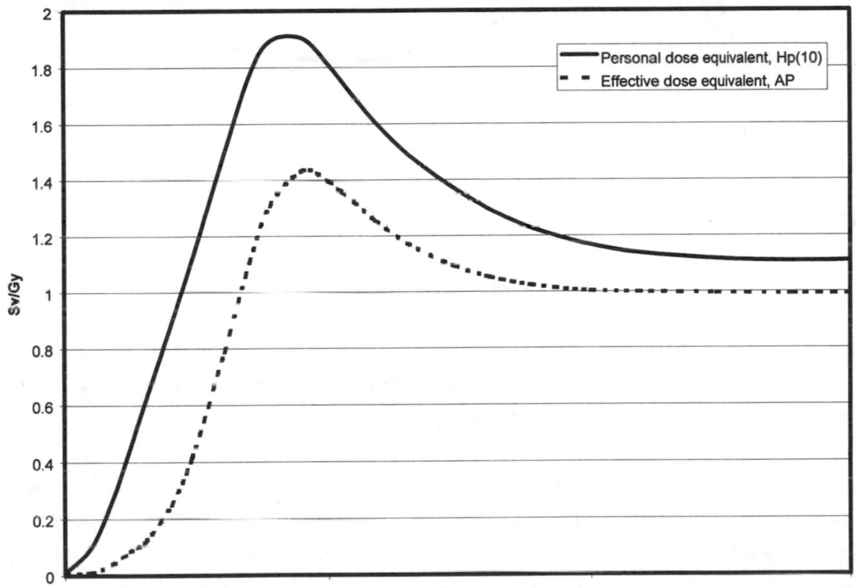

Fig 13.3 – The effective dose for AP geometry and *Hp(10)* compared to air kerma.

compliance. The response of a typical air ionization chamber in air is shown in Fig. 13.6 compared to $H_p(10)$ and the effective dose from a plane source at the ground surface that is infinite in extent. The location on the ordinate for the ionization chamber response is approximate, since the calibration is against free air ionization. Free air ionization measurements yield roentgen units and must be converted to air kerma. The ion chamber response is R/R (exposure in air) whereas the effective dose response is Sv/Gy. This particular instrument responds similarly in all orientations to a beam of radiation, except the lower energies. The electronics unit and housing wall differentially attenuate the lower energies. It is possible that such an instrument can be successfully incorporated into a measurement program, if care is taken in the interpretation and calibration. Most hand-held ionization chambers do not have sufficient sensitivity for cleanup applications.

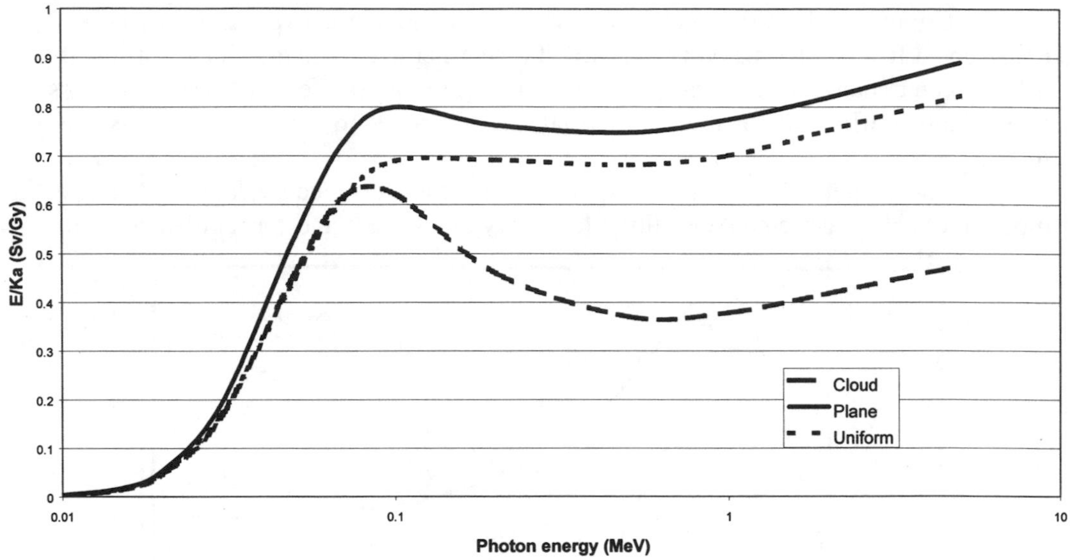

Fig 13.4. Energy and geometry dependence for effective dose for several possible environmental exposure conditions of a radioactive cloud, radioactivity on an infinite plane, and radioactivity uniformly distributed in soil.

A particular instrument, the pressurized ion chamber, was designed for uniform directional response and response at low doses. The common name for this instrument is the PIC. The PIC was developed by Environmental Measurements Laboratory and is well characterized (Beck 1972a). The energy response of this instrument is shown in Fig. 13.7 and compared to $H_p(10)$ and the effective dose from an infinite plane source. A PIC is calibrated in μR h^{-1}. The PIC ionization chamber is a spherical steel shell containing argon at 18 atmospheres. The PIC measures neither air kerma nor effective dose, nevertheless, because it is well characterized it is often accepted as a standard for human exposure. The PIC can also be usefully employed in a measurement program as long as the limitations are recognized and accounted for.

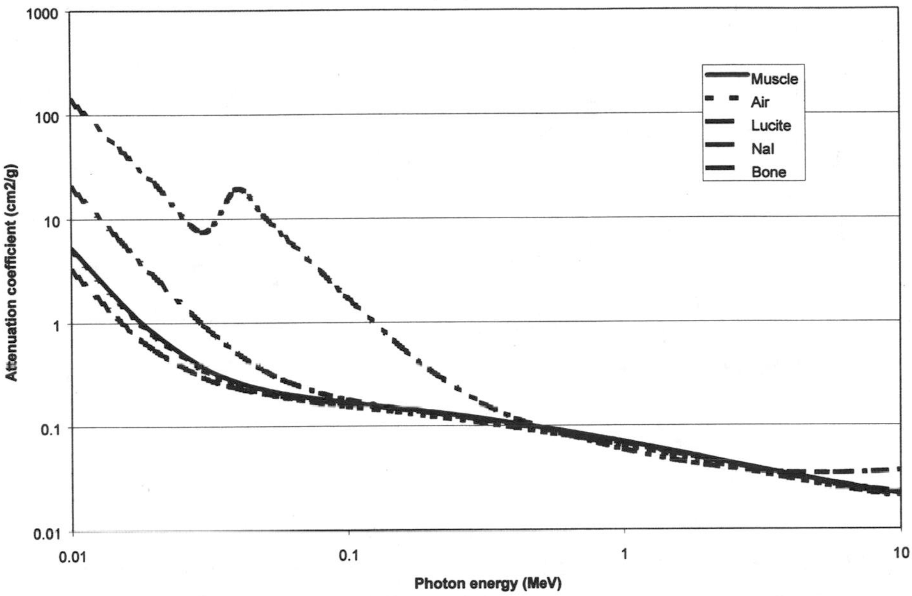

Fig 13.5 – Mass attenuation coefficients of several common detector materials compared to muscle.

Air ionization is used in other types of instruments. GM type detectors have an energy response typified in Fig. 13.8 and compared to $H_p(10)$ and effective dose from an infinite plane source. This type response is generally unacceptable for dose measurements for unknown energies, but can be useful for dose rate measurements in situations where the calibrating conditions closely match the measuring conditions. Generally, GM counters are more useful for locating and measuring radioactivity than for dosimetry. The compensated GM detector was developed to give better dose rate response. The energy response of a compensated GM detector is, also, shown in Fig. 13.8.

A popular instrument for environmental gamma exposure measurements is the NaI scintillation detector calibrated to $\mu R\ h^{-1}$. The high-density detector has excellent event response to gamma and, therefore, good sensitivity to background gamma levels. This detector does not have an energy response that allows general calibration to effective dose because it counts events and does not have the same energy response as tissue. The energy response of a μR meter is shown in Fig. 13.9 and compared to $H_p(10)$ and effective dose from an infinite plane source.

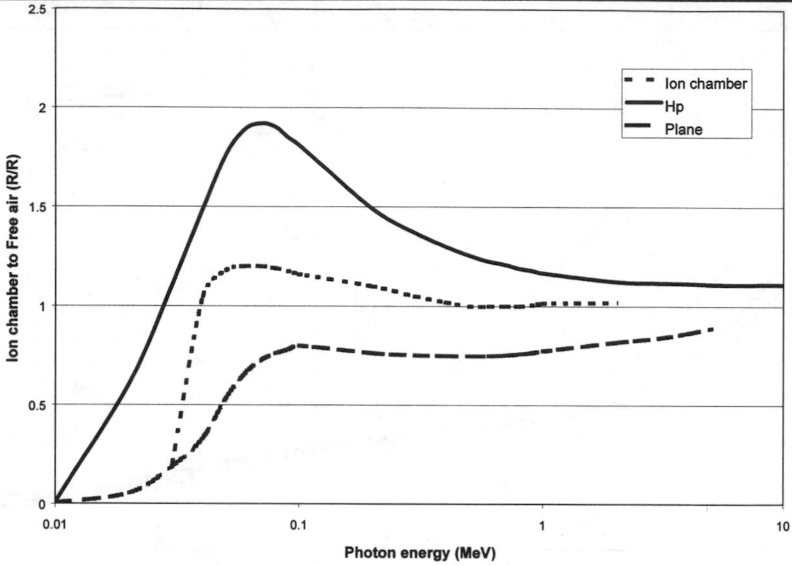

Fig 13.6 – Energy response of a hand-held ionization chamber compared to a free air ionization chamber. Further comparison is made to $H_p(10)$ and effective dose from an infinite plane source.

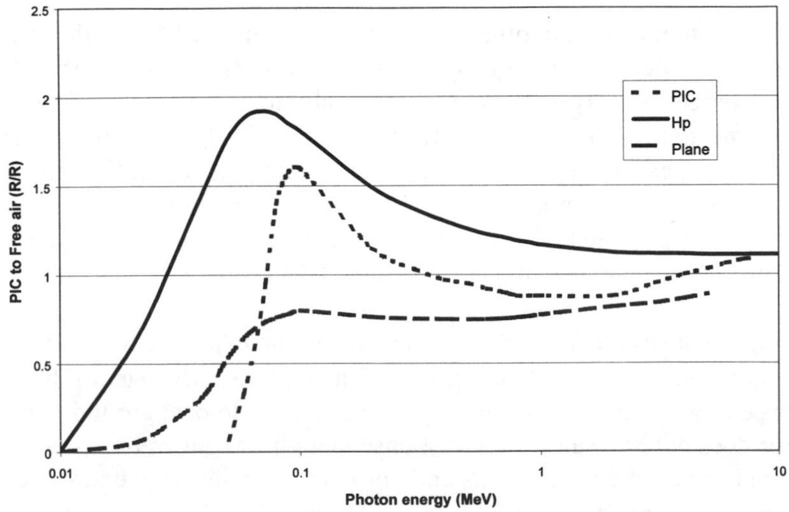

Fig 13.7 – Energy response of a pressurized ionization chamber (PIC) compared to a free air ionization chamber. Further comparison is made to $H_p(10)$ and effective dose from an infinite plane source.

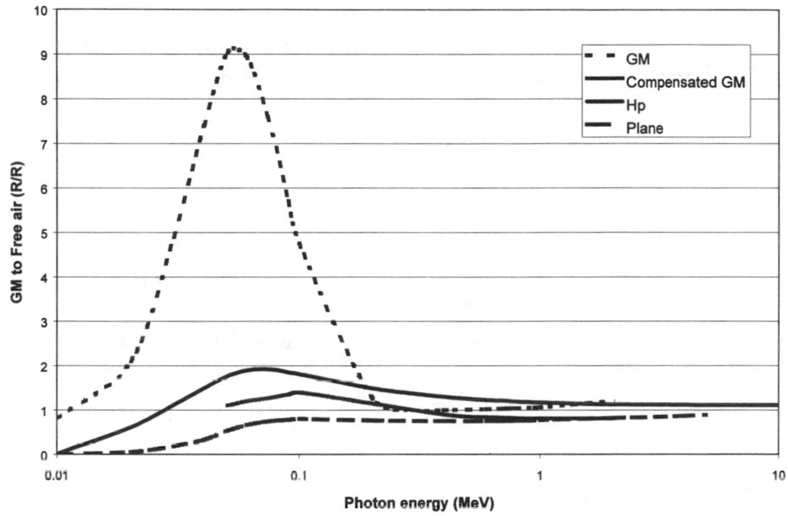

Fig 13.8 – Energy response of a GM detector and a compensated GM detector compared to a free air ionization chamber. Further comparison is made to $H_p(10)$ and effective dose from an infinite plane source.

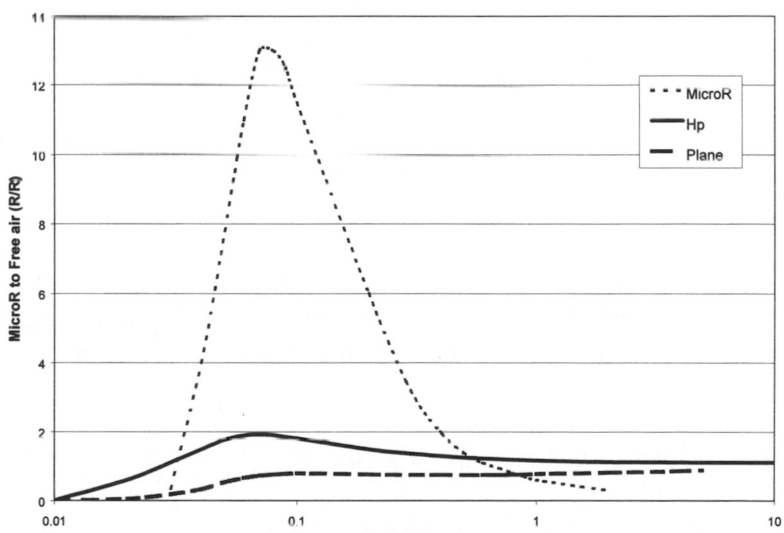

Fig 13.9 – Energy response of a microR meter compared to a free air ionization chamber. Further comparison is made to $H_P(10)$ and effective dose from an infinite plane source

A common field calibration when a NaI μR meter is used is to correlate the μR meter to a standardized PIC. Figs. 13.7 and 13.9 show that neither instrument correctly measures effective dose, but the PIC is acceptably close and is accepted as a standard. Fig. 13.10 shows the results of such a correlation.

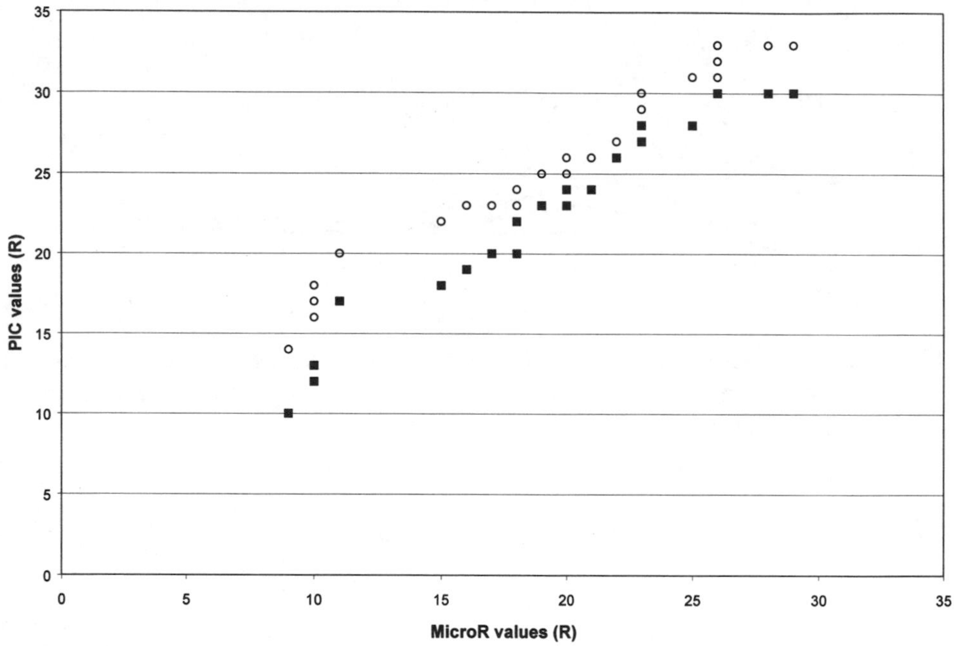

Fig 13.10 – Correlation of a μR meter to a PIC for purposes of correcting the NaI bias.

Fig. 13.10 shows two sets of data that are almost parallel but with different intercepts. The measurements were made at two locations similar in soil radionuclides but at different altitudes. The major difference between the data sets is the amount of cosmic radiation.

The fit to both data sets was

$$PIC = A + 0.51 microR \tag{13.6}$$

The constant A is nearly equal to the cosmic dose rate. The NaI response, a single pulse, and calibration to ^{137}Cs (low energy relative to cosmic radiation) ensures that there is little response to the cosmic radiation.

Another instrument for environmental gamma exposure measurements is the plastic scintillation detector calibrated to μrem/h. The low-density detector has poorer event response to gamma than the NaI detector and, therefore, poorer sensitivity to background gamma levels. This detector has very good energy response that allows general calibration to effective dose because it has an energy response similar to tissue. The response of a microrem meter is shown in Fig. 13.11 and compared to $H_p(10)$ and effective dose from an infinite plane source.

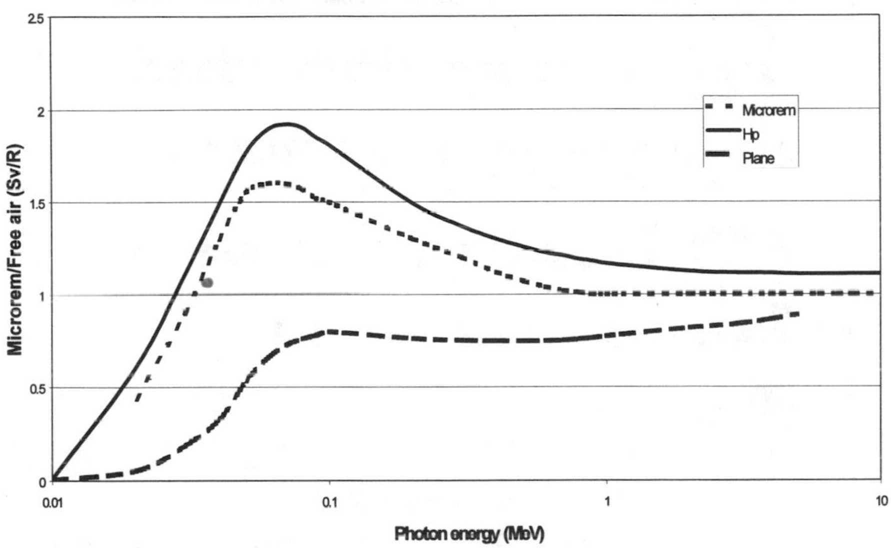

Fig 13.11 – Energy response of a microrem meter compared to a free air ionization chamber. Further comparison is made to $H_p(10)$ and effective dose from an infinite plan source.

13.3.2 Calibrating Conditions

Under good calibrating conditions instruments can approximate the various ICRU 57 definitions. Depending upon the type of instrument, the energy range of the approximation may be limited. Therefore, if the field conditions are known, the calibration could allow a good approximation of dose under the given field conditions. Nevertheless, calibration is usually inappropriate for field conditions. $H_P(10)$ is the usual calibrating geometry, it overestimates effective dose in the best calibration configuration. $H_P(10)$ does not approximate any other exposure geometry, but is always conservative.

A particular problem for calibration is that the calibration spectrum (point source) does not follow a field spectrum (extended source). Fig. 13.12 shows the energy spectral difference between a point ^{226}Ra source and ^{226}Ra distributed evenly in soil (Beck 1972b, NCRP 1976).

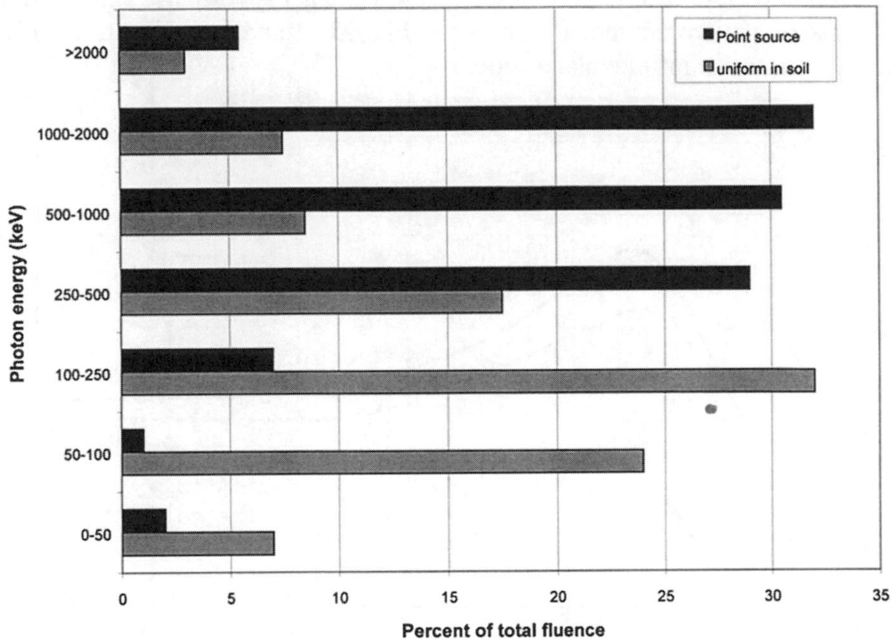

Fig 13.12 – The energy spectral difference between a point ^{226}Ra source and ^{226}Ra distributed evenly in soil. There is nearly twice the total energy in a point source spectrum of an equal number of photons.

The difference between the two spectra will deliver a quite different air kerma for an equal number of photons. In some cases, instrument characteristics may compensate for the air kerma difference, but it should be obvious, that an instrument is only correctly calibrated for the calibrating conditions.

Calibration to effective dose for a given environmental situation should be possible, but tedious. Placing a source or sources at sufficient positions in the plane and recording the instrument response can simulate an infinite plane. The measurements of response are repeated and summed for enough configurations to approximate a planar source. Such a calibration must be accomplished for enough photon energies to ensure that the calibration acceptably fits the environmental spectrum. The final calibration would be an iterative process.

A more likely calibration would be to use a computer model of the instrument to predict the response. A single or a few calibrations to a parallel beam could be made based on the model to effect an acceptable calibration. Nevertheless, it is most likely that calibration is possible only for limited types of detectors. The best technique is to recognize the approximate magnitude of over response.

13.4 MEASURING CONCENTRATION FOR CALCULATION OF TOTAL DOSE

Total dose requires knowledge of radionuclide concentration and a model such as RESRAD. The usual method for obtaining soil concentration is to collect samples and analyze them in a laboratory. Laboratory analysis has limited accuracy because time and cost usually restrict the number of samples. The accuracy is affected by the representativeness of a few samples compared to the area sampled. Radionuclide concentration can be determined using *in situ* gamma spectrometry, count rate, or dose rate. These field techniques are superior in representativeness because they cover a larger area (field of view) than a single soil sample. It is usually possible to acquire many more field analyses than laboratory analyses. Assumptions of radionuclide distribution in soil or the level of detail concerning the distribution affects the accuracy of field measurements of soil concentration. All field techniques require a known or presupposed distribution in the soil with depth and a lateral distribution that does not vary appreciably across the field of view of the detector.

The measurement of dose rate does not distinguish from natural background but presumes an average background and considers all dose rates above the average as an increase due to contamination. Nevertheless, a given location would be considered as natural background as long as it was within the uncertainty of background. The uncertainty of background requires that an LLD of other level of detection be developed to distinguish elevated dose rate from background dose rate. The measurement of concentration requires that an LLD be established for purposes of design of the investigation. The LLD should have a definite relationship to a parameter of the investigation such as the clean up level. A recommended LLD goal is 10% of the parameter. The LLD developed for concentration measurement involves the detector, detector size, count time, detector height and shielding, and distribution of the radionuclides.

Dose rate instruments can estimate soil concentration if the ratio of soil concentration to dose rate is known. Dose rate instruments may have more utility locating areas of contamination rather than quantifying concentration, nevertheless, it is possible to model the dose rate based on concentration. The difficulty with dose rate instruments is in obtaining sufficient resolution and differentiation from background.

Gamma spectrometry is the most obvious method for concentration measurement since it can be related to concentration and geometry. The calculation of the relationship may not be simple, but the principles are well known. In many cases calibration can be used instead of calculation and is usually required at some point to verify the process.

Gamma spectrometry can be accomplished with a multi-channel analyzer or with only one or two gamma energy windows using discriminators. NaI detectors are often used for *in situ* spectrometry because they are relatively inexpensive and rugged. A method must be devised to calibrate a NaI system to a distribution in soil. There are numerous published reports using NaI that can be used to estimate field conditions and the ability of a system to achieve acceptable sensitivity for given requirements.

Based on information from Beck 1972, 37 Bq kg^{-1} (1 pCi g^{-1}) of ^{137}Cs uniformly distributed laterally and vertically in the soil will yield 0.022 photopeak counts per second (cps) per cm^2 of detector face placed 1 meter above the ground surface. A 10 cm x 10 cm NaI detector should collect 1.8 cps for each 37 Bq kg^{-1} of ^{137}Cs. A background of about 19 cps is found in the photopeak area for a NaI detector of this size. The background level requires 23 photopeak cps for a 95% lower limit of detection (LLD) for this particular detector in a one second count. The LLD for different count times (1 and 30 seconds) and detector configurations are given in Table 13.1. Doubling the background will increase the LLD by approximately 40% while lowering it by half will decrease the LLD by roughly 25%.

Table 13.1 - LLDs for NaI detectors and detector arrays for 1- and 30-s (WHC 1993).

Detector Diameter (cm)	No. of Detectors	Total cps per 37 Bq kg^{-1} of ^{137}Cs	LLD in 1-s (Bq kg^{-1})	LLD in 30-s (Bq kg^{-1})
5	1	.045	1069.3	162.8
10	1	1.78	477.3	77.7
20	1	7.1	225.7	40.7
20	4	28.5	107.3	18.5
20	6	42.8	88.8	14.8
20	10	71.3	66.6	11.1

Gross counts are an option if there is no need to identify the radionuclide specifically. Gross counts allow a 5 cm x 5 cm NaI detector to have an LLD of about 703 Bq kg^{-1}. Kocher (1985) showed that the flux from 1 Bq kg^{-1} of ^{137}Cs uniformly distributed through soil of density 2 was equivalent to 82973 Bq/m^2 distributed on the surface. Assuming 30% efficiency for intercepting the emitted photons the cps are

0.038 per Bq kg^{-1}. A background count rate of 2000 cpm for the detector would allow a 703 Bq kg^{-1}1-s LLD. The gross count LLD is about 2/3 of the photopeak LLD suggesting that gross counts are more sensitive than photopeak counts, but without discrimination between contaminant and background.

The estimated LLDs for photopeak and gross counts are for simplified geometries. Actual field conditions will not involve the face of the detector only. Table 13.1 ignores the volume of the detectors (the data pertains to 10 cm x 10 cm detectors only. Smaller detectors will have a smaller photopeak to background ratio, while larger detectors will have a larger ratio. The LLDs depend upon counts as taken by a scaler. If detection depends upon count rate by sound or visual observation, the LLDs will likely be double or more. If detection depends upon computer analysis and mapping, the LLD may be lower. The LLDs were estimated using an assumed background. The LLD will vary with the square root of the background at a location and the effective LLD will depend upon the variability of background. The effective LLD is a function of the mean background plus the propagated variability in background. Computer analysis along with measurements of background through pulse height analysis can allow continuous calculation of the LLD.

The scanning of a large area using NaI can be accomplished several ways, depending upon the terrain, ground cover, and required sensitivity. Single detectors have a limited field of view so swinging the detector usually increases coverage. Multiple detectors with overlapping fields of view can increase the sensitivity and quality of the survey. The observation that gross counts can be more sensitive than photopeak counts suggests that a large array of gross counting detectors may offer the best sensitivity. A single large plastic scintillator may substitute for a large array.

A plastic scintillator of equivalent thickness to a 5 cm NaI detector for 600 keV photons is 15 cm if the attenuation coefficients are 0.3 cm^{-1} for NaI and 0.1 cm^{-1} for plastic. A plastic scintillator 100-cm x 100-cm x 15-cm will have a background of 20,000 cps (compared to the 2000 cpm of a 5-cm x 5-cm NaI). The LLD would be 660 cps. Assuming 2 π efficiency for a surface deposit, then 370 Bq m^{-2} of ^{137}Cs on the surface would produce 120 cps m^{-2} [=370 Bq m^{-2} x 0.5 x 0.85 (abundance) x 0.78 (detector efficiency)] for the detector. The 1-s LLD is 2035 Bq m^{-2} (=370·660/120). The Kocher 1985 relationship of 1 Bq kg^{-1} uniform is equivalent to 82973 Bq m^{-2} surface gives and LLD of 0.025 Bq kg^{-1} uniform (=2035/82873).

The gross counting analysis shows that if it is not necessary to identify the radionuclide, then photopeak analysis does not offer an advantage. When radionuclides must be identified, even better resolution than NaI should be considered. Germanium detectors can be the solution to higher resolution and large, high-efficiency Ge detectors allow short acquisition times (Helfer 1988). Small, all-orientation cryostats allow easy handling and arrays of detectors. Cost and handling restrict Ge detectors to the more critical applications that cannot be effectively handled by other means. Germanium detectors can be very useful in assessing environmental concentrations of fission products; it is also proving very useful for measuring concentrations of plutonium. The measurement of plutonium depends upon the 60 keV gamma from ^{241}Am. The accuracy of the measurement is a function of the ratio of plutonium to ^{241}Am and the distribution of each in the soil.

The original method for using ^{241}Am to detect plutonium in field surveys was the Field Instrument for Detecting Low Energy Radiation or FIDLER (Schmidt 1966, Tinney 1969). The detecting element of the FIDLER is a 12.5 cm diameter NaI(Tl) crystal that is 0.16 cm thick. The thin detector has a high efficiency for low-energy photons and low efficiency for high-energy photons. The FIDLER detects the ^{241}Am gamma well, while discriminating against most environment photons. This technique has its limits and generally does not work if the contamination includes mixed fission products or if the natural background is high (NCRP 1976). The FIDLER has uses where the concentration of plutonium is high, on the order of 1000 Bq m^3 and when rapid scanning is adequate to map an area of contamination. Cleanup levels of less than 1000 Bq m^3 usually require Ge detectors.

The measurement of plutonium by measuring ^{241}Am involves several steps and consequent propagation of error. Assume that plutonium is present at 8 times the americium, since the 60 keV gamma has an abundance of 36%, then the plutonium concentration is 24 times the inverse of the efficiency of counting the photons, which assumes a distribution of the americium. If the count is near the LLD, then there is a 30% error in counts. The prospect of determining the plutonium concentration is daunting. Nevertheless, if sufficient care is taken, *in situ* measurement of plutonium concentration using germanium spectrometry can be very superior to soil sampling. Soil analysis for plutonium involves a very small volume of soil, often about 10 g, and multi-step, harsh chemistry with low yields.

Measurement of plutonium in soil using *in situ* germanium spectrometry is usually more accurate and less expensive than soil sampling, especially if the area is large. The accuracy of *in situ* spectrometry is dependent upon soil sampling to determine the soil concentration profile. The concentration profile is usually expressed as

$$S(z) = S_0\, e^{-(\alpha/\rho)\rho z} \tag{13.7}$$

where $S(z)$ is the activity per unit volume of soil (Bq cm^{-3}) at depth z (cm), S_0 is the activity per unit volume (Bq cm^{-3}) at the surface ($z=0$), α is the inverse of the relaxation length of the exponential depth profile (cm^{-1}) and ρ is the soil density (g cm^{-3}).

The depth profile is expressed in terms of the mass depth, ρz (g cm^{-2}) with the degree of penetration into the soil characterized by the depth parameter α/ρ (cm^2 g^{-1}). As

$\alpha/\rho \rightarrow \infty$, the depth profile approaches an infinite plane atop the ground. As $\alpha/\rho \rightarrow 0$, it approaches a uniform source distribution with depth. The effect on accuracy for assuming the ^{241}Am is distributed uniformly with depth is shown in Fig. 13.13.

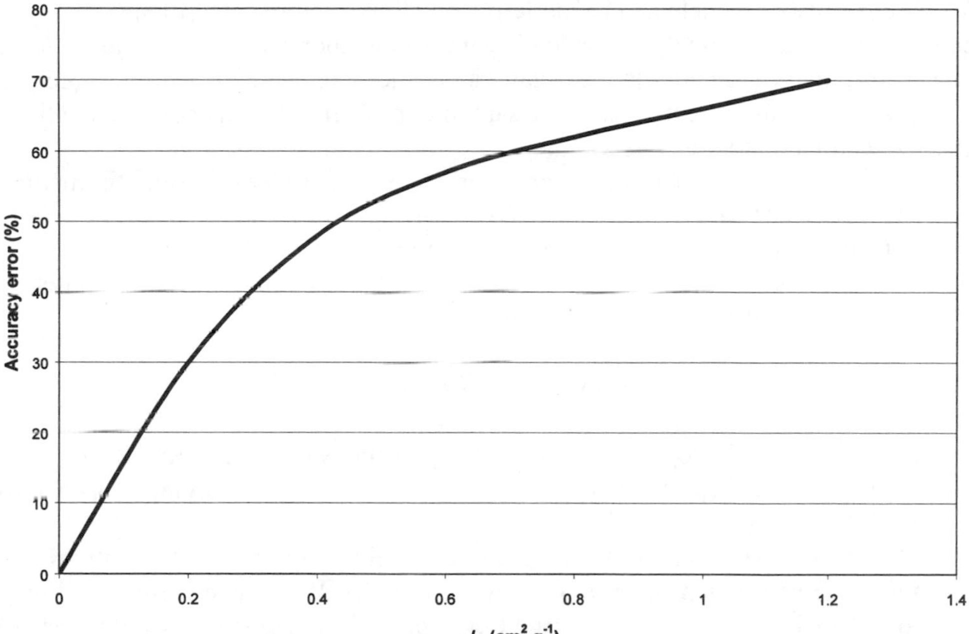

Fig 13.13. The accuracy error as a function of α/ρ in ^{231}Am concentration estimates when the nuclide's depth profile is misidentified as a uniform distribution (Fong 2997).

The depth profile will be the largest source of error if the *in situ* sampling design properly accounts for other sources of error from count time, field of view, and calibration (Fong 1997). The depth profile will be less of a source of error as the energy of the photon increases. It will always be the largest source of error if the radionuclides are laterally uniform. The energy of the photon limits the depth to which the soil can be sampled. An *in situ* detector will not collect information from ^{241}Am ($E_\gamma = 70$ kev) from more than about 8 cm depth of soil and no deeper than about 25 cm for ^{137}Cs ($E_\gamma = 661$ kev) and the information decreases exponentially from the surface.

The discussion concerning plutonium used ^{241}Am as an indicator for plutonium. Gamma emitters can be used for alpha and beta emitters when there is a mixture of radionuclides if a relationship is established and reasonable uniformity of the mixture can be demonstrated.

13.5 CONCLUSIONS

Environmental measurements of radiation tend to be oriented to risk. Radiation risk is usually expressed as annual or lifetime dose as TEDE. It is not possible to measure TEDE. It can only be modeled. The modeling of TEDE requires environmental concentration of the radionuclides, which is generally a laboratory measurement. *In situ* gamma spectrometry is an alternative to laboratory measurements. A well-designed *in situ* measurement program can over come sampling problems associated with small sample size and laboratory analysis.

Portable, dose rate instruments are generally unable to measure environmental dose rate correctly and cannot measure TEDE. Dose rate instruments should be able to measure a response that can be interpreted based on model predictions. Dose rate instruments, therefore, may be able to indicate TEDE if variation of background response is not larger than the response to be measured.

13.6 REFERENCES

Beck, H. L., Decampo, J; Gogolak, C. In situ Ge (Li) and NaI Gamma-Ray Spectrometry, New York: U. S. Department of Energy; Environmental Measurements Laboratory; Report No. HASL-258, 1972.

Beck, H. L. The physics of environmental radiation fields, page 101 in The natural Radiation Environment II, Adams, J. A. S., Lowder, W. M. and Gesell, T., Eds., Report CONF-720805 (U. S. Energy Research and Development Administration, Washington); 1972.

Fong, S. H.; Alvarez, J.L. Data quality objectives for surface-soil cleanup operation using *in situ* gamma spectrometry for concentration measurements. Health Physics 72: 286-293, 1997.

Fong, S. H.; Alvarez, J.L. When is a lower limit of detection low enough, Health Physics 72(2): 282-285; 1997.

Helfer, I. K.; Miller, K. M. Calibration factors for Ge detectors used for field spectrometry. Health Phys. 55:15-29; 1988.

International Commission on Radiological Protection. Recommendations of the International Commission on Radiological Protection, ICRP Publication 26, Pergamon Press, Oxford; 1977.

International Commission on Radiological Protection. Limits for Intakes of Radionuclides by Workers, ICRP Publication 30, Annals of the ICRP 19, Pergamon Press, Oxford; 1988.

International Commission on Radiological Protection. 1990 Recommendations of the International Commission on Radiation Protection, ICRP Publication 60. Annals of the ICRP 21(1-4) Pergamon Press, Oxford; 1991

International Commission on Radiological Protection. Conversion Coefficients for use in Radiological Protection against External Radiation, ICRU Report 57, International Commission on Radiation Units and Measurements, Bethesda, Maryland; 1998.

Kocher, D. C.; Sjoreen, A.L. Dose rate conversion factors for external exposure to photon emitters in soil, Health Physics, 48:193-205; 1985.

National Council on Radiation Protection and Measurements. Environmental Radiation Measurements, National Council on Radiation Protection, Report No. 50, Washington, D.C.; 1976.

Schmidt, C. T.; Koch, J.J. Plutonium survey with x-ray detectors, page 1 in Hazards Control Progress Report No. 26, Report UCRL-50007-66-2 (Lawrence Livermore Laboratory, Livermore, CA.) 1966.

Tinney, J. F., Koch, J.J.; Schmidt, C.T. Plutonium survey with an x-ray sensitive detector, Report UCRL-71362 (Lawrence Livermore Laboratory, Livermore, CA.) 1969.

U.S. Environmental Protection Agency. Technical Support Document for the Development of Radionuclide Cleanup Levels for Soil, Office of Air and Radiation, EPA 402-R-011A; 1994.

U.S. Environmental Protection Agency. Environmental Radiation Protection Standards for Yucca Mountain, Nevada, 40 CFR Part 197 Subpart F; 1997.

U.S. Nuclear Regulation Commission. Final rule on radiological criteria for license termination, 10 CFR 20, US. Nuclear Regulatory Commission, 1997.

U.S. Nuclear Regulation Commission. MARSSIM Multi-Agency Radiation Survey and Site Investigation Manual (MARSSIM), Final, NUREG-1575, 1997.

Westinghouse Hanford Company. Large-scale Remediation Analytical Program Plan, Westinghouse Hanford Company, WHC-SD-EN-AP-148, 1993.

Yu, C.; Zielen, A.J.; Cheng, J.J; Yuan, Y.C.; Jones, L.G.; LePoire, D.J.; Wang, Y.Y; Loureiro, C.O.; Gnanapragasam, E.; Faillace, E.; Wallo, A.; Williams, W.A.; Peterson, H. Manual for implementing residual Radioactive material guidelines using RESRAD, Version 5.0, ANL/EAD/LD-2; 1993.

Zankl, M., Petoussi, N.; Drexler, G. Effective dose and effective dose equivalent – the impact of the new ICRP definition for external photon irradiation, Health Physics 62:395-399; 1992.

Chapter 14

EMERGENCY RESPONSE INSTRUMENTATION

Carson A. Riland

14.1 INTRODUCTION

This chapter is shorter than most, as the majority of instrumentation is the same used for everyday health physics activities. The major difference is the circumstances under which the instruments are used. Health Physics instrumentation needs for emergency response operations can vary greatly with the arrival time, the unique situation and the mission of responding elements. Determination of appropriate instrumentation requires assessment of several factors, including; the skill level of instrument users, conditions of instrument use (radiation levels, environmental conditions, etc), storage and transport of instruments, and costs. This chapter discusses portable instrumentation used to respond to emergency situations. The chapter begins by examining potential radiological emergency response scenarios, then discusses the skill level of response personnel. Potential hazards are also identified, before instrument selection is discussed. The chapter concludes with a summary of issues involved in the use of instruments during an emergency response situation and how to maintain equipment readiness.

14.2 EMERGENCY RESPONSE SCENARIOS

An emergency response may bring to mind different pictures dependent upon the perspective of the individual. A health physicist employed at a medical center may consider a broken cable on a high-dose rate irradiator (brachytherapy unit), a spill of radioactive material, or receiving contaminated patients an emergency. Whereas an Aerial Measurement Survey Team may consider a lost or stolen source, a terrorist threat to use radioactive material, or the dispersal of radioactive material an emergency. While many may agree in principle that a general level of emergency response may exist for a multitude of scenarios, the importance or severity level of each scenario may differ from person to person. For the purpose of establishing a common ground for this chapter, any "unplanned event" will be considered an "emergency." Generally speaking, emergency response scenarios can be divided into two categories; those involving the dispersal of radioactive material (posing an internal dose hazard, e.g. – plutonium release) and those which only pose an external dose hazard (e.g. – a lost source). A discussion of a variety of emergency response scenarios follows, with attention on the unique circumstances that need to be considered when selecting and maintaining instruments for responding to these events.

Let's first consider events that only pose an external dose hazard. A stuck source may be due to a broken cable, a dented tube, or pneumatic problems. An example could be a broken or snagged cable for a radiography source. These conditions could include a wide range of exposure rates. Responders will typically want to determine the location of the radioactive material, and the type, energy and intensity of radiation emitted (which should allow identification of the material). Response personnel will also require dose assessment, indicating the need for personnel dosimetry. In the case of a lost source, responders will want to use instruments with high sensitivity and be able to rapidly search large areas. The same is true when looking for a radiological or nuclear device based upon a terrorist threat. An additional scenario could be a criticality event (this may also cause dispersal).

The other category of events involves the release and/or dispersal of radioactive material. These scenarios may involve either the combination of internal and external dose hazards, or internal dose hazards, only. Examples include reactor accidents or the spill (or intentional dispersal) of radioactive material that emits both beta and gamma radiation. A spill is from a breach of containment possibly due to fire or a transportation accident. Other dispersals may be the intentional act of a terrorist. For these types of scenarios, again the location and identity of the dispersed material is of primary interest. In addition, the extent of radioactive material dispersal (size of and concentration in the effected area) and its impact on the environment and population are also items of concern. This would indicate that monitoring to determine the type and concentration (ground deposition and airborne) of contamination would need to be performed. In addition, personal contamination surveys of both the public and response personnel will be needed. Response personnel will again require dose assessment, indicating the need for personnel dosimetry. In addition to dispersal of beta/gamma emitting radioactive material, beta only and alpha emitting radioactive material (e.g. – weapon accident, Weapon of Mass Destruction {WMD} device, or satellite accident/re-entry) can also be released. These materials present primarily an internal dose hazard. Extensive contamination surveys will be required, as well as deposition measurements, identification of the material dispersed and airborne contamination measurements. Dispersal scenarios may require extensive sampling activities, increasing in emphasis as the response evolves. Samples will typically be screened in the field and transported to a laboratory for analysis. Laboratory instrumentation has been covered previously, however attention should be given to transportation and field operation of laboratory instrumentation used for mobile laboratories.

For the above mentioned emergency response scenarios, the radiation hazard is often assumed to be the primary hazard. Typically this is NOT the case. A variety of additional hazards will be present. These include general safety hazards (e.g. – driving safety or electrical hazards) and also physical and chemical hazards. These hazards must not be overlooked when developing safety plans or choosing instruments for responding to radiological emergencies.

14.3 EMERGENCY RESPONDERS

Emergency response personnel will have a varied background of expertise, dependent on the type, level, and phase of the response and the responder's individual duties. For example, a fire involving radioactive material will include fire, police and medical personnel, typically with limited radiation training. A Hazardous Material unit, that has slightly better radiation training, may also be present. At some point, a state team, Environmental Protection Agency (EPA) Radiological Emergency Response Team (RERT), or Department of Energy (DOE) Radiological Assistance Program (RAP) Team may also be on-scene. The skill level of each group of responders will differ greatly. Instrumentation needs for each response organization will also differ greatly.

First responders typically have a very limited knowledge of radiation hazards. They will usually refer to the Department of Transportation's "Emergency Response Guidebook" and follow whatever directions are found. They will also call for assistance from the local HazMat Team and possibly the state. The state radiological response organization may handle the situation themselves or ask for additional assistance. Additional regional assistance may come from EPA RERT, DOE RAP, and/or a National Guard Civil Support Detachments (CSDs). These organizations may then request assistance from a variety of national level response assets. An overview of assets follows beginning with local responders, progressing to regional responders and ending with the national level response organizations.

- First Responders (Fire, Medical and Police) – Minimal radiological knowledge. Will require significant explanation of radiological risks. Must have simple instrumentation (Go or No Go).
- HazMat Teams – Some limited knowledge of radiation physics and risks. Will typically be able to use slightly more complex instruments.
- Facility Teams – Usually well trained and knowledgeable of local resources and hazards. Typically adequately trained in radiation protection. Often on-site only.
- CSD – The National Guard's WMD response element. Similar knowledge of radiation protection as a HazMat Team. Often more knowledge of WMD issues and availability of additional resources.
- State/City/County/Tribal Teams – Radiological knowledge varies by location. States with a reactor facility typically are reasonably well trained for reactor response. Often unfamiliar with non-reactor scenario response.

- EPA RERT – Well trained with significant radiological knowledge, especially as it pertains to radiological response. Usually a small team of approximately six response personnel.
- DOE RAP Team - Well trained with significant radiological knowledge, especially as it pertains to radiological response. Usually a small team of seven response personnel. Three to four teams are available for each of the eight RAP regions.
- DOE Radiological Emergency Assistance Center and Training Site (REAC/TS) – DOE's experts on biological effects from accidental exposure to radiation. Available to provide advice to local medical facilities and personnel responding to accidental exposures.
- DOE Aerial Measurements System (AMS) – DOE's remote monitoring capability. Can assist in performing both aerial searches for lost or stolen sources, or for performing measurement of a ground deposition of a plume from dispersal of radiological material.
- DOE Accident Response Group (ARG) – Provides on-site response during the recovery of a nuclear weapon.
- DOE Nuclear Emergency Search Team (NEST) – DOE's response team for radiological or nuclear weapons' threats. Can also be useful in searching for lost or stolen radiological sources.
- Department of Defense (DoD) Air Force Radiological Assistance Team (AFRAT) – A deployable team prepared to respond to a variety of emergency scenarios on both a national and international level.
- Federal Radiological Monitoring and Assessment Center (FRMAC) – Compilation of local, state and national level assets into one organization for coordinating all off-site monitoring and assessment activities. Limited local knowledge, but familiar with large-scale radiological emergency response organization and needs.

This is a sampling of potential radiological emergency response organizations. There are a variety of other organizations, but the skill levels will be similar to the organizations listed.

14.4 HAZARDS

In order to identify instrumentation required for a radiological emergency response, an understanding of the potential hazards and environmental conditions is required. As previously mentioned, the radiological conditions may include external exposure, personnel and equipment contamination and airborne radioactive material. While these conditions might be expected, additional hazards could be present and environmental conditions could vary greatly. Additional hazards could include biological hazards, such as insects (ticks, bees, etc), animals (wild or domestic), or bacteriological (natural or from a WMD). Many of these hazards must be observed, as with general safety hazards. Exceptions are certain bacteriological hazards, which can be sampled or possibly field evaluated (WMD). A variety of chemical hazards need to be considered. Oxygen deficiency which may be an immediate threat to life, is a concern. Additionally, combustible and explosive levels of contaminants need to be evaluated. Other chemical toxins may also need to be considered. Carbon monoxide or toxic gases are some examples of chemical hazards from typical response activities. A WMD response could include a variety of chemical hazards intended to harm both the public and/or response personnel.

Environmental conditions can present physical hazards to response personnel and can affect instrument operation. Many hazards can be a result of local weather, such as heat or cold, rain, snow, ice, floods, or lightening. Manmade physical hazards will also exist, such as, fire, noise, etc. A WMD scenario may also present a hazard of collapsing buildings or additional devices. While these conditions may present physical hazards, they also can have a severe impact on instrument operation. These factors may include temperature, barometric pressure, humidity, magnetic or electric fields, and/or competing contaminants.

14.5 INSTRUMENT SELECTION

A wide variety of instrumentation may be employed during an emergency response. Instruments used are generally the same types that are used for routine operations, with a few exceptions. Emergencies still require dose rate measurements, contamination monitoring, airborne radioactivity level measurements and perhaps other physical and chemical hazards (heat, cold, oxygen deficiency, etc). Accidents that involve dispersal of radioactive material may also require ground deposition measurements. The basic types of instrumentation include;

- Geiger-Muller detectors – for beta-gamma contamination and gamma radiation (teletectors)
- Ionization chambers – exposure rate, beta dose rate
- Proportional counters – contamination monitoring, neutron dose
- BF_3 tubes (Bonner sphere or snoopy) – for neutron dose
- Scintillation detectors – sensitive detectors for exposure rate, alpha and/or beta contamination, ground deposition
- Semi-conductors detectors – sensitive detection of charged particles and photons, ground deposition, spectroscopic identification of unknown radionuclides (*in situ* spectroscopy)

Proper selection of instruments to be used for emergency response is important not only to respond to a radiological accident, but also to ensure instruments are operable and accurate in a variety of environments. Typical types of instruments should include those able to measure exposure rate (both high and low range) and those able to measure contamination levels (airborne or deposited on a surface). All instrumentation used should comply with minimum operational standards for radiological detection instruments (ANSI 1992a, 1992b).

Other factors in instrument selection include familiarity of users with the instruments and ease of operation. Users need to be trained on the instruments they may be expected to operate. The simpler the instruments are to use, the less training time is required. Cost is always a factor in selection of equipment, as well as customer service. Developing a good, ongoing relationship with a local vendor may help reduce costs. Combining purchases with other organizations may also help in reducing costs. When considering costs, consideration of calibration, maintenance and transportation costs should be included. It should be determined if special calibration techniques are involved which are beyond your facilities capability. Also, determination of special calibration and/or check source requirements is desired. Finally, it should be determined if additional transportational costs will be incurred.

14.5.1 Personnel Dosimetry

Personnel Dosimetry needs will vary with the phase of the response. First responders will not typically have dosimetry, except perhaps Pocket ion chambers. It's typically not practical to provide personnel dosimetry monitoring for first responders, unless responders are dedicated to a specific facility. It is possible to provide Self Reading Dosimetry (SRD) that can be issued on-site to the responders who need it. Most response personnel deployed on a state, regional, or national level, are radiological workers and will have dosimeters. Consideration should be given to using a separate dosimeter for the response or whether the individual's routine dosimeter should be worn. In addition to personnel dosimeters, SRDs should also be worn. Pocket ion chambers can be used, but electronic dosimeters may be preferred, as they can have dose and dose rate alarms set. Care should be taken when selecting an electronic dosimeter to ensure insensitivity to radio frequency and other non-ionizing electromagnetic interference.

Internal dose must also be monitored. Intake may be verified by bioassy methods similar to those used for normal Health Physics dose assessment. Field methods can be used to provide estimates of internal dose through air sampling. Both area air samplers and Personal Air Samplers (PAS) can be used to estimate an individual's intake (through inhalation) and should be considered.

14.5.2 Non-Dispersed Material Instrumentation

A wide variety of dose/dose rate measurements may be required for measuring non-dispersed material. Dose rates may be due to gamma, beta or neutron radiation. During local emergency response (First Responders), instruments will need to be able to measure higher dose rates. An exposure rate instrument capable of higher ranges would be a basic first instrument of choice. An additional resource would be a telescoping detector for remote measurement. A gamma camera may also be useful for identifying the location of the source from a distance. This device uses a position sensitive detector to super-impose a "radiation map" onto a video picture of an area. This allows for remote identification of where a source may be located while minimizing dose to response personnel. Later phase deployable instruments may be required to have greater sensitivity and often will not be required to monitor higher dose rates. Auto-ranging instruments can be particularly useful during the initial response, when responders need to be able to readily see changes in dose rates without having to change scales. Also, initial responders may be wearing personal protective equipment (PPE) which may make changing scales more difficult. Instrumentation is also unique to the response situation. Instruments for first response personnel should be as simple and easy to use as possible. The ideal instrument would only have an "on/off" switch and something similar to a "green/yellow/red" readout. Many "off the shelf" instruments are appropriate when responding to a known location or when performing life-saving activities. The search for a lost or stolen source often will require unique non-standard equipment, capable of covering a large area in a timely manner. This is especially true for terrorist threats.

In addition to measuring dose rates and locating sources, identification of the source may also be required. Some indication is given by the type of radiation present (e.g. – neutron). Additional methods will include *in situ* gamma spectrometry. Current spectrometers vary in complexity. Handheld units using NaI(Tl) detectors are available from several vendors with built in simple analysis routines. More complicated High-Purity Germanium (HPGe) detector systems are also available, but are higher in cost and more difficult to maintain, deploy and operate.

14.5.3 Dispersed Material Instrumentation

Dispersed radioactive material will require a variety of measurements to be taken using many different instruments. Many of the same instruments used for non-dispersed materials may be used for these scenarios. Those radioactive materials that emit gamma radiation will still require the use of exposure rate instruments and may also require identification. Some of these scenarios may require HPGe detector based *in situ* counting systems or collection and analysis of field samples to determine the radioisotope mix. An example is reactor release scenarios, where the isotopic mix is important in determining dose to the public. Identification of beta emitters will require sample collection and analysis. Liquid Scintillation Counters (LSC), now available in a smaller, more portable package, are generally the instrument of choice for identification of beta only emitting radioactive material. The same method can be used for alpha emitting identification, but spectroscopy of associated photons is preferred. Another recent development to aid in alpha emitting radioactive material identification is a handheld alpha spectroscopy instrument that can perform gross spectroscopy measurements in the field.

Alpha and beta contamination monitoring equipment utilizes the same types of detectors used for non-emergency operations. It can be useful, from a packaging and transportation standpoint, to use a readout unit that can accept a variety of probes. "Smart" probes, which store calibration and settings information, are the ideal and are available from several vendors. Ground deposition measurements may include *in situ* measurements using a HPGe detector based system, FIDLER (Field Instrument for Detection of Low-Energy Radiation) instruments or the update to the FIDLER, the Violinist system. The Violinist system uses the same basic detector as the FIDLER, but includes a 256-channel spectroscopy system. These instruments are generally used and maintained in the same manner as for non-emergency operations. The main difference is the age or weathering of the deposition. *In situ* systems need to be calibrated for surface deposition (the most likely case for accident plume deposition).

In addition to ground monitoring, contamination monitoring of personnel, equipment, vehicles, buildings, etc will need to be performed. Beta/gamma survey instruments will use GM, proportional or scintillator detectors. Alpha survey instruments will use proportional or scintillator detectors. Both probes can be easily damaged. Scintillation detector probes are sensitive to light leaks and magnetic fields. Proportional detectors can become insensitive with the loss of counting gas. In some cases, indirect surveys (swipe or smear surveys) may be desired. Instruments need to be able to count the swipes in a reproducible manner. Counting jigs can be purchased or manufactured for using contamination survey instruments in the scaler mode of operation. Portable alpha/beta survey instruments with reproducible geometries are available from several vendors. Most can operate under battery or AC power.

14.5.4 Air Monitoring

Air monitoring would include the same type of air monitoring equipment used for non-emergency operations. As an example, Personal Air Samplers (PAS) are often used to estimate a person's intake during operations. When planning air-sampling activities, care should be taken to achieve the best sensitivity possible. This is especially true for radioisotopes with low Annual Limit of Intakes (ALIs). High-volume air samplers may be used for initial response to monitor plume concentrations (non-iodine). High-volume air samplers are later used to monitor resuspension of radioactive material. Low-volume portable grab samplers are often used for airborne plume monitoring, especially for reactor accidents. Low-volume continuous flow air samplers can be useful for either long-term air sampling in the field or performing health and safety air sampling for mobile laboratories or at a hotline. Air samples are often screened in the field to obtain initial results. Whenever field screening or analysis is performed, care should be taken to account for radon and thoron progeny. Some progress has been made in using Continuous Air Monitors (CAMs) for outdoor air monitoring, but with limited applications.

14.5.5 Industrial Hygiene Monitoring

In addition to radiation measurements, some radiological accidents may also involve monitoring for other non-radiological hazards, such as oxygen deficiency, explosiveness, carbon monoxide, and/or other contaminants. Noise levels, high heat or other physical hazards may also need to be monitored. The need to monitor for other physical and chemical hazards will vary by situation. Some common hazards to be monitored include; Lower Explosive Limit (LEL), oxygen levels, toxic gases, noise levels, and heat/cold levels. Useful Industrial Hygiene monitoring instrumentation usually starts with a multi-gas meter. These instruments typically have sensors for monitoring: the percentage of oxygen present, the Lower Explosive Limit (LEL), carbon monoxide levels, and one or two additional user selected sensors. Colorimeter tubes are also a good addition. Colorimeter response kits stocked with a variety of commonly used tubes can be purchased from several vendors. Photo or flame ionization detectors and a gas chromatograph could also be useful in identifying chemical hazards, but are more costly and the gas chromatograph can be difficult to deploy. Sound level meters and Wet Bulb Globe Temperature (WBGT) meters are used for monitoring noise levels and heat stress, respectively.

14.6 EQUIPMENT READINESS AND USE

Emergency response to radiological accidents presents unique circumstances for instrument use. Maintaining readiness of instrumentation for radiological emergency response also presents unique problems. While difficulties exist for maintaining instrument readiness for local emergencies, the problems intensify for instrumentation that may be deployed to a variety of locations and climates (Riland 2000).

14.6.1. Quality Assurance / Quality Control

One of the primary issues in maintaining detection instruments is quality assurance / quality control (QA/QC). Typical QA/QC for instrumentation include annual calibrations and routine response checks. The foundation for a routine QA/QC program is outlined in a variety of references (ANSI 1997, ANSI 1994, DOE 1999, NCRP 1991). Response checks typically occur at routine intervals (weekly, monthly, etc.) and may use a variety of radiation sources. Sources may include National Institute of Standards and Technologies (NIST) traceable sources, "button" check sources, sources built into the instrument (in the probe cover, along the side, etc.) or even consumer products (e.g. lantern mantles). These check sources may be stored at a designated location or with the instrument, depending upon the organization's source accountability procedures. Whichever type of source is used, it should have an expected response for a properly calibrated instrument. Additionally, industrial hygiene monitoring equipment requires calibration and operational or "bump" checks. These instruments are typically calibrated and maintained in accordance with manufacturer's procedures.

Air sampling equipment also requires routine calibration and maintenance. Air samplers should be calibrated to the flow rate typically used for field air sampling operations. Calibration equipment should be field deployable in order to recalibrate in the field. Field calibration is desirable in order to adjust airflow for changing environmental conditions (e.g. – atmospheric pressure) and to recalibrate if different flow rates are desired.

In situ instruments should be calibrated in accordance with manufacturer's instructions. For FIDLER and Violinist systems, this typically means using the calibration software from the Hotspot codes available from Lawrence Livermore National Laboratory. HPGe system calibration will vary by vendor. One example is Canberra's calibration based upon an initial characterization of a detector and computer modeling to determine a series of calibration geometries.

14.6.2 Packaging, Storage and Transport

Instruments will generally be packed as part of a "kit." A typical "kit" will usually contain personal protective equipment (PPE), self-reading dosimeters and forms for logging measurements. Kits may also include flashlights, pens, personal air samplers, sampling supplies, decontamination supplies, etc. The kit may be stored at the location to be used; making the packaging used less important. However, if the kit will be transported, it will need to be of rugged construction and equipment would need to be packaged carefully to avoid damage.

Storage locations depend upon expected need. Some kits may be needed at the location where the risk associated with operations may require emergency response equipment to be stored in the immediate area. Most facilities have emergency response kits distributed at strategic locations about the facility, much like fire extinguishers. If instrument kits are designed to be transported, they must be readily available to the transport method, e.g. vehicular transport requires location near loading dock or loading area. Air transport would require access to an airport, airbase or other landing area. Some instruments may have rechargeable batteries, requiring storage in a location with AC power to keep batteries charged. Equipment should also be stored in an "environmentally safe" area. Temperature and humidity extremes may adversely effect instruments and other kit contents.

Transportation needs for kits may define packaging requirements. Local emergencies may only require hand carry or short vehicular movement of kits. While these modes of transport may require ruggedness of packaging, weight and size limitations are minimal. When deploying to remote locations, weight and size limitations become more important. Ground transport has limits on total load, but is still typically less restrictive than air transport. Military and charter air travel require more restrictive size and weight limitations, with air cargo even more restrictive and check baggage being the most restrictive. Unfortunately, availability of transport is generally from the most restrictive size and weight limitations to the least (for air transport). Checked baggage limitations are 45.4 kg (100 lbs.) and 178 lineal cm (70 lineal inches) (length x width x height). In addition, sources can usually be transported via ground, military air and charter air transport. Air cargo and checked baggage source transport is at the pilot's discretion (approval is not anticipated).

14.6.3 Emergency Response Instrument Use

Generally, instrument operation does not change significantly during an emergency response from non-emergency situations. The instrument will still turn on or off in the same manner. What may often change are the environmental conditions in which the instrument is used, the user of the instrument and often the frame-of-mind of the user. The responders must be able to understand the effects of non-normal radiation fields on their instruments. Older instruments may not operate properly when overranged. Some scintillation and GM detectors will over respond at lower photon energies. Instruments can also respond differently in varying environmental conditions. Scintillator based instruments may have stability problems in varying temperatures. Temperature can also effect Liquid Crystal Displays (LCDs). Humidity can effect ion chamber instrument response and several instrument systems could be difficult to operate in the rain. Rain or moisture will also effect alpha survey capabilities and could make indirect surveys difficult. Radon progeny can effect both air monitoring and direct contamination survey results (particularly in dry environments for the latter). Altitude and temperature have significant effects on instruments that measure airflow.

Response personnel need to be aware of what their results actually mean. As an example, instruments will almost always report a result. The validity of the result (e.g. – above or below the Minimum Detectable Activity {MDA}) has to be determined by the user. Response personnel must also recognize the limitations of each instrument and ensure the proper instrument is used. For example, a GM detector based instrument designed to measure exposure would not be the instrument of choice for performing alpha contamination surveys. Choosing the proper instrument also requires evaluation of the sensitivity required by the measurement and capability of available instruments to meet sensitivity requirements. This may require longer count times, slower scanning speeds or selection of a different instrument.

14.7 SUMMARY

Instruments used for emergency response are typically the same as used for everyday health physics activities. The manner in which instruments are used, maintained, transported, and how results are evaluated are key differences. Instrument selection and use is based upon the intended user, the hazards present, the environmental conditions, maintaining equipment readiness, transportation issues cost and results required. A little preplanning, focusing on emergency response requirements will allow for a more successful emergency response instrumentation program.

14.8 REFERENCES

American National Standards Institute 1989a - American National Standards Institute, American National Standard Performance Specifications for Health Physics Instrumentation – Portable Instrumentation for Use in Normal Environmental Conditions. New York, NY. N42.17A-1989.

American National Standards Institute, American National Standard Performance Specifications for Health Physics Instrumentation – Portable Instrumentation for Use in Extreme Environmental Conditions. New York, NY N42.17A-1989.

American National Standards Institute, American National Standard Performance Specifications for Direct Reading and Indirect Reading Pocket Dosimeters. New York, NY. N13.5-R 1989.

American National Standards Institute, American National Standard for Calibration – Calibration Laboratories and Measuring and Test Equipment – General Requirements. New York, NY. Z540-1-1994.

American National Standards Institute, American National Standard Radiation Protection Instrumentation Test and Calibration, Portable Survey Instruments. New York, NY. N323A-1997.

Department of Energy, Portable Monitoring Instrument Calibration Guide. DOE G 441.1-7. Washington, D.C. 1999.

National Council on Radiation Protection, Calibration of Survey Instruments used in Radiation Protection for the Assessment of Ionizing Radiation Fields and Radioactive Surface Contamination. NCRP 112-1991.

Riland, C. A.; Tighe, R. J.; Bowman, D. R. Readiness Issues for Emergency Response Instrumentation, In; Instrumentation, Measurements and Dosimetry, proceedings of the 33rd Health Physics Society Midyear Topical Meeting. Madison, WI: Medical Physics Publishing; 2000:119-122.

Chapter 15

RADIATION INSTRUMENTS IN UNIVERSITY / RESEARCH LABORATORIES AND ACCELERATORS

Raymond H. Johnson, Jr. and Shawn W. Googins

15.1 INTRODUCTION

Radiation safety programs in university / research laboratories and accelerators depend upon radiation instruments for information to verify regulatory compliance and safety program performance. Since radiation does not produce a response by any of our five senses, we must rely on radiation measurement instruments to detect the presence of radiation or radioactive materials. Radiation instruments become our eyes and ears to tell us the type and the amount of radiation that may result in exposures to workers or members of the public.

15.1.1 Instrument Needs In University and Research Laboratories

The radiation instrument needs in university and research laboratories vary greatly depending upon the scope and complexity of the radioactive materials and radiation producing equipment present. This review is intended to provide practical guidance to someone, such as a Radiation Safety Officer (RSO), who is responsible for choicc and operation of radiation instruments, as well as understanding their limitations and interpreting radiation measurements. This review is not intended to include detailed theories or principles of radiation instrument operations. Instead, this review seeks to cover the practical aspects of major radiation instrument needs that will be experienced by the majority of university and research radiation safety programs. Such programs may range from: 1) the use of a few MBq (microcuries) of materials as innocuous as ^{3}H and ^{14}C and a few diagnostic x-ray machines, up to 2) large programs involving multi-curie use or production of radioactive materials by accelerators or a research reactor, and 3) full diagnostic radiology and nuclear medicine departments.

In university and research laboratories the need for radiation measurements is often a matter of human nature and the fact that spills of radioactive material occur in the normal course of ongoing research. Monitoring is required to detect unknown spilled material, confirm proper cleanup, assess contamination levels for possible dose assessment, and to meet a myriad of regulatory requirements. These requirements are derived from either Agreement State or Nuclear Regulatory Commission (NRC) regulations for contamination limits for areas and equipment that may be occupied by occupationally exposed persons (restricted areas), or by members of the public (unrestricted areas), or equipment and facilities that need to be decommissioned or released for unrestricted use.

It is important to draw a distinction between restricted and unrestricted areas for determining radiation instrument requirements. These two types of areas differ in terms of sensitivity needs for radiation measurements. The required sensitivity, whether driven by regulatory or safety program requirements, will determine which instruments to use and how to use them. For example, sensitivity needs may determine the speed of probe movement for contamination monitoring and, for laboratory instruments, sensitivity determines factors such as sample counting time.

15.1.2 Practical Radiation Instrument Challenges

Because radiation instruments are such a primary tool in radiation safety programs, the training of RSOs, radiation safety staff, and other occupationally exposed personnel must have a significant emphasis on radiation instrumentation. Such training should address several challenges for practical use of radiation instruments (Johnson 2000). The first challenge is to decide which instrument(s) to purchase or choose for particular needs. The second challenge is to know how to properly use the chosen instrument for acquiring the desired data. The third challenge has to do with defending the acquired radiation data. Users of radiation instruments should be able to answer the question, "How do you know whether the data are any good?"

15.1.3 Approach for this Review

In addition to addressing the challenges described above, this review will include practical information on portable and laboratory instruments for:

A. Contamination monitoring
B. Exposure rate measurements
C. Radioactive materials shipping and receiving
D. Radionuclide identification
E. Area monitoring
F. Personnel monitoring and bioassay

15.2 INSTRUMENTATION COMMONLY USED IN UNIVERSITY AND RESEARCH LABORATORIES

15.2.1 Portable Instrumentation

15.2.1.1 Contamination Monitoring. The most commonly used radiation instruments for contamination monitoring, or detecting and quantifying radioactive materials on laboratory surfaces and equipment, include: Geiger-Mueller (GM) detectors, either pancake or end window, and sodium iodide (NaI) crystals. The design and operation of GM detectors are described earlier in this book. Let it suffice to say here that GM detectors are best suited for medium to high-energy beta emitting radionuclides, although GM detectors may detect alpha, beta, and gamma emitters. NaI detectors are best suited for gamma emitters. Thin crystal NaI detectors designed to measure low-energy gammas will also measure high-energy betas. Monitoring for alpha contamination may best be done with a zinc sulfide (ZnS) detector or a proportional counter.

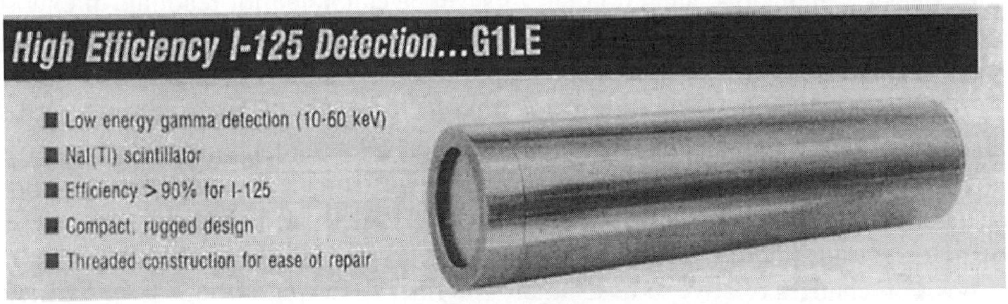

Fig 15.1 - Thin window, thin crystal sodium iodide (NaI) detector.

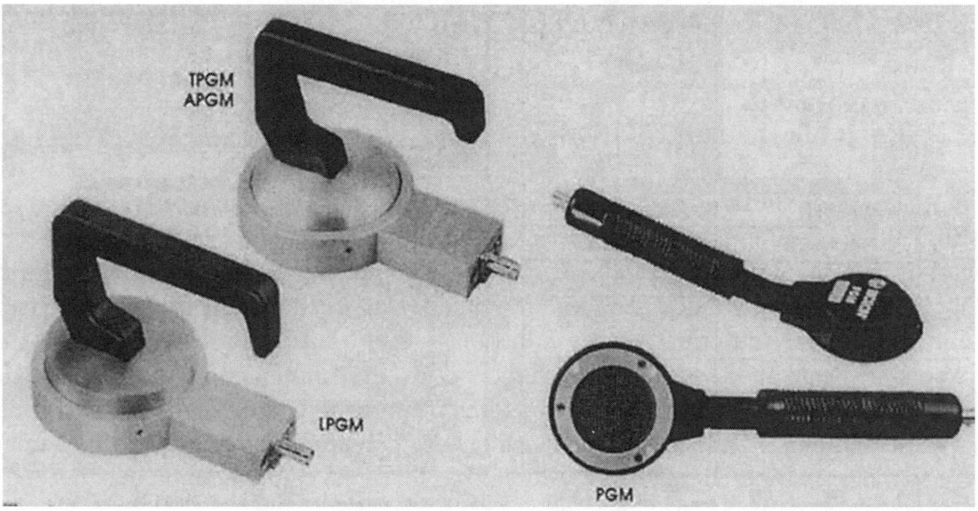

Fig 15.2 - Shielded (left) and unshielded (right) pancake GM probes.

15.2.1.2 Exposure Monitoring. Ion chambers are the commonly used instrument for exposure monitoring. However, ion chambers operating at ordinary atmospheric pressure are not very sensitive. More sensitivity can be achieved by using a pressurized ion chamber or other detectors calibrated for exposure readings. For example, NaI detectors, plastic scintillators, and GM detectors can be calibrated for readings of exposure or ionization in air. However, care should be taken with their use due to possible energy dependent responses.

15.2.1.3 Pocket Ion Chambers and Digital Dosimeters. Direct reading pocket ionization chambers, or pocket dosimeters, function on the simple premise of gas ionization. A central electrode is given a positive charge using a simple battery. Negative ions are produced by ionization in the gas surrounding the electrode and are attracted to the positive electrode, thereby neutralizing the charge. A fiber is attached to the electrode and its position changes relative to its charge. The fiber is viewed through a lens which allows the user to see the position of the fiber against a calibrated scale (typically from 0-200 mR) (Cember 1983). While reasonably accurate (±15 to 20 %), the instrument's electrode can be easily shorted by dust fibers inside a shirt pocket and is susceptible to leakage. Dropping the dosimeter can also change the reading. In each of these cases, the instruments mode of failure will result in a higher reading and a complete discharge to a full scale reading is possible.

These dosimeters are sensitive primarily to photons, but will respond, although poorly, to high energy beta radiation. Because the dosimeter is usually within a shirt pocket, the dosimeter is further shielded from beta radiation and its use for beta measurement is not recommended.

Electronic versions of the pocket dosimeter or Electronic Personal Dosimeters (EPDs) are also available with digital readouts, typically liquid crystal displays. These dosimeters serve the same function as pocket ion chambers, but yet their functionality is much extended. EPD's are battery powered, use a semi-conductor based detector, often provide dose rate information, and respond to photons in the 30 keV to 3 MeV range with ±30% accuracy. Many devices, as seen in the figure below, have nearly linear energy response from 50 keV to 3 MeV.

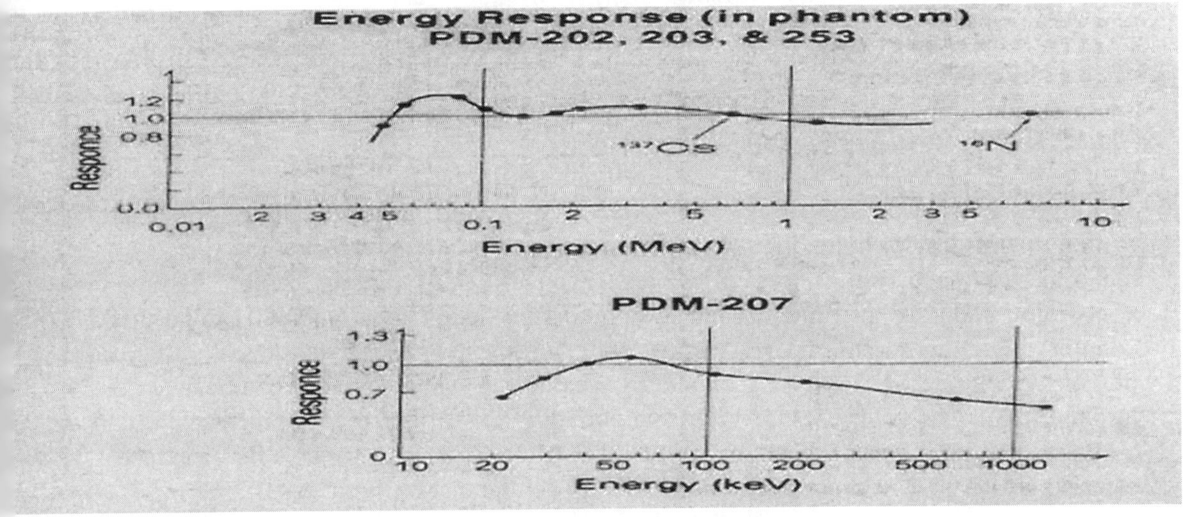

Fig. 15.3 - EPD Energy Response Graphs from Bicron, Inc.

Since EPD's span multiple dose ranges, it is no longer necessary to have multiple devices. An EPD can range from 10-10,000 μSv (or 10 to 10,000 mrem). They are considerably lighter, and are not subject to discharge failures like a standard pocket dosimeter. Cost for an EPD device is approximately $300 each. A picture of these devices is seen in Fig. 15.4.

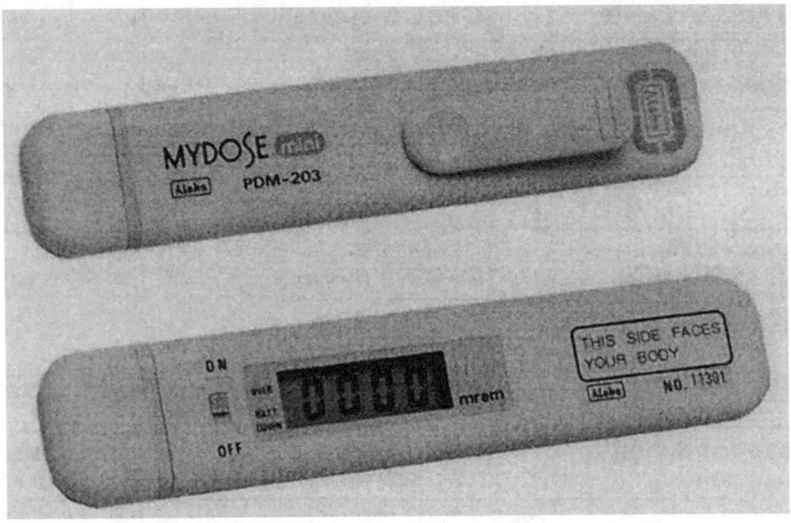

Fig 15.4. - Picture of an EPD from Bicron, Inc.

15.2.1.4 Analysis of Removable Activity. Normally, measurements of removable contamination are determined first by collecting a sample of removable radioactive material on a swipe, wipe, smear, or leak test. Secondly, this sample is usually analyzed in a laboratory radiation instrument, such as a liquid scintillation counter (LSC), proportional counter, or gamma spectrometer. Most recently, however, small (about the size of a bread box or large toaster) portable LSC counters have become available which can be taken into the field. Due to their light-weight and portability (lack of shielding), the background on these systems is greater, but they can easily function to perform surveys for clearance of packages for shipment and for basic laboratory contamination surveys.

15.2.2 Laboratory Instrumentation

15.2.2.1 Liquid Scintillation Counters: The liquid scintillation counter is perhaps one of the most versatile laboratory instruments available in the university/biomedical research setting. A liquid scintillation counter has the capability to analyze virtually all types of samples containing alpha, beta, and low energy photon emitters in virtually any media with the exception of soil and other similar "not easily dissolved " materials. A LSC can separate and quantify alpha from beta emitters; measure very low-energy beta emitters, such as ^{55}Fe, ^{3}H, ^{63}Ni; as well as measure high-energy beta emitters such as ^{32}P and $^{90}Sr/Y$. It can even identify low-energy photon emitters, such as ^{125}I, and moderate-energy photon emitters, such as ^{51}Cr (using the 4.38 keV auger electron). For beta emitters, a LSC has the capability to provide an energy spectrum printout to be used for visual nuclide identification.

The principle behind the liquid scintillation counter is the transfer of energy from radioactive decay to a scintillation media. The energy absorbed by the media moves molecules into an excited state within the media solvent. This absorbed energy is then released by the scintillation media in the form of visible light photons that are detected simultaneously by two photomultiplier tubes (PMTs) operating in a coincident mode. The magnitude of the light emitted by the scintillation media is directly proportional to the radiation energy released during a radioactive decay. Typically a laboratory LSC output is sorted into three "energy channels." A low-energy channel accepts low-energy events (beta decay) from ^{3}H (tritium). A medium-energy channel records the signal from ^{14}C and ^{35}S and a high-energy channel records the signal from ^{32}P. LSC equipment may contain several different programs or memory settings to customize channel energy ranges for particular radionuclides.

Complications with LSC analyses may arise from chemical and color quenching. Quenching results in loss of the light signal produced by the scintillation media and a loss of counting efficiency. This loss of signal can cause the energy spectrum to shift to a lower energy channel (to the left) and can be severe enough to result in the complete loss of signal (when the light output is less than the electronics lower level discriminator). Physical absorption of the beta energy in the sample itself, or absorption of the light output by the sample, can also result in a reduced signal output. Other factors can result in an increased light output (false activity), such as chemical luminescence. Usually LSCs are designed to allow correction for quenching effects and to notify the operator of excessive quench/chemical luminescence. Thus, these problems are not typically severe.

LSCs have minimum detectable activity (MDA) and lower limit of detection (LLD) levels in the range of a fraction of a Bq/L to several Bq/L above background for alpha and beta emitters. However, LSC equipment is limited for high-count rates and the amount of radioactivity should be limited to less than about 37 kBq (about a microcurie). This may require that small samples (in the order of 10's of micro-liters) be analyzed for radioactive waste samples. Detection efficiency of LSC equipment ranges from 30% for ^{3}H to almost 100% for ^{90}Sr.

High-energy beta emitters such as ^{32}P and $^{90}Sr/^{90}Y$ can be measured in water by Cerenkov radiation counting without the use of any scintillation media (although at a lower detection efficiency). Another drawback of LSC analysis is that while detection sensitivity is very high, when liquid scintillation fluids are used they must be disposed of as radioactive waste. In some cases, scintillation media may also be considered chemically hazardous as well.

15.2.2.2. NaI Gamma Spectrometers. Used in both the laboratory and field, the NaI detector is perhaps the most ubiquitous detector used in radiation safety programs. The basis of a NaI gamma spectrometry system consists of a NaI detector that may be either a solid cylindrical crystal or a cylindrical crystal with a coaxial hole in the top whose depth is about ¾ the crystal length. This latter design is called a well crystal and a sample may be placed inside the crystal for increasing detection efficiency. Gamma rays deposit energy in the crystal that then releases the absorbed energy as visible photons of light. The light photons are amplified through a photomultiplier tube and the output is connected to a multi-channel analyzer/PC combination where the energy spectrum can be displayed and analyzed by gamma spectrometry software.

Quantitative analysis can be performed on samples containing radionuclides with simple gamma ray spectra. However, NaI based systems are not suited for detailed spectrometry on samples containing may different radionuclides with gamma emissions that have similar energies. When the radionuclides are known, however, NaI based systems perform exceedingly well, are highly portable, and do not require detector cooling. A number of portable handheld gamma spectrometry systems are available from companies such as Exploranium and Bicron. Because of the typical crystal housing thickness, the low-energy efficiency suffers, and the lower energy limit of detection may typically be greater than 15 to 20 keV. Thus these type of detectors are better suited for use with low-energy gamma emitters, such as ^{125}I, when modified thin widow versions of the NaI crystal are used. NaI detectors with thin aluminum or beryllium windows are commonly used as portable instruments for low-energy gamma contamination monitoring.

15.2.2.3 Solid-State Semi-Conductor Gamma Spectrometers. The basis of these systems is a solid-state semi-conductor material (lithium doped silicon (sili) or high-purity germanium (HPGe)) in which gamma rays deposit energy and create free electrons that are collected as pulses. These detectors are much better at distinguishing gamma rays of similar energy (high resolution) than NaI systems. The reason for these detector's high resolution is that the units (or "packets") of gamma energy measured are in units of single electron volts. In contrast, in a NaI detector individual energy packets are a function of the energy of the light pulses produced by the crystal when it scintillates and the intrinsic photomultiplier tube resolution. The packet size for NaI light pulses and variations in the PMT signal make the size of these packets many times greater than a HPGe detector. Thus gamma rays of similar energy produce overlapping signals from a NaI detector that do not allow individual gamma rays to be distinguished from each other (Knoll 1979). Typical counting times for samples on semi-conductor systems range from a few minutes (for higher activity samples) to thousands of minutes for samples at environmental levels. HPGe based systems can detect and quantify very low energy photon emitters with energies as low as several keV using beryllium end windows.

15.2.2.4 Gas and Gas Flow Proportional Counters. A gas flow proportional counter can be used to measure either beta or alpha emitters. This type of instrument consists of a detector chamber in which a mixed gas such as propane and argon (P-10 gas) flows through the chamber at a slow rate. The detector chamber is operated at a voltage where the output pulse is proportional (hence its name) to the energy deposited in the detector. Since alpha particles have much higher energies than beta particles, their pulse sizes are much greater than a beta would generate.

Typically this is a counter where the sample is placed on a planchet (a circular metal tray) made of aluminum or steel. The planchet is placed in very close proximity to the detector window, or in the case of the windowless counter, the sample is placed within the actual chamber itself. Using pulse height discrimination, the pulses can be easily separated into "alpha" or "beta" signals and counted separately for the same sample.

The efficiency obtained from a proportional counter is comparable to a portable ZnS detector for alpha particles (although typically slightly higher because the window can be made thinner). Dramatic increases in efficiency can be achieved using windowless counters with a 4π geometry where the sample is actually placed within the detector itself. The efficiency of a windowless counter may range from 40 to 50 percent for alpha particles and greater than 60 percent for medium to high-energy beta particles. A windowless proportional counter is one of the only ways to detect and measure ^3H without using a LSC.

While proportional counters are very efficient detectors and they result in very little radioactive waste generation, they do not provide the spectrometry capabilities of the LSC. Also the handling of hundreds or thousands of planchets with paper smears is quite cumbersome in comparison to liquid scintillation vials. Laboratory proportional counters can be equipped with automatic sample changers. In the field, these instruments require a source of counting gas in order to function.

To get around the necessity of having a counting gas, an air proportional counter is available which functions with air as the counting gas for conducting surveys of alpha particle emitters. Air proportional counters function very well in a dry climate like Nevada or New Mexico, but they may not function at all in a high humidity climate like Florida. A survey team using air filled proportional counters from New Mexico, not understanding this limitation, could arrive in Florida and find that they cannot perform alpha surveys.

15.2.2.5 Alpha Spectrometers. Alpha spectrometry equipment uses solid-state semi-conductors which are thinly coated with either gold or aluminum in a ruggedized detector. Because of the very limited range of alphas in air, the detector and sample are placed within a vacuum chamber and the air is removed from the chamber. The vacuum level in the chamber is typically in the order of 10^{-4} torr. The output of each detector is sent to a multi-channel analyzer equipped with spectrometry software for identifying radionuclides by alpha energy.

Again, because of the short range of alphas in any medium, extensive chemistry and extraction of the radionuclide by electro-deposition on metallic plates or thin membrane filters must be performed to obtain adequate results. And even in these cases, correction for detection efficiency as a function of sample mass (density thickness in mg/cm^2) must be performed. If samples are too thick, no measurements can be made at all due to sample self absorption of the alpha particles. But with proper chemistry and extraction techniques, isotopic identification and quantification of elements such as Pu, Th, U, Am and Ra can be performed. Crude measurements can also be performed on raw samples to grossly identify and quantify radionuclides with no chemical preparation, although the alpha spectra will not have very good resolution.

Background count rates on solid-state alpha spectrometer systems are exceedingly low. This allows for detection limits in such systems as low as 0.001 Bq/g (0.03 pCi/g) with counting times for samples and backgrounds exceeding a couple of days.

It is possible to contaminate a solid-state alpha detector if the activity of the sample is high and the radioactive material is not properly plated to the sample substrate. If alpha spectrometry is performed in the field on raw samples such as pulverized soil or paper/cloth smears, contamination of the detector is a very real possibility. In addition, the quality of the spectrum obtained is very poor.

15.2.3 Fixed Instrumentation

15.2.3.1 Area Monitoring. Area monitors are typically deployed in locations with irradiators that may not be entirely self-shielded, or other areas in which higher levels of radiation may be present intermittently. Such conditions may occur at a radiation instrument calibration range, a radiation therapy room, or an accelerator facility. Area monitors might also be placed in an area to prevent the removal of sources or to alert individuals to the movement or arrival of radioactive materials in transport tubes or dumbwaiter systems.

An area monitor is typically an energy compensated Geiger-Mueller detector with circuitry that activates an audible and/or visual alarm to alert personnel about radiation levels in the area. Other types of detectors, such as NaI and plastic scintillators, as well as neutron monitors, are also available for area monitors. Area monitors are typically wall mounted and have options for either internal or external detector probes. The range of these instruments can be from 0.1 $\mu Sv\ h^{-1}$ (0.01mrem h^{-1}) to 100 mSv h^{-1} (10 mrem h^{-1}) and higher with external remotely mounted probes. Normally such monitors require electricity, but can be powered by battery if necessary. Alarm points may be set by a digital keypad or by a screwdriver adjustment on the inside of the monitor itself.

NRC regulations may require an area monitor as an alarm system in accordance with Title 10 CFR 20, Subpart G, 20.1601(a), "Control of access to high radiation areas" (NRC 1999). The area monitor serves as a control device to energize a conspicuous visible or audible alarm signal. Thus, an individual entering the high radiation area will be aware of the potential for significant radiation exposure.

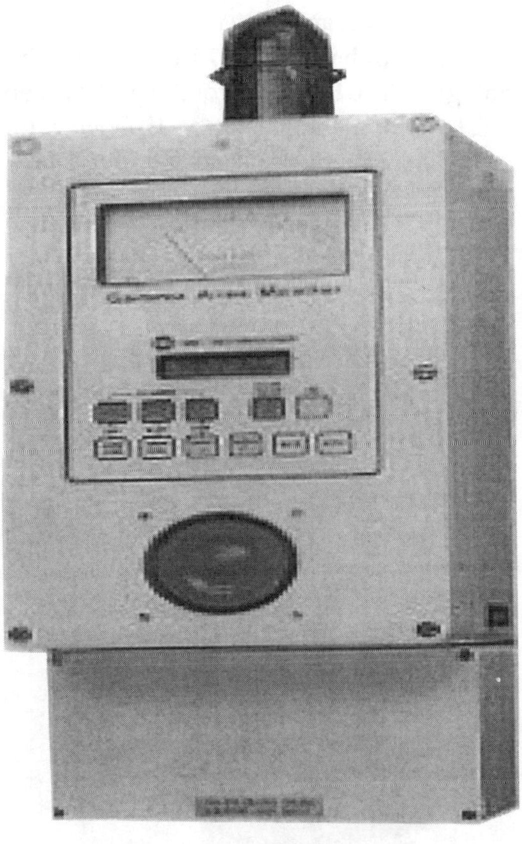

Fig 15.5. - NE Technology's Ga4 Area Monitor (NE 2001).

15.2.3.2 Hand and Foot Monitors. These devices are often used to rapidly screen individuals leaving areas of potential contamination such as a cyclotron facility, radioactive materials area, or radiopharmaceutical preparation area. A hand and foot monitor would be placed at the exit location from any of these areas and serve as an official survey point to detect radioactive materials on a person's hands or feet. These monitors target areas of high probability for contamination. The detectors for hand/foot monitors are typically proportional counters or GM detectors. In some cases a removable GM probe on the device is also available to "frisk" the remainder of the individual. Sensitivity for hand and foot monitors, such as the NE Technology's HFM7, for a 5 second count time and a 0.1 $\mu Sv\ h^{-1}$ (0.01 mrem h^{-1}) background, is said to range from approximately 4 Bq (10^{-4} μCi) for hands and 7 Bq (2×10^{-4} μCi)for feet for alpha contamination 20 Bq (5×10^{-4} μCi)for hands and 37 Bq (10^{-3} μCi)for feet for beta contamination. (NE Technology 2001). Efficiency of the probes is approximately 41% for ^{14}C on the hand detector and 16% for the foot detector. Efficiency for other high energy radionuclides is upwards of 64 and 37% for the hand and foot detectors, respectively.

Fig 15. 6 - NE Technology's HFM7 Hand and Foot Monitor.

15.2.3.3 Portal Hand-Foot Monitors: Often a hand-foot or portal monitor is setup to facilitate rapid and thorough screening of personnel for contamination. This may be an elaborate portal device in which the individual stands on grates suspended over proportional counters or GM detectors and places his/her hands inside a chamber which has similar detectors monitoring the backs and palms of the hands. Elaborate portal monitors can cost in excess of $25,000.

A more typical portal monitoring setup would consist of a large area probe (50-150 cm^2 or more) filled with argon-methane or standard GM gas. These detectors have high efficiencies of typically 40+ percent for ^{14}C and other beta-gamma emitters such as ^{60}Co, ^{137}Cs, ^{67}Ga, etc.

15.2.3.4 Laundry-Waste Monitors. Many research facilities that use outside laundry services and incineration for their medical pathogical waste (MPW) screen any outgoing materials for radioactivity prior to release. In addition, carts of regular waste may be screened to prevent release of radioactive materials to normal sanitary landfills. Radioactive materials may be introduced into these "outbound" streams of materials through the careless disposal of radioactive materials by researchers, patients who have had nuclear medicine procedures, and other research activities.

Typically, facilities accomplish laundry and solid waste screening by using a handheld thin window NaI probe or a micro-R meter (Erdman 2001). However, a number of biomedical research facilities have implemented a fixed waste monitoring system where either the boxes of MPW, or carts of laundry or waste are passed by stationary NaI detectors. These systems incorporate either 1"x1", 1"x3," or "3x3" NaI probes that are attached to a rate meter with a visual and audible alarm system (Erdman 2001, Austin 2001)*. The detectors should be strategically placed at a loading dock where materials enter and leave the facility.

Alarm points (or points considered to be detectable) are typically set on scaler instruments for laundry-waste monitors at approximately 2-3 times the background rate. Sensitivities that have been achieved are approximately 0.41 MBq (11 μCi) for ^{99m}Tc inside a standard loaded trash or laundry cart (Erdman 2001) and 0.37 MBq (10 μCi) for ^{137}Cs at a distance of approximately 1 meter (Bicron 1999). Additional testing of a "standard" MPW waste disposal system showed that it was able to successfully detect the following radionuclides and activities (Austin 2001): ^{201}Tl, 0.007 MBq (0.18 μCi); ^{123}I, 0.009 MBq (0.24 μCi); ^{125}I, 0.014 MBq (0.38 μCi); ^{111}In, 0.016 MBq (0.44 μCi); ^{67}Ga, 0.028 mBq (0.77 μCi); ^{99m}Tc, 0.033 MBq (0.88 μCi); ^{131}I, 0.05 MBq (1.34 μCi); and ^{51}Cr, 0.559 MBq (15.1 μCi). Commercial systems are available like Bicron's RWS-4 or LFM-2 systems which can detect photon emitters from 20 keV and up (see photo below).

---

* Austin, Sean. Personal Communication, 2001.

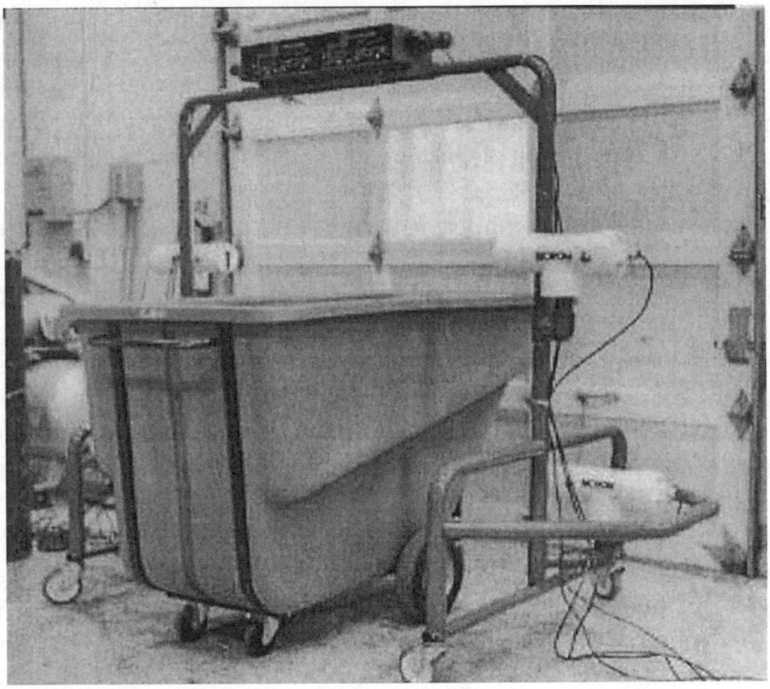

Fig 15. 7 - RWS-4 Laundry/Waste monitoring System.

15.3 CHOOSING THE RIGHT INSTRUMENT AND/OR PROBE FOR EACH MEASUREMENT NEED

The first question to consider in choosing the right instrument and/or probe is ***"What is my need for radiation data?"*** What decisions for radiation safety or regulatory compliance will require measurements of radiation? Ideally, choices on radiation instruments and probes should be governed by the decisions to be made on the basis of radiation measurements. Choosing the best instrument for the job is a big factor in assuring that the best data are acquired for the necessary decisions. Unfortunately, RSO's often find themselves inheriting instruments from a predecessor and then having to do the best they can with what they have. In some cases the inherited instruments may be marginally or even completely inadequate for the required decisions. For example, an RSO who uses a low-energy gamma emitter, such as ^{125}I, may inherit an end window GM detector and belatedly discover that it will not detect contamination levels of this nuclide.

15.3.1 Instruments for Exposure Measurements

Two types of radiation safety decisions are routinely based on radiation measurements. The first has to do with detecting the presence of radiation that may represent a potential for exposure or radiation dose to workers or the public. Only one type of instrument is designed to provide a true measure of "exposure" defined in terms of ionization in air from x-rays or gamma rays. This is the ion chamber or pressurized ion chamber that contains ordinary air as the radiation detection medium and gives a readout in units of roentgen (R hr^{-1}), milliroentgen (mR hr^{-1}), or microroentgen (μR hr^{-1}). Thus, whenever a measurement of exposure is needed in units of roentgen, the first instrument of choice should be an ion chamber. It should be noted that for health physics purposes, an exposure of one mR corresponds to a dose equivalent of one mrem.

Other instruments may also give readings in units of roentgen and may be adequate for exposure measurements, but do not actually measure "exposure." These include instruments that do not use air as the detection medium and only give surrogate or "apparent" exposure readings by an independent calibration procedure. For example, count rate instruments, such as sodium iodide (NaI) or GM detectors, can be calibrated to give readings in roentgen units. This is done by observing the instrument responses when placed at points of known R/hr on a calibration range and creating a meter scale in units of R hr^{-1}. Again, care must be exercised when using such instruments, especially NaI detectors, due to energy dependent response.

Although, count rate meters are responding to radiation events, rather than exposure (radiation energy absorbed in air), they will provide different readings at different distances from a calibration point source according to the inverse square law (the radiation intensity varies as the square of the distance from the source). At defined distances from a calibration source, the mR hr^{-1} (exposure) can be determined by using a NIST traceable source . A count rate meter can then be placed at each distance and "apparent" exposure readings recorded to create a meter scale in units of R hr^{-1}. Such count rate instruments may provide an adequate measure of exposure, if properly calibrated and properly used as discussed in the following section on instrument procedures.

15.3.2 Instruments for Contamination Measurements

The second type of decision that will be frequently required in radiation safety programs deals with the detection of unwanted of unknown radioactive materials. The presence of radioactive materials on surfaces or objects that may be handled by radiation workers or others would be considered contamination. Such materials may result in direct exposure to workers from external radiation or they could be inhaled or ingested and result in internal exposures. The RSO needs to know the amounts of radioactive materials (contamination) that may be found on work surfaces, equipment, floors, clothing, or even on a worker's body.

Measurements of the amount of radioactivity are best made in units of counts/minute (cpm) because radioactivity is defined in terms of disintegrations (radiation emissions or decays) per unit time (dpm). Instrument responses in cpm can be converted to dpm by appropriate calibration, efficiency, and energy response corrections .

The type of radiation to be measured should guide the choice of meter and/or probe for contamination measurements and the sensitivity needed to satisfy decision requirements. Decision criteria for contamination usually are specified in terms of whether the activity is removable or fixed. Removable activity is determined by rubbing 100 square centimeters of contaminated surface with a sampling material to collect a sample of the radioactive material. This procedure is called a swipe, smear, or wipe sample. Whatever cannot be rubbed off by this procedure is considered fixed activity. Guidelines for contamination decisions are given in Regulatory Guide 8.23 (U.S.NRC 1981). For example, in an area restricted (posted) for use of radioactive materials, removable activity is allowed up to 37 Bq (2,200 dpm) / 100 cm^2 for beta and gamma emitters. Five times this level, or 180 Bq (11,000 dpm) / 100 cm^2), is allowed for fixed activity.

15.3.2.1 Choice of Instrument for Low-Energy Beta Emitters. The type of radiation and the decision criteria will determine the best choice of instrument. For example, low energy beta particles from tritium (18 KeV maximum beta energy) are not measurable at all by GM or NaI detectors (Dunkelberger 1999). Therefore, the preferred method for measuring tritium contamination is to sample the contaminated surface with a swipe and then to analyze the removable tritium with a liquid scintillation counter.

At slightly higher beta energies (50 to 200 KeV), the instrument of choice would be a proportional counter or GM detector. For example, for C-14 or S-35 with maximum beta energies at about 160 KeV, there are two GM detector designs that could be adequate. These are the end window GM and the pancake GM. Whenever possible, the pancake GM detector is preferred because it has a detection efficiency about three times greater than the end window GM. For C-14/S-35, the detection efficiency of a pancake GM is about 4 to 5%, while the end window GM detects only 1 to 2%. Detector size is also important and the pancake GM has a much larger detector area which allows faster probe speed during a survey and a greater likelihood of detecting the presence of radioactivity. Because of the low detection efficiencies for C-14 and S-35, it may be preferable to sample for removable activity of these nuclides with swipes for analysis by liquid scintillation counting, where detection efficiency may exceed 60%.

15.3.2.2 Choice of Instrument for Medium and High-Energy Beta Emitters. Medium-energy beta emitters include isotopes with energies from about 200 to 500 keV. Examples include ^{33}P with beta energy of about 250 keV and ^{45}Ca with beta energy of about 260 keV. Beta energies above 500 keV would be considered high energy. Isotopes in this category include I-131 with beta energy of about 600 keV, ^{32}P with beta energy of about 1.7 MeV, and $^{90}Sr/^{90}Y$ with beta energy of about 2.2 MeV. The detection efficiencies for end window and pancake GM probes for each of these nuclides are shown in Table 1. In each case, it is notable that the pancake probe gives about three times greater efficiency or sensitivity than the end window GM. Therefore, whenever there is a choice, the pancake GM is always preferred. Although end window GM detectors are somewhat less expensive at about $125 compared to about $225 for pancake GM detectors, the pancake is well worth the extra cost.

Table 15.1 Instrument response for medium and high-energy beta emitters

Nuclide	Energy	EndGM - %	PanGM - %
^{33}P	248 keV	5	15
^{45}Ca	257 keV	5	15
^{131}I	606 keV	8	25
^{32}P	1.71 MeV	10	30
$^{90}Sr/^{90}Y$	544 keV 2.245 MeV	13	40

15.3.2.3 Choice of Instrument for Gamma Emitters. Although GM detectors will measure many gamma emitters, GM detectors are best for medium or high-energy beta emitters. Therefore, to detect gamma emitters, a better choice is to use sodium-iodide (NaI) scintillation crystals. The energy of the gamma emitter dictates the choice of the size of the NaI crystal. For low energy gamma, such as ^{125}I at about 27 and 35 keV, the best choice is a thin window, thin crystal, NaI detector, which has a detection efficiency of about 20%. Low activity ^{125}I may not be detectable at all with GM detectors.

For higher gamma energies above 50 to 100 keV, larger NaI crystals may be even more efficient (the larger the crystal, the greater the probability of intercepting a higher energy gamma ray). Higher energy gamma emitters may also be detected with a GM detector, but the NaI will always have a much higher sensitivity. However, there may be some penalty with the higher sensitivity. Namely, NaI detectors also have a much higher sensitivity to background radiation. Therefore, a high background signal may mask measurements of small amounts of contamination. In areas of very high background, such as near stored gamma emitting sources, small areas of contamination may be easier to find with the less sensitive lead or tungsten shielded pancake GM probe.

15.3.2.4 Choice of Instrument for Alpha Emitters. Portable survey instruments for alpha particle emitters typically use either a zinc-sulfide (ZnS) scintillation detector or a proportional gas ionization detector. Both of these detectors are usually covered with a very thin aluminized mylar film that absorbs some of the alpha particle energy. Also, because the distance of travel of alpha particles is usually less than two inches through air, these alpha detectors must be held very close to the source in order to be in the range of alpha particles. Detection efficiencies for alpha emitters for these detectors are about 25%. Removable activity for alpha emitters is usually determined by sampling with swipes and analyzing the removed material with a liquid scintillation or laboratory proportional counter.

15.3.2.5 Instrument Choices for Analysis of Removable Activity. Removable activity is determined by swiping surfaces to collect a sample of the radioactive material. Liquid scintillation counting will work well for most nuclides, however, it may not always be the best method of choice. For example, gamma emitters may be analyzed with greater sensitivity in a well-crystal NaI detector. These types of gamma detectors also have the capability with a gamma spectrum analyzer to determine the gamma energy or fingerprint for identifying particular gamma emitting radionuclides. For determining specific gamma energies, a detector with even better capabilities uses a high purity germanium (HPGe) detector.

Removable activity for alpha or medium/high energy beta emitters may also be measured with internal proportional counters. The advantage a proportional counter has over a liquid scintillation counter is that no scintillation solution is required and therefore the samples are easier to prepare for analysis. This is also true for gamma emitters that can be counted dry in a NaI well-crystal detector.

15.4 RADIOACTIVE MATERIALS PACKAGE SHIPPING AND RECEIVING

The surveys conducted for shipping and receiving radioactive materials are not considered to be unique. Surveys for radioactive material packages require both exposure rate and removable contamination measurements. The instruments used to accomplish these types of measurements are the same as used in laboratory monitoring. Exposure rates can be measured by GM counters, preferably an energy compensated GM tube, or an ion-chamber. Measurements of removable activity can be done by a LSC, gamma counter, or proportional counter. Some facilities may use a ZnS scintillator or a thin end-window NaI detector with a single drawer counting chamber attached to a simple scaler and printer to record results.

Requirements for instrument performance are driven by the Department of

Transportation regulations contained in Title 49 Code of Federal Regulations parts 100-185. Of particular importance for radiation safety programs is part 173 subpart H (DOT 1999), which contains the basic requirements for the shipment of radioactive materials. Radiological measurements include the following:

Measurement Need	Measurement Method
Transportation Index ("exposure" rate at 1 meter)	Ion chamber, GM
counter,	
	μR-meter
Package/vehicle surface "exposure" rate	Ion chamber, GM
counter,	
	μR-meter
Removable Contamination (α,β,γ)	LSC, Proportional, or Gamma counter.
Activity to be shipped	LSC, Gamma counter,
HPGe.	

We talk loosely about "exposure" rate, but the DOT regulations specify "dose equivalent" rate on the package surface and at one meter for the transportation index. We make the assumption that the quality factor is equal to one (only gamma photons are present) and that the exposure rate is equal to the dose equivalent rate.

A meter that is calibrated for exposure, typically using ^{137}Cs as the calibration source, is absolutely required. When surface exposure rates are measured it is essential that the instrument be calibrated to respond to the known or true exposure reading *"at the surface of the detector probe or meter body."* Many detectors or probes may be calibrated at a point that is at the center of the detector, which in the case of an ionization chamber may be several inches from the surface of the meter body. Caution must be exercised to make sure the survey is measuring the dose equivalent rate at the surface and not several inches away. Instruments available should be able to measure exposure rates down to below 0.005 mSv h^{-1} (0.5 mrem per hour) and have an upper range of up to 10 mSv h^{-1} (1,000 mrem per hour).

Instrument sensitivity is not generally a problem for measuring removable activity. This is because the limits for removable contamination are high with respect to instrument sensitivity. Instruments should easily be able to detect activity at the limits of: 0.4 Bq/cm^2

(22 dpm/cm^2) beta-gamma and 0.04 Bq/cm^2 (2.2 dpm/cm^2) alpha. When these limits are combined with the required surface area to be surveyed (300 cm^2) this equates to 120 Bq (6600 dpm) and 12 Bq (660 dpm) per swipe sample respectively for beta/gamma and alpha contamination(DOT 1999).

15.5 PERSONNEL BIOASSAY AND "FRISKING"

15.5.1 Instrumentation for Bioassay

In a typical university and biomedical research laboratory, internal contamination is determined by personnel bioassay using urine analysis and a LSC for most alpha, beta, or low-energy gamma emitters. Internal contamination with radioactive iodine can be determined by thyroid counting using a NaI detector attached to a single or multi-channel analyzer. Intakes are then calculated using recognized models for intake and excretion such as an ICRP30 based model (ICRP1987). A typical thyroid counting system would consist of a 2"x2" lead shielded NaI probe mounted on a stand and attached to a simple scaler device and perhaps a printer. A device such as this would cost approximately $6,000.

What is often overlooked by radiation safety personnel is the use of air sampling equipment. For example, airborne radio-iodine can be sampled with activated charcoal, as an adsorbent sampler, followed by analysis of these samples using a NaI gamma counter. These measurements make possible a determination of the concentration of radioactive material in the breathing zone air, and thus an estimate of the intake of radio-iodine. This is an excellent compliment to regular thyroid counting.

LSC counting of urine is initiated by obtaining a urine sample from the individual monitored. The urine sample is often preserved using thymerosol tablets and refrigerated to prevent growth of bacteria/fungus. The preservative does not interfere with the LSC analysis and prevents radioactive material concentration by bacteria and fungal growth. Such growth could cause urine sample aliquots to be non-representative of actual urine concentrations.

Typically, a one mL sample of urine is placed in 5 to 15 mL of scintillation cocktail, and analyzed in both a LSC and a gamma counter. Both gross alpha/beta screening of the urine, as well as spectral analysis/display can be employed to determine the presence and identity of any radioactive material that may be present. Count times for urine samples are in the range of 10 to 20 minutes, and can be shortened to 1 minute for rapid screening of many samples in an emergency situation. If alpha spectrometry is available, the urine can be chemically prepared and analyzed for alpha emitters.

In a pinch, gross beta analysis can be performed on evaporated and dried aliquots of urine using a gas proportional counter or even a thin window NaI probe. Caution must be exercised in this case because the process of drying/evaporation will result in the loss of volatile radionuclides.

15.5.2 Instrumentation for Frisking

After the use of unsealed sources of radioactive material, or after working in areas of suspected removable contamination, it is necessary to check (frisk) for contamination on one's person (hands, body, tops of shoes, clothing) as well as survey for contamination throughout the laboratory or work area. In university facilities, instrument selection for these purposes should be the same as the equipment used to monitor the research laboratory for contamination. Instrumentation choices are the pancake GM (preferred over an end window GM tube because of its higher surface area and sensitivity) for beta emitters such as ^{14}C, and ^{32}P. If the surveyor uses poor technique (probe movement too fast or detector held too far above surfaces), the use of a GM or proportional counter will be insufficient for low energy beta emitters, such as ^{14}C and ^{35}S.

For very low energy beta emitters, such as ^{3}H, the only acceptable survey method is to collect a sample of removable activity on a smear. This sample is then analyzed by a liquid scintillation counter or windowless proportional counter. Tritium is not detectable by any instrument with a window, because the beta energy of tritium is too low to penetrate the window. Windowless proportional counters for portable use are not good alternatives for personnel monitoring or frisking for ^{3}H because a good seal between clothing and skin surfaces cannot be obtained, and the process takes too long for practical application.

Where low-energy photon emitters like ^{125}I are involved, care must be taken to provide a detector that is capable of detecting the low-energy emissions. Typical NaI probes have thick detector casings, poor efficiency for low energy photons, and may not be able to detect energies in the 10 to 30 keV range. Thin aluminum or beryllium end windows are required to extend the energy response of the NaI detector down to approximately 10 keV. Even with this extended lower range, problems may exist because of the small probe size and the high background. As noted elsewhere in the literature and this chapter, probe size and background are two important variables that affect detection ability.

15.6 VERIFICATION OF PROPER INSTRUMENT FUNCTION

After determining the best radiation measurement instrument, the next challenge is knowing how to use it properly to acquire the needed data. Many new RSO's think that using a meter is simply a matter of turning on the power, pointing the instrument at the source of the radiation signal, and observing the needle or digital readout. It is much more involved. There are many factors that may cause a meter reading to be inaccurate and unacceptable.

15.6.1 Checking Instrument Operation and Simple Troubleshooting

Before taking any radiation measurements, one should first determine that the meter or laboratory instrument is operating properly (Dunkelberger 1999). For portable meters, usually the first step to verifying instrument operation is to turn the power switch to the battery setting and determine that the battery is adequately charged.

15.6.1.1 Battery Check. - If the meter fails the battery test, the first thing to check is whether the meter has batteries installed. Since batteries are prone to discharge and leak if left in the meter for several months, it is a good idea to remove the batteries when meters may not be used for several weeks or longer. If there are batteries in the meter, but they are discharged, then they should be replaced with fresh D cells (preferable alkaline). Unfortunately, dead batteries left in a meter may have leaked or corroded with resulting damage to the meter. If this happens, remove the corroded batteries and attempt to clean the battery terminals with a flat blade screwdriver or a small wire brush. Be sure the spring loaded contacts in the base of the meter are free to move. If the meter fails to respond to new batteries, then it may have a damaged circuit board or broken battery terminal connector from battery corrosion and it will need to be sent to a repair shop.

15.6.1.2 Check Source Response - After determining that the battery reading is acceptable, verify the check source reading. Each meter should have available a relatively long-lived check source (with a half-life of four years or longer) preferably with an emission of the type and energy of radiation similar to the radioactive material which you plan to measure. For example, for an alpha meter a good check source could be ^{241}Am. For medium/high energy beta probes, such as an end window or pancake GM, a good check source is ^{204}Tl. For low energy gamma detectors, a good check source is ^{129}I. For higher energy gamma or beta detectors, a good check source is ^{137}Cs.

After determining that the proper check source is available for the particular detector, the detector should be held in contact with the source and the reading compared with the corresponding reading recorded on the calibration form or sticker on the side of the meter. You should expect to get a reading close to the recorded value (within 10 to 20%). If you get a significantly lower reading or the meter does not respond, you should check to be sure you have the right check source for the probe. If you still get a low reading, it may be due to radioactive decay of check source. This could be a factor for ^{204}Tl, which has a relatively short half-life of about 4 years. If the recorded check source reading is less than 1,000 cpm at the time of calibration, it may have decayed to a level near background.

15.6.1.3 Possible Cable Failure - If the check source is verified as suitable, but the probe still does not respond, the reason may be cable failure. You should check that the cable is attached tightly to both the meter base and the probe. Also check the cable for visible signs of damage, such as broken insulation. If the cable appears sound and is tightly attached, check to see whether the meter responds erratically when the cable is stretched or flexed. Often cables fail from internal damage that is not visible to the eye. If the meter still does not respond, or responds erratically, then you should replace the cable. Next to batteries, cables are the weakest component of portable survey instruments. It is always a good idea to keep an extra cable or two on hand for replacements. When you order extra cables be careful to specify the proper connectors for both the meter base and the probe. Survey meters customarily use either BNC or Type C connectors. You can also buy adapters to convert between the two connector types, if needed.

15.6.1.4 Possible Probe Failure - After replacing the probe cable, if you still get no response from the meter, it may be due to probe failure. First, check for obvious signs of probe damage. For instance, if the mica window is broken on an end window or pancake GM probe, then the probe will not function. If your NaI probe rattles, the photomultiplier tube could be broken. Dropping a NaI probe can easily break the photomultiplier tube. GM tubes are very susceptible to puncture fractures of the mica windows. For either of these types of damage, the probes can be repaired in a repair shop.

If there is no obvious indication of probe damage, the only way to check a probe is to replace it with one from another meter that you know is working. If the replacement probe works on your meter, that is a good indication of failure of the original probe. If the meter still fails to respond, that is an indication of meter base failure. Both probe failure or meter base failure will require repair in a meter repair shop.

15.6.2 Exposure Meter Check Out

The operation of exposure meters can be verified in a similar manner to contamination meters. The first thing to do is to verify battery condition and replace batteries or clean battery contacts if needed. Next, you can verify a check source reading with an appropriate check source. Useful check sources may be either beta or gamma emitters. It does not matter what the source may be as long as it results in a reproducible reading for verifying meter operation.

15.6.3 Laboratory Instrument Check Out

Laboratory instruments are usually verified in a manner similar to portable instruments, although there are usually no batteries to check. The normal procedure when starting up a radiation counting laboratory each day is to run check source and background readings on each instrument. These readings are then compared, not to a single check source reading, but to a series of previous readings plotted as control charts.

15.6.4 Laboratory Instrument Trouble Shooting Tricks

15.6.4.1 Chemical or Photo Luminescence. This malady strikes the LSC counter and can usually be attributed to either true chemical luminescence or light stimulated luminescence. Light related problems can be cured by dark adaptation of the vials before counting. Usually 10 to 15 minutes in the counter with the lid closed cures this problem. For very low level counting applications this problem has been dealt with by placing the counter in a room setup as a darkroom and with actual refrigeration of the samples. A good rule of thumb would be to avoid strong overhead lights immediately above the counter and to keep the vials and counter out of direct sun light and away from windows to minimize photo-luminescence problems.

Chemical luminescence is induced by the presence of chemicals in the samples which react with the scintillant and produce light flashes. This can happen with urine samples or with mixed waste samples containing numerous chemicals. Chemically stimulated luminescence can often be treated with a 10 to 100 micro-liters of acetic acid available at any photo store. Both the chemical and photo luminescence phenomena produce false activity in the first, lowest energy, channel and can be severe enough to produce count rates in the thousands of cpm.

If your LSC has the ability to display or print out a spectrum, you will observe that chemical and photo luminescent peaks have a characteristic shape that is unlike the standard broad peak from a beta emitter. Most often these peaks appear to be narrow and shaped much like one might expect when viewing a good NaI spectrum. This recognition will help you to immediately identify the problem of chemical or photo luminescence.

15.6.4.2 Static Electricity Induced Activity. Static electricity is usually only a problem in the winter or in very dry climates. With plastic LSC vials, static discharge can yield false counts, again primarily in the first channel. Many instruments have static eliminators to reduce this effect. It can be cured in the laboratory by adding a humidifier or something as non-scientific as wiping plastic vials and the sample rack with a dryer sheet before placing the vial into the LSC itself.

15.6.4.3 Cerenkov Radiation in High Energy Beta Samples: Even if a sample has LSC fluid in it, it is still possible to have cerenkov radiation produced in the sample. This appears as a very broad peak which can be mistaken for ^{3}H present in the sample, as the energy of the peak typically shows up in the upper portion of the ^{3}H window of an LSC

15.6.5 Factors for Proper Instrument Usage

Since all radiation measurements are made by comparisons, it is important to know the conditions of the meter calibration. Radiation measurement instruments are ideally calibrated for the specific nuclide and conditions intended for actual measurements. The closer the similarity between the conditions of calibration and use, the better the quality of the measurements. This means that a meter calibrated in reference to a ^{137}Cs source (with gamma energy of 662 keV) will be most accurate when measuring ^{137}Cs exposure or contamination. If the meter is used to measure another radionuclide with a gamma energy that is either much lower (100 keV) or much higher (1 MeV or greater), the instrument may not respond in the same way as it did for ^{137}Cs and the meter will read either higher or lower than it should. This variation in meter response with energy is called energy dependence. Preferably, you either calibrate for exactly the same energies that you wish to measure, or you choose an instrument that is designed to give a reasonably level response (in reference to the calibration source) over the range of energies that you will be measuring.

RSOs are encouraged to prepare a table of the nuclides allowed on their license and list the properties of each nuclide including: half-life, emission of alpha, beta, or gamma rays, and the percent abundance and energy of each emission. This information should be provided to the instrument sales representative when you order an instrument, in order to choose the best instrument for your needs. This information should also be provided to the calibration laboratory for best calibration of your instrument.

When you send your meter for calibration it is also important to tell the calibration laboratory how you plan to use the meter. For example, most end window GM probes are calibrated with the mica window pointed at the source and held about 1/8 to 1/4 inch away from the source. However, this same probe could be calibrated for exposure measurement, but for that purpose the probe is normally held sideways to the source. The orientation of the probe and source is called geometry and the best readings are achieved when the geometry of calibration and use are the same.

Concern for reproducible geometry means that you cannot monitor floors with a pancake GM probe by swinging the probe over the floor while holding on to the cable. This is actually bad practice for two reasons. First, it violates probe geometry (pancake probes are calibrated flat and parallel with the source) and secondly, carrying the probe dangling at the end of the cable will likely result in damage to the cable insulation and cable failure.

Geometry is also important for exposure meters. Depending on the meter design, the calibration beam may be directed at the bottom or side of the meter case. Again, best readings are achieved by holding the meter in the same orientation, when taking actual readings, as the meter was held for calibration. Exposure meters are designed to give accurate readings when the entire detector is in the radiation field. Thus, a narrow "pencil" beam of intense radiation will give a lower reading than the actual radiation intensity within the beam.

Other factors could also cause a meter reading to be in error. These include resolving or dead time loses, saturation of GM detectors, detection efficiency and sensitivity problem , use of the wrong probe or meter for the particular source, background interference, and variations in standards and calibration conditions as explained above.

15.7 CALIBRATION OF INSTRUMENTS

Laboratory instrumentation calibration requirements begin with five essential elements;

1. The purchase or attainment of sources which are ultimately traceable to the National Institutes of Standards and Technology (NIST) as either standard reference materials or traceable through secondary standards;
2. The initial and periodic calibration of laboratory equipment using NIST traceable sources;
3. The tracking of instrument performance/constancy checks on a routine basis;
4. Compensation for instrument response due to sample geometry, sample size and the physical condition of the sample;
5. A written set of procedures for laboratory analysis, documentation, and data review and verification.

NIST traceable sources can be obtained from NIST itself or from any number of commercial laboratories who are considered "reference" laboratories. These laboratories participate in cross checks and "certification" by NIST. Once obtained, the sources may be used directly or may be aliquoted volumetrically of gravimetrically into smaller quantities and placed in the geometry of the samples that will actually be counted.

Table 15.2 Typical Calibration Sources for Gamma Spectrometry (with Low E option)

Radionuclide	Energy(keV)	Half-Life
^{241}Am	59.5	432 y
^{109}Cd	88	462.6 d
^{57}Co	122	271.79 d
^{139}Ce	166	137.64 d
^{203}Hg	279	46.59 d
^{113}Sn	392	115.09 d
^{137}Cs	662	30.0 y
^{88}Y	898	106.63 d
^{60}Co	1173	5.27 y
^{60}Co	1332	5.27 y
^{88}Y	1836	106.63 d

Calibration of laboratory instrumentation should be conducted: 1) when the instrument is new, 2) when changes or repairs are made (for example to detector amplifiers, pre-amps or associated electronics) which change the detection efficiency, and 3) periodically (perhaps annually) if the tracking of instrument response indicates a statistically significant change in detector function. Often, modern day instrumentation (particularly laboratory instrumentation) is so consistent that the efficiency, gain, and energy calibration might not change in years and will necessitate only periodic checks against known standards over the energy range used.

Portable instrumentation may be as stable as laboratory instruments, but often due to the use/abuse of instruments in the field more frequent calibration, typically semi-annually or annually is required. This is a change from historical guidance recommending quarterly calibration for field instrumentation.

15.8 INTERPRETING RADIATION MEASUREMENTS

Part of the challenge for RSOs is to recognize that radiation is a statistically random phenomenon. Consequently, if a radiation measurement is repeated twice, ten times, or one hundred times, each reading will be different. This can be demonstrated in the classroom in several ways. First, turn on a GM meter and hold it in the air away from any known source. Ask the class to listen to the frequency of the clicks, which represents the background signal. Ask them to notice how many clicks they might hear in a ten-second interval. Then ask what that number might be if they repeated the exercise for several ten second intervals? As the students listen to the clicks on the meter they will quickly realize that they are listening to random events. There are clearly groups of clicks together separated by short or long intervals without clicks. To further dramatize the randomness, you could ask if anyone could write music to that beat? Once students recognize the randomness of the GM signal, it is important that you also ask them whether that randomness is due to some artifact of the meter? Health physics students must understand that the randomness is due to the radiation signal and not the response of the instrument.

At this point, students may agree that radiation is a statistically random phenomenon, but they do not yet have any real world experience of what that means when conducting measurements. This experience may be gained by a laboratory exercise to demonstrate the variations not only with multiple readings on a check source by a single instrument, but also with repeated measurements with a variety of instruments.

15.8.1 Laboratory Exercises with Exposure Instruments

This laboratory exercise allows intercomparison of 6 to 8 meters, all of which are calibrated to give exposure readings in units of R/hr. The meters include analog and digital ion chambers, a pressurized ion chamber, a NaI microR meter, a plastic scintillator microrem meter, and a side wall GM probe and an NaI probe attached to a meter base with a cpm readout. Intercomparisons of these meters provide students with the opportunity for hands-on observation of differences, in terms of response time, readout variability, size, weight, ease of use, etc. Each meter is used to read exposure at distances of 5, 10, 15, and 20 cm from a ^{137}Cs check source with an activity of about 10 microcuries. Thus, four readings are taken with each instrument as a basis for students to observe instrument response and to consider which meter they would choose, if they had to choose only one. Taking readings at four distances also allows students to observe the effect of distance.

They are asked to determine whether the inverse square law is verified by their readings, and if not, why not.

The exercise also includes one additional task. Namely, students are asked to convert all of the exposure readings into activity measurements. They are shown how to make this conversion by means of the exposure rate constant or specific gamma ray constant for ^{137}Cs. This illustrates the inter-relationship of exposure and activity units. This also helps students understand that a screaming count rate detector with a high cpm reading (10,000 to 100,000 cpm) does not mean that the exposure is high (the exposure in contact with the source is less than 1.5 mR h^{-1}).

After they have converted all of the exposure readings to activity units, students are then asked to determine their best estimate of the activity of the ^{137}Cs source. They are not given any guidance on how to make this estimate. They usually have about 25 different data points to work with. Some of the meters, such as the microR meter and the NaI probe, read off scale at the 5 cm distance and therefore cannot provide data at that point. Also, the analog ion chamber gives essentially the same reading at 15 and 20 cm. The range of calculated activities is usually from about 6 to 14 microcuries due to variations between meters and the different distances.

Students are then asked to come up with a single number that they believe represents their best estimate of the ^{137}Cs source activity. Usually students will take a simple numerical average of all data points to come up with a single number. They are reluctant to throw out any data points or to use any judgment on data selection. Sometimes, however, students will notice that two or three of the meters seem to give readings that are in close agreement and they will use only those readings for their best estimate. They may also notice that some of the readings do not follow the inverse square law and they may discard those data points.

After students report their best estimate of the ^{137}Cs activity, they are asked to explain their process and reasons for their answer. It is pointed out that usually RSO's will not have the benefit of multiple instruments, but only one. Therefore, they are also asked, if they had to choose only one of the various meters, which one would they chose and why? Quite often students will choose a digital meter over an analog. Somehow, the display of real numbers gives them better assurance of accuracy, compared to the swinging fluctuations of an analog needle movement. They usually dislike the slow response of ion chambers. They also tend to like the more steady readings achieved with scintillation detectors.

15.8.2 Determining the Precision of the Reported Best Estimate

Students are asked to discuss the precision of their reported best estimate. They are shown how to use normal statistics for calculating the mean and standard deviation for their 25 measured data points. They are then asked how would they determine the quality of their estimate if they had only one reading, which is the usual case in actual practice?

Students are then introduced to normal distributions by illustrating the results of measuring a source's activity with 100 different instruments, or taking 100 readings with a single instrument. In a normal(Gaussian) distribution 68 of the 100 readings should fall within a range of +/- one standard deviation and that 95 of the readings should be within a range of +/- two standard deviations. Assuming the instrument was properly calibrated, properly operating, and properly used, then the mean of the 100 readings should be a good estimate of the true mean value.

Students are further shown how to estimate the precision of a single measurement, where one standard deviation can be estimated by the square root of the total counts collected for a particular counting time. Since most laboratory instruments provide an output in terms of count rates, the students are shown how to estimate one standard deviation by the formula:

$$\sigma = \sqrt{\frac{N}{T}}$$

Where N = the sample count rate, and
T = the sample counting time
N and T are in the same units of time

Also, since radiation measurements always include a background signal, the students are shown how to estimate one standard deviation to include the uncertainty of the sample

$$\sigma = \sqrt{\frac{N_{s+b}}{T_s} + \frac{N_b}{T_b}}$$

count and the background count as follows:

Where N_{s+b} is the count rate of the sample plus background
N_b is the count rate of the background
T_s is the sample counting time, and
T_b is the counting time for the background

15.8.3 Reporting Measurement Results

Once an estimate is made of the uncertainty of a sample measurement, the result must be reported in a way that makes the uncertainty clear. The usual method is to report the measured value +/- one standard deviation. Another option is to report the measured value and its coefficient of variation (CV). The CV is defined as the measured value divided by one standard deviation (multiplied by 100 to convert to percent).

The reporting of measurement data also needs to consider how many significant figures to include. Some may find it difficult not to carefully report all six or eight digits that may come from the printout of a measurement instrument. However, it is important to consider how many significant figures can be defended. Since no measurement can have an uncertainty less than the original calibration standard (typically 5%), then reported values with more than two significant figures may not be defensible. For example, a reported value of 10 (which is two significant figures) says that the uncertainty is +/- 1, i.e. the real value is not a 9 or an 11 (or it would have been reported as a 9 or an 11). A reported value of 10 +/- 1 therefore gives an estimated uncertainty of +/- 10%. However, if the value is reported with three significant figures, such as 10.0 (which says the real value is not 9.9 or 10.1), then the implied uncertainty is now +/- 1%. Since this is less than the expected calibration error, it cannot be defended.

Other sources of measurement uncertainty also need to be understood by RSOs. These include not only counting uncertainty (which accounts for the randomness of radiation), but also uncertainties that may be due to calibration errors, counting time, amount of radiation, background variations, geometry, uniformity of samples, sample selection bias, sample preparation errors, and volume or weight errors. Usually, reported measurement uncertainties are only based on counting errors, they do not include all other sources of uncertainty. Therefore, when defending the quality of measurements, it is important to realize that reported measurements probably have much greater uncertainty than given with reported values.

15.8.4 Defending the Quality of Reported Measurements

RSOs should be able to answer the question, ***"How do you know whether the data are any good?"*** The defensibility of measurements that may be reported and used for expensive decisions should be a matter of concern for all RSOs. We have reviewed many factors that contribute to the quality of measurements. These include using the right instrument or probe appropriate for the radionuclide and material being measured. The measurement instrument also must be used properly. This usually means using the instrument in the same way it was calibrated. Or conversely, getting the calibration done in the same way that the instrument will be used. Once the choice is made for the proper instrument, and its proper operation has been verified, then results can be reported with an estimate of uncertainty based on the randomness of radiation. However, RSOs should be able to address questions about all of the other sources of uncertainty and the normal standards of practice for minimizing their effects on the overall precision of a measurement.

The accuracy of measurements (closeness to the true value) is verified by the evaluation of readings from known sources. The disintegration rates of standards, or the exposure rates of known sources, are usually verified by reference to NIST. The reproducibility of measurements (precision) is verified by duplicate samples or split samples. Reproducibility can also be evaluated by repeating measurements of the same sample. A basic rule of practice is to never report a measurement value, especially one that may lead to expensive decisions, without confirming the value by a repeat measurement. The primary basis for defensibility of measurement data is careful documentation of instrument calibration and instrument operational verification. It is also important to document all quality control measures for verifying precision and accuracy of measurements.

15.8.5 Measurement vs. Calibration Conditions

15.8.5.1 Geometry Considerations. Instrument calibrations are performed on calibration ranges or using setup gigs to assure reproducibility in instrument calibration, source homogeneity, and geometry in a standard configuration. A standard configuration is important to measure instrument performance as well establish a "norm" against which to base validity of instrument calibrations and possible changes in instrument performance.

However, once the calibration is completed, the geometry may not be held a constant. In the field instruments may not be used in fixed "source to detector" geometry. In fact the geometry will vary during the same survey, and even more greatly with different personnel. Also, 1) contamination found in the field is usually not homogeneous, 2) the contamination may not be the same energy as the calibration source, and 3) the surfaces present in the field may have a dramatic effect on efficiency. You might be tempted to ask at this point "Why did I bother to calibrate the instrument in the first place?" The point of this review is not to cast aspersions on the calibration procedures, but to point out one of many conditions that may affect the efficiency of the measurement and hence its sensitivity.

A typical pancake GM detector will have an efficiency of approximately 20-30% for ^{90}Sr in a standard calibration setup at a few mm's to approximately 5 mm from a plated source. If the detector position is raised by a cm or more above the source, the efficiency can easily drop to 15–25%. Considering that an individual may change the probe to surface distance during use in the field, the efficiency can be easily reduced to 10-20% or lower simply due to survey technique.

Simple inverse square relationships that are valid for photon emitting sources cannot truly be used for beta emitters. At best such a relationship may only be considered approximate for beta emitters (NCRP 1991 Report 112). For alpha meter calibrations this inverse square relationship cannot be applied at all due to the extremely short range of alpha particles in air.

When instrument surveys are conducted other "geometric" factors should also be considered . An example is the surface characteristics of the field source as compared to the calibration source, particularly for measurement using alpha monitoring equipment. When using ZnS , gas proportional, and air proportional counters, the surface "roughness," dust, moisture content, and "probe covering" (used in an effort to prevent probe contamination) are highly influential. A slight covering of surface dust can completely mask significant contamination of an alpha emitter. This has been personally observed by one of the authors at a radium source/luminous paint manufacturer and distributor. Attempts made by emergency response personnel as a "standard procedure" to protect alpha scintillation and gas proportional equipment from contamination by bagging instruments in plastic makes the instruments completely useless. Similar and complete nullification of instrument effectiveness was observed at a remedial site when alpha scintillation equipment was used to detect contamination on newly sodded and watered areas of a park area. This same effect may occur in a laboratory due to the washing and waxing of floors or the good intentions of a staff member using a plastic cover over a GM pancake probe, alpha meter, or other instrument to prevent contamination or damage from sharp objects. This is one good reason to perform visual audits of surveyor performance, conduct extensive training, and have standard operating procedures that consider such factors.

15.8.5.2 Energy Dependence. Due to the limited range of alpha, beta, and low energy photon radiation, the energy of the calibration source in comparison to the energy of the radiation in the field measurement conditions must be assessed and accounted for. Because the measurement of these three types of radiation is dependent upon a thin detector window to allow entry of the radiation into the detector, the effect of this window on the efficiency of measurement is considerable (NCRP 91). Because of this effect, the NCRP recommends that instruments used for beta measurement be calibrated over multiple energy ranges from less than 0.3 MeV, from 0.4 MeV to 0.8 MeV, and at energies greater than 1.5 MeV (NCRP 1991). This is amply demonstrated by observing the difference of efficiencies between ^{14}C-^{35}S and ^{32}P - ^{90}Sr which range from as low as 10 % to as high as 40%.

15.8.6 Monitoring Technique and Speed: Effects on Sensitivity

The method (speed of probe movement) during a survey using portable instrumentation can have a significant effect upon whether or not contamination is detected. If the movement of the probe is too slow, the maximum sensitivity (with respect to speed) is obtained but the area surveyed may be so small that an effective survey program may not be completed. If probe movement is too fast, large areas can be quickly covered, but sensitivity suffers to the point that significant contamination may be missed.

It is essential that a proper balance of speed and sensitivity be obtained to maximize both sensitivity and area coverage. The magnitude of the effect is dependent upon many factors and has been examined in detail within two publications: NUREG-1507, Minimum Detectable Concentrations with Typical Radiation Survey Instruments for Various Contaminants and Field Conditions (NRC 1997) and MARSSIM (EPA 1997).

Speed of probe movement by itself will affect the ability of the instrument to detect contamination due to the characteristic electronic response of the instrument. Typical speed of probe movement is between: 1) 0.5 m per second for wide area wheel mounted scanning probes used on floors and large surfaces to 2) perhaps 5 to 15 cm per second for a typical field/laboratory survey. As an example, the ability to detect contamination of 500 dpm/100 cm^2 using an alpha scintillation probe starts at over 90% for probe speeds below 5 cm per second and drops to less than 50% for a probe speed of over 10 cm per second. When the probe size (dimensions of the probe in line with the direction of scanning) drops to 5 cm, the probability of detection drops to less than 15 percent (EPA 1997). These data show that probe speed can defeat the intentions of the best calibrations and the calculated MDA's.

Other human factors can affect the sensitivity of the instrument survey in more profound ways then actual physical factors. Two of the major human factors are surveyor fatigue and noise.

15.8.6.1 Surveyor Fatigue: When an individuals are tired, they are less able to discern differences in the audible count rate of an instrument. Another result of surveyor fatigue is boredom that may lead to apathy and inattention to detail. Tasks such as repetitive surveys of laboratories and corridors of a typical research facility are classic setups for apathetic fatigue. It is essential to keep individuals motivated by rotating assignment of personnel so that this type of fatigue does not become a factor.

15.8.6.2 .Noise Since the surveyor often relies on an audible signal (increase in count rate) from a portable field instrument, the level of background noise can mask the ability of a surveyor to hear the increased count rate. In very noisy areas surveyors may require headsets or ear pieces to discern changes in count rate. It has been noted that the use of a headset increased the ability of a surveyor to discern contamination by a factor of ten (EPA1997). Likewise, if surveys are being conducted in quiet ancillary office areas/laboratories, the surveyor may reduce the volume of the instrument to avoid disturbing or upsetting office and laboratory workers and thus to decrease the sensitivity of the survey by a factor or ten.

15.8.7 Background Radiation: Effects on Sensitivity

Several similar equations have been published to derive the minimum detectable activity, or MDA, in a sample. Regardless of the equation used, the MDA is inversely related to the magnitude of background radiation levels. In addition to this simple concept, the possibility of interference with alpha contamination measurements can occur in elevated radon atmospheres where charged radon daughters affix themselves to the mylar or mica detector faces artificially raising the background over time, and therefore reducing the MDA. In a hospital/research environment, the background may also change in both portable and fixed (laboratory instrumentation) due to patients containing radioactive materials, the movement of sources, and the temporary storage of radioactive waste.

One might think of the background radiation level as analogous to a noisy room where the surveyor is attempting to hear an audible increase in count rate from a survey meter. It is important to select the typical background radiation levels carefully to enable the detection of the presence of true radioactivity and reduce false positives. A "typical" MDA quantity used for surveys with portable instruments might be set as 4 times the standard deviation of background. Application of such a calculation is not in itself complicated. What is complicated is determining which "background" to use given the change in background from other sources present in building materials, nearby radiation sources, patients, etc. It is recommended that a thorough evaluation of instrument background be made at survey locations and the results of these surveys recorded and maintained. Prior to the conduct of each survey, a measurement of instrument background should be conducted and compared to stored values for a given instrument and physical location.

For laboratory-based analyses, one would obtain samples (air, water, grass, vegetation, rain water, soil, urine, removable contamination smears, etc.) to establish the "environmental" background. In addition, samples of these representative media, as well as "blanks," which contain everything but the radioactive analyte, should be analyzed to set the MDA and Lower Level of Detection (LLD) for the laboratory analysis. This can result in a significant change from the "a priori" LLD/MDA. For example, the gross LLD for ^{32}P in a 1mL urine sample might be in the range of less than 3 cpm based upon instrument background alone. Once one takes samples of urine from a background population and performs the same analysis, the LLD/MDA is in the range of 11 dpm per ml due to the large variation in the urinary ^{40}K levels of the background population.

With this knowledge it will be possible to use accurate backgrounds to assist in establishing amore defensible and realistic MDA. A thorough discussion of the topics touched upon here can be found in: MARRSIM (EPA 1997) and NUREG-1501, Background as a Residual Radioactive Criterion for Decommissioning (NRC 1994).

15.8.8 Backscatter and Self-absorption:

Effects due to both backscatter and self-absorption could easily be considered in the category of geometric/physical effects. Self-absorption is an important consideration when lower energy beta emitters and alpha emitters are involved because the radioactive material may be imbedded in the surface of the material to be monitored. Conversely, backscatter is important when the materials are non-porous and of a high atomic number that facilitates the scatter of betas back into the detector media. The self- absorption problem is by far more important because it results in the underestimation of contamination.

These two factors play a part in both laboratory and field instrumentation calibration and efficiency calculations. In the laboratory, self-absorption is most important with low energy beta emitters, low energy photon emitters, and alpha emitters. In the measurement of high-energy beta emitters like ^{90}Sr, self-absorption is not typically a severe problem. An increase in effective absorber thickness from 0 (just the detector face) to almost 10 mg/cm^2 results in a count reduction of less than 10 % for ^{90}Sr, while an increase in 10 mg/cm^2 of sample thickness for a low energy beta or an alpha emitting radionuclide may result in an almost total loss of detection efficiency.

15.9 TRACKING AND MAINTAINING INSTRUMENT PERFORMANCE

15.9.1 Quality Assurance and Quality Control

Laboratory instrument function is typically evaluated either daily or prior to each use. The evaluation of instrument function is accomplished by a background count and a constancy source check. It should be noted that the source used for the constancy check does not have to be NIST traceable, although the same source should be used every time for each instrument. The sources used should be able to detect changes in efficiency and energy calibration.

Since most laboratory instruments use an amplifier and analog to digital converter, it is quite possible, and has been observed, that either the low or high energy calibration might be correct, while the other may have drifted sufficiently to result in non-detection of lower or higher energy photons from radionuclides that are indeed present. For example, a daily check of a HPGe detector system had been conducted for years with only a ^{60}Co source (1173 and 1332 keV photons) for constancy and energy calibration. But it was discovered that the energy calibration for photons supposed to appear at the energy of 240 KeV had shifted sufficiently to be missed. This points out the importance of instrument performance checks on a periodic basis at both the low and high energy regions. This holds true for NaI detectors and LSCs as well.

A daily background, and/or blank sample check should be run with each analysis group. This will help to identify changes in instrument background due to both electronic noise and background from stored sources, patients, as well as cross contamination of samples.

15.9.2 Control Charts

All of the quality control data are plotted on a control chart (which can be an Excel™ or Quattro Pro™ spreadsheet). Data are compared not to a single check source reading, but to a series of previous readings plotted graphically. With a little work, control charts can also be used: 1) to evaluate instrument performance according to action points such as ±2 and 3 standard deviations from the expected mean, 2) to identify departures from the expected range, and 3) to compensate for the radioactive decay of instrument check sources.

15.9.3 LLD and MDA

Much has been written on the topic of the LLD and MDA parameters. Two of the classic references for calculating these quantities are Currie, L.A., "Limits for Qualitative Detection and Quantification Determination", Analytical Chemistry, 40(3):587-593, 1968 and Altshuler, B., Pasternak, B., "Statistical Measures of the Lower Limit of Detection of a Radioactivity Counter," Health Physics 34(9):293-298. 1963.

These two papers examine the effects of instrument background on the establishment of LLD and MDA. The important thing to remember is that the LLD for a particular measurement should always be a small fraction of the appropriate regulatory limit. With modern day, normally functioning, laboratory analytical equipment with modest counting times in minutes, LLDs and MDAs are typically hundreds and thousands of times lower than regulatory limits for surface contamination, air concentrations, and "free" release limits for equipment. If you maintain your LLD's at low levels, there will be no problems being able to detect quantities of concern.

The challenge does not present itself in the control of instrumentation performance, but in the control of factors such as bioassay frequency and the time elapsed between radioactive materials use and the performance of measurement. As the time from use increases, the corresponding LLD required to detect an excursion above or near regulatory concern decreases. It is essential that these factors be examined to assure that the LLD/MDA that is chosen will meet those criteria.

15.10 RADIATION INSTRUMENTS AT ACCELERATOR FACILITIES:

Measurement of radiation at accelerators must be viewed as two separate tasks: 1) the measurement of prompt radiation, both neutron and photon, which is present at the time the accelerator is operating and 2) measurement of radiation produced by induced radioactivity. Measurement of prompt photon radiation can be successfully accomplished using ion chambers operated at high collection potentials and shielded from radiofrequency and magnetic fields. High collection potential instruments, as compared to commonly available survey instruments, may be required to reduce recombination of ions within the chamber to insignificant levels of 1-2% (ICRU 1971, Knoll 1979). The problem of recombination effects is usually experienced at locations within the shielding of the facility and is typically not a problem outside shields unless the shielding is inadequate or severe streaming through voids is encountered. An ionization chamber is a popular instrument less prone to recombination effects than other instruments such as GM detectors (ICRU 1971).

15.10.1 Radio-Frequency (RF) and Magnetic Interference

Commercially available ion chamber survey instruments typically operate at collection potentials that range from 27 to 67 volts (Keithley 1981, Bicron 1984). It is also important that the ion collection times are long compared to the radiation pulse duration (ICRU 1971). If it can be determined that the ion collection times are greater than the pulse spacing or frequency, then the measurement device should be adequate for use in a pulsed field (ICRU 1982).

15.10.1.1 RF Characteristics. The characteristic frequency of radiation pulses for a cyclotron typically range from 5 to 47 MHz (CYC 1984). It is essential to determine pulse frequency for your particular cyclotron or accelerator. Thus, collection times for accurate measurement are met if the ion collection times are much greater than 2.0E-7 seconds, representing a frequency of 5 MHz. Satisfactory results may be obtained using typical ionization chambers and exposure rate meters at a cyclotron since the conditions approximating continuous radiation sources exist. This is due to the fact that the ion collection times for most ion chambers are on the order of several milliseconds and radiation pulse durations are on the order of microseconds (Knoll 1979, ICRU 1971, Boag 1966).

15.10.1.2 Interference from Magnetic Fields. Magnetic field interference is of particular concern with older instruments that use feed switches to change scale magnitudes. This is due to unwanted auto-ranging when the magnetic field sets the range of the instrument, while the physical range switch does not appear to change. It is possible for magnetic interference to even turn the instrument off (Broseus 1985)†. The severity of malfunction depends upon the orientation of the instrument in the magnetic field. For example, rotating the instrument 90 degrees may cause the instrument to read zero, where in its original position the instrument indicated an exposure rate of 500 mR h^{-1}. Radio-frequency (RF) interference may also induce electrical currents in some survey instruments causing inaccurate and/or sporadic response (Bradley 1970). Both sources of interference can result in the misinterpretation of meter output and prevent potentially high radiation exposures from being identified. While you may assume that these problems only exist when the accelerator is "on" (producing beam), it is possible to have magnets active and the RF source off and encounter these problems. This interference could mask potentially significant sources of exposures due to induced activity in close proximity to the accelerator.

15.10.1.3 Magnetic Shielding of Instruments. Shielding for magnetic interference can be accomplished on instrumentation by using mu-metal that may allow for normal operations in magnetic fields (Rank 1985). Mu-metal, composed of nickel, iron and copper, is used for magnetic shielding because of its high magnetic permeability and low hysteresis losses. Prompt photon spectrometry accomplished using NaI detectors coupled to photomultiplier tubes (PMTs) may be severely affected due to gain shift induced by magnetic fields (Broseus 1985, Finn 1985). Recalibration of this equipment is necessary to compensate for gain shifts. In fact, changes in location and orientation of the detector relative to the magnetic fields may result in the need to recalibrate the system for each change in detector placement (Broseus 1985). Above a given magnetic field strength, PMT- based instruments can yield severely reduced or even null readings (Schlapper 84). Such instrumentation may operate normally in magnetic fields of 1 gauss and achieve extended operation to approximately 2 gauss with mu-metal shielding

---

† Broseus, R.W. Personal Communication, 1985.

(Sclappler 1984). Because of the susceptibility of PMT-based instrumentation to magnetic fields, solid state detectors are the equipment of choice around accelerators due to their resistance to such interference (ICRU 1971). The magnitude of magnetic fields very significantly with the type of accelerator, but it is not uncommon to see fields in the range of 0.006 Tesla (T) (60 gauss) to 0.03 T (300 gauss)at 5 cm from typical cyclotrons and fields of 0.0025 T (25 gauss) at a distance of one meter.

15.10.1.4 RF Radiation Effects on Instrumentation. Intense radio frequency (RF) fields can be a source of false readings if RF shielding is not provided (NCRP 1978a). The effect of RF on survey instruments is quite variable. Bradley found commercially available scintillation detectors to be unaffected; however considerable instability of ion chambers and Geiger counters was noted (Bradley 1970). Erroneous readings induced by RF appear to be most significant above 60 MHz for all instrumentation except GM counters, in which false readings of 100 mR h^{-1} were induced by 10 MHz RF fields (Bradley 1970). The frequency of RF fields typically found around a cyclotron is generally below 47 MHz, so ionization chambers will not be severely affected (CYC 1984).

The use of GM detectors to measure exposure rates is further discouraged due to their well-known dead time characteristics. Because of the side photon energy spectrum around a cyclotron and the tendency of pulse oriented instrument, like the GM detector, to recombination effects and energy dependence, GM counters have the possibility to introduce large measurement errors in such an environment (ICRU 1971, NCRP 1978a). In addition, due to the very high exposures possible due to induced radiation, a specialized instrument commonly called a "Teletector^{tm}" can be successfully used to measure exposure rates in excess of 1000 R h-1. A teletector consists of a small GM detector (usually a high and low range detector) at the end of a pole extension of about 3.5 meters long. This enables the health physicist to measure high exposure rates without getting close to the source of exposure.

15.10.2 Use of Ion Chambers for Accelerator Surveys

15.10.2.1 RF Shielding of Instrumentation. :Since ionization chambers are most frequently used and available to measure exposure rate, the remainder of this review will focus on instruments of this type. Commercially available instruments incorporating RF shielding are generally more expensive and not commonly available. The HP may be forced to modify an instrument to enable measurements in a severely affected environment. Inexpensive methods, such as the use of metalized marquisette fabric (a metal coated plastic mesh cloth), have been found to provide attenuation of up to 40 db for RF radiation of 350 to 1000 MHz (Reynolds 1961). A bag of the metalized cloth material is created and the instrument is placed inside. This is a very workable solution, as the instrument may be viewed and operated within the cloth bag without difficulty. Even though the frequencies used to test the cloth's shielding effectiveness were much higher than the RF encountered at a typical cyclotron, instruments shielded with the metalized cloth behaved in a manner comparable to commercially shielded instruments (Properzio 1970). One advantage of using the metalized cloth, is that energy response of the instrument is not affected. The effect of the shielding material on instrument response must be considered and accounted for.

15.10.2.2 Energy Dependence. Other factors to consider in the measurement of prompt photon radiation include energy dependence and angular response. The accurate measurement of exposure rate requires that electronic equilibrium be established. Due to the high energy of prompt photons at accelerators, it is often necessary to correct for non-equilibrium conditions to correctly assess exposure and absorbed dose (ICRU 1971). Thick buildup caps must be used to yield accurate measurement (ICRU 1964). A buildup cap of 1.5 g cm^{-2} should be sufficient to establish equilibrium for photons in the 1 to 5 MeV range, but may result in under response if a substantial component of the field is of lower energy. It should be noted that commercial ion chambers are supplied with buildup caps that are substantially thinner than 1.5 g cm^{-2}.

Photons detected outside the shielding barriers will generally be of less energy than that of incident photons and the energy of the accelerated particles. In most cases, behind thick shields, the necessity for such buildup caps is not required due to attenuation and buildup within the shield (Patterson 1973). However, in the proximity to shielding penetrations and voids, equilibrium conditions may not exist and caution must be exercised in interpretation of readings.

Photon radiation associated with induced activity at accelerators falls within the range of measurement that the ICRU considers to be "not normally a problem"(ICRU 1971). Characterization of delayed or induced photon radiation around accelerators, when not complicated by RF or magnetic fields, is relatively simple. Activation products at accelerators are generally positron emitters that have characteristic annihilation photons of 511 keV, as well as photons having energies between 0.04 and 2.0 MeV. Within this energy range a commercially available instrument with a wall thickness of 0.4 g cm^{-1}, sufficient to establish electronic equilibrium, will usually be accurate to within 10 to 15% (Keithley 1981).

Measurement of exposure rates from induced activity are made when the accelerator is typically off. One might assume that RF or magnetic interference would not be a concern. But even with the equipment off, residual magnetic fields can still affect measurement accuracy. It is suggested that instrument performance be thoroughly tested and measurements taken with the accelerators magnet on and off and with the instrument in different spatial orientation (rotations of angle and position) and the behavior documented. Instruments have been observed to fail completely, giving null, zero or significantly reduced readings at beam ports, or along beam lines near strong sources of magnetic fields.

15.10.3 Radiation Environment at Accelerators

15.10.3.1 Radiation Due to Induced Activity. It is important to recognize what the potential magnitude of exposure rates surrounding an accelerator may be, so that instrumentation with the proper range is selected. The following table lists some exposure rates that may be encountered outside the accelerator itself as well as removed components/devices associated with the accelerator.

Component	Exposure Rate C/kg h^{-1}	Exposure Rate R h^{-1}
Targets	0.258 +	Up to 1000+ on contact (Barbier 1969)
Deflector/Septum	0.009 to >0.0258	3.5 to > 100 (Jacobi 1982)
Beam Probes	0.0001 to 0.0006	0.5 to 2.5
Hill Sector Edge	5.13 E-3	20 (Smith 1985)
Ion Source	1E-4 to 6E-4	0.5 to 2.5
Dees/Dee Stem	8E-4 to 2.58E-3	3.0 to 10 (Burgenton 1985)
Faraday Cups	1.81E-3	7 or Greater (Jacobi 1982)
General Area outside Vacuum Tank		
TRIUMPF	5.1E-6 to 2.58E-4	0.02 to 1 (Burgenton 1985)
Naval Research Lab	1.29E-6 to 2.58E-6	0.005 to 0.01
Johns Hopkins	1E-4 to 2.54E-4	0.5 to 1 (Broseus 1985)
Mt. Sinai	1.0E-4 to 1.1E-4	0.5 to 4.4 (McLeod 1982)
UCLA	1.29E-6 to 2.58E-4	0.005 to 0.1 (Bi82)
NIH, CS-30	5.1E-6 to 1.3E-2	0.02 to 5.0 (Googins 1986)
NIH, JSW	2.58E-7 to 2.58E-5	0.001 to 0.1 (Googins 1986)

Some common beta emitting radionuclides found in accelerators:

Radionuclide	Avg BetaEnergy (keV)	% Emission	Photon Energy Max (keV)	% Emission
^{11}C	386	99.7	511	198
^{13}N	491	99.8	511	200
^{15}O	735	99.9	511	200
^{18}F	249	96.7	511	193
^{22}Na	215	89.8	511, 1274	179, 99.9
^{65}Cu	278	17.8	511,1345	35.6, 0.49
^{65}Zn	143	1.4	511, 1115	2.8, 50.8
^{60}Co	95	100	1173, 1332	100,100
^{56}Co	607	19.7	846, 1238, other	99.1, 67
^{183}Re	Many conversion electrons.	100	59,67,162, 291,many.	59, 25, 23,3, etc.
^{184}Re	Many conversion electrons	100	59, 216, 252, many.	14.8, 9.6, 10.9, etc.

15.10.4 Beta Monitoring

15.10.4.1 Importance of Beta Radiation Measurement. Monitoring of beta radiation is especially important around a cyclotron or accelerator. This is due to the presence of significant beta radiation from targets, activated accelerator components, and radiopharmaceuticals (Wells 1985‡, Plascjak 1985§). Severe radiation burns can result if the individual is unaware of the hazard.

Measurement of beta radiation fields in and around an accelerator facility can be complicated by magnetic fields, RF and geometric conditions. The RF and magnetic concerns are the same as previously mentioned for photon measurements and will not be discussed further.

Additional complications are introduced by the misinterpretation of ionization chamber readings. Ion chambers are generally supplied with an end wall shield to establish electronic equilibrium, and such caps/shields are often used to determine penetrating/non-penetrating (photon/beta) components of a radiation field. The problem arises because ion chambers have been shown to underestimate the beta radiation dose by easily as much as a factor of 100 (Selby 1983).

Specialized beta measurement instrumentation is now available, but expensive. The alternative is to determine a calibration factor for beta dose so that available ion chambers can be used to sufficiently estimate dose rates. Thermoluminescent dosimeters can be deployed in conjunction with ion chamber readings to develop calibration factors for this purpose. Calibrations for a variety of commercially available instruments have been conducted (Walker 1983). It is essential when these, and other, calibration factors are used that they are applied with caution and with consideration of possible geometry corrections.

15.10.5 Neutron Monitoring and Instrumentation

Measurement of the prompt neutron dose equivalent around accelerators and outside of shields may be accomplished through a variety of methods. However, each device has specific measurement limitations that must be known.

Two commonly used instruments are known as a neutron rem-meter, designed by Hankins, and the multi-sphere method developed by Bonner. The rem-meter can be used to measure dose equivalent directly with a BF_3 tube inside a polyethylene sphere of 12 inch diameter (Hankins rem-meter) (Hankins 1963).

---

‡ Wells, K. Personal Communication, 1985.

§ Plascjak, P. Personal Communication, 1985.

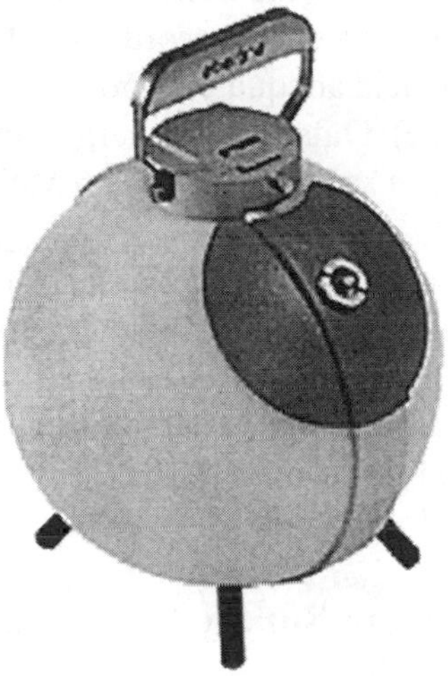

The LiI(Eu) based instrument can be used for the determination of neutron energies and intensities by placing the detector inside of different diameter polyethylene balls (Bonner Spheres). This information can be used for the calculation of dose equivalent. The principle behind neutron detection using a ^{10}B enriched BF_3 tube is the capture of thermal neutrons via the high thermal neutron cross section $^{10}B(n,\alpha)^7Li$ reaction. The alpha particle created in the reaction ionizes the gas within the detector and creates a pulse that is measured. The BF_3 tube is surrounded by a moderating material (polyethylene sphere) which thermalizes fast neutrons so that they may be detected by the boron loaded gas in the tube. Because the detector functions by detecting a pulse created by an alpha particle, this instrument exhibits excellent photon rejection in radiation fields of up to 200 Rh^{-1} (ICRU 1971). In addition, this instrument has excellent resistance to magnetic fields and appears to function normally at magnetic field strengths of up to 70 Gauss (0.007 Tesla).

The Hankin's type of rem-meter is intended to give a reasonable dose equivalent response for neutrons from thermal to 7 MeV neutrons (Hankins 1963). In addition, the BF_3 based instrument demonstrates very little directional dependence, unlike a long counter or Snoopy type neutron monitor. Combination of several types of errors can easily result in the overestimation of dose equivalent from 3-15 times the actual values. It may be thought that over response is better for protection purposes, but it may result in the imposition of undue work restrictions, shielding, and undue cost.

Another factor to consider is the energy of the neutron field being measured compared to conditions present during instrument calibration. Because of the energy dependence of neutron dose equivalent calculations, the instrument readings may not adequately reflect the true dose equivalent. According to Hankins (Hankins 1963), most rem-meters have been shown to yield adequate response from 100 keV to 6 MeV (with a maximum of + 65% overestimate). Outside this energy range at accelerator installations where large portions of the neutron spectrum exceed 6 MeV, the rem-meter response is no longer proportional to dose.

The use of the $^{6}Li(n,\alpha)$ reaction is used similarly to the reaction in the BF_3 tube except that the ionization created by the alpha particle is converted to light and then to an electronic pulse in the photo-multiplier tube. A Li based system will not function as well as a BF_3 tube based instrument around an accelerator due to the susceptibility of photo-multiplier tubes to magnetic fields. However, Li-based systems can be operated successfully in magnetic fields of up to 1.0E-4 Tesla (1 gauss) (Sc84).

15.10.6 General Contamination Survey Instruments

The instrument of choice for contamination surveys at an accelerator facility is a pancake GM probe equipped meter. Typically any standard GM detector system is appropriate for use at an accelerator facility. Because of the possible high background due to the presence of either activated accelerator structures, or target material, or radiopharmaceuticals in production, it may be necessary to use a probe which is shielded by either Pb or W (HP210T probe for example). The use of NaI-based survey meters will not be successful, particularly within the cyclotron vaults, due to the high background that can mask surface contamination.

15.11 SUMMARY AND CONCLUSIONS

This chapter has attempted to provide a broad review of radiation instruments needs in university and research laboratories and in accelerators. The information provided is intended for practical guidance for all radiation safety staff. In particular, we have addressed challenges involving choices of instruments, how to verify proper instrument function, how to use instruments properly and how to evaluate the quality of instrumental data. Both portable and laboratory instruments were reviewed for contamination monitoring, exposure rate measurements, measurements for package shipping and receiving . area monitoring, and personnel monitoring. For each type of instrumentation, we have also reviewed the limitations that could affect the quality of measurements.

For reliable radiation measurements, the appropriate instrument should be used or the limitations of available equipment should be recognized. Instrument users need to understand the effect of energy dependence and geometry on measurement results. Other factors that should be known include interferences such as quenching effects for liquid scintillation counting or the effects of radio-frequency and magnetic fields on portable instruments used in accelerators. The best radiation measurements are achieved by reproducing all of the calibration conditions as closely as possible.

5.12 REFERENCES

Altshuler, B,. Pasternak, B. "Statistical Measures of the Lower Limit of Detection of a Radioactivity Counter", Health Phys 34:293-298. 1963.

Barbier, M.M. Induced Radioactivity, Amsterdam, North Holland, 1969.

Bicron Corporation, RSO-50 Ion Chamber Survey Meter Specifications, Bicron Corporation, Newbury, OH, 1984.

Bicron Corporation, RWS-4 Specifications Sheet, Bicron Corporation, Newbury, OH, 1999.

Boag, J.W. 1966, Ionization Chambers, In Radiation Dosimetry, Vol II, Chapter 9,Academic Press, New York, NY.

Bradley, F.J., Jones, A.H. RF Response of Radiation Survey Instruments, In : Electronic Product Radiation and the Health Physicist, BRH/DEP report 70-26, Proceedings of The Health Physics Society Fourth Annual Midyear Topical Symposium, 1970.

Brinkman, G.A. Dose Rates During Isotope Production, International Journal of Applied Radiation and Isotopes, 33, 109-115, 1982.

Burgenton, J.J. Cyclotrons for Radionuclide Production, Proceedings of the Eight International Conference of Accelerators in Research and Industry, Denton, TX, Nuclear Instruments and Methods in Physics Research, B10/11, 956, 1985.

Cember, H. Introduction to Health Physics, 3rd edition. McGraw Hill, NY, NY, 1996.

Currie, L.A. Limits for Qualitative Detection and Quantification Determination, Analytical Chemistry, 40(3):587-593, 1968.

Dunkelberger, II, R.O. Some tips for Operational Health Physicists, Operational Radiation Safety, Health Phys, 76(Supplement 2): S67-S70, 1999.

Environmental Protection Agency 402-R-97-016, Multi-Agency Radiation Survey and Site Investigation Manual, MARSSIM, 1997.

Erdman, M.C., Miller, K.L., Achey, B.E. Experience With a Medical Waste Portal Monitoring System, Operational Radiation Safety, Supplement to Health Physics, Vol 80, S13-S15, 2001.

Googins, S. Cyclotron Health Physics and a Proposed Radiation Safety Manual for a

Cyclotron Facility used for Radionuclide Production, Masters Thesis, Georgetown University, 1986.

Hankins, D.E. New Methods of Neutron Dose Rate Evaluation, Neutron Dosimetry, Vol II, International Atomic Energy Commission, 1963.

Hidex Inc, Specifications for the Triathler Multilabel Tester, http://www.hidex.com/moreinfo.html., 1999.

International Commission of Radiological Protection, Limits on Intakes of Radionuclides by Workers, ICRP Publication 30, Part I and II, Pergamon Press, New York, NY, 1979.

International Commission of Radiation Units and Measurements, The Dosimetry of Pulsed Radiation, ICRU Report 34, International Commission on Radiation Units and Measurement, Washington, DC, 1982.

International Commission of Radiation Units and Measurements, Radiation Protection Instrumentation and Its Application, ICRU Report 20, International Commission on Radiation Units and Measurements, Washington, DC, 1971.

Jacobi, L.R., Sandel, P.S. Health Physics Considerations of Cyclotron Maintenance, Communication, L.R. Jacobi to R.W. Broseus, 1982.

Johnson, R.H., Jr., Radiation instrument training for new RSOs – defending the quality of measurements. In: Instruments, Measurements, and Electronic Dosimetry. Proceedings of the Health Physics Society 33rd midyear topical meeting, Virginia Beach, VA , Jan. 30 – Feb. 2, 2000. Medical Physics Publishing, Madison, WI, 2000.

Keithley Instruments, Survey Meter Users Manual, Keithley Instruments, Inc., Cleveland, OH, 1981.

Knoll, G.F. Radiation Detection and Measurement, Wiley and Sons, New York, NY, 1979.

McLeod, T.F. Communication, T.F. McLeod to R.W. Broseus, 1982.

NE Technology Inc, Specifications for the NE HFM7 Hand and Foot Monitor and Ga4 Area Monitor, http://www.netechnology.co.uk., 2001.

National Council on Radiation Protection and Measurements, 1991, Report 112, Calibration of Survey Instruments Used in Radiation Protection for the Assessment of Ionizing Radiation Fields and Radioactive Surface Contamination.

National Council on Radiation Protection and Measurements, 1978, Report 58, A Handbook of Radioactivity Measurement Procedures, NCRP, Washington, DC.

NUREG-1507, Minimum Detectable Concentrations with Typical Radiation Survey Instruments for Various Contaminants and Field Conditions, U.S. Nuclear Regulatory Commission, 1997

NUREG-1501, Background as a Residual Radioactivity Criterion for Decommissioning, U.S. Nuclear Regulatory Commission, 1994.

NUREG/CR-4884, Interpretation of Bioassay Measurements, Brookhaven National Laboratory, Upton, LI., NY. BNL-NUREG-52063.

U.S. Nuclear Regulatory Commission, Regulatory Guide 8.23, Radiation Safety Surveys at Medical Institutions, Washington, D.C., 1981.

Properzio, W.S. An Inexpensive RF Shield for Use with Radiation Survey Instruments, Health Physics, 19, 442, 1970.

Patterson, H.W., and Thomas, R.H. Accelerator Health Physics, Academic Press, New York, NY, 1973.

Reynolds, M.R. Development of a Garment for Protection of Personnel Working In High Power RF Environments, Biological Effects of Microwave Radiation, Plenum Press, NY, 71, 1961.

Schlapper, G.A., Kay, D.C., Neft, R.D. Dose Equivalent Measurements in The Area of Reduced Shielding at the Texas A&M Variable Energy Cyclotron, Radiation Protection Management, 1(4) 57-46, 1984.

Selby, J.M. Field Measurement and Interpretation of Beta Doses and Dose Rates, In Proceedings of the International Beta Dosimetry Symposium, Washington, DC. Feb 15-18, 1993. Available as NUREG/CP-0050, USNRC, 1983.

Smith, L., 1985, New England Nuclear Inc,. Personal Communication.

Tenth Annual International Conference on Cyclotrons and Their Applications, 1984, Conference Proceedings, Michigan State University, Ann Arbor, MI, 1985.

U.S. Department of Transportation, Title 49 Code of Federal Regulations, Parts 100-185, Transportation, Subpart H. 1999.

Walker, E., Jacobs, R., The Response of Selected Survey Instruments to Various Types and Energies of Beta Radiation, Proceedings of the International Beta Dosimetry Symposium, Washington, DC. Feb 15-18, 1993. Available as NUREG/CP-0050, USNRC.

Chapter 16

RADON MEASURING TECHNIQUES

Bernard L. Cohen

16.1 INTRODUCTION

^{222}Ra decays with a 3.8 day half life by 5.5 MeV alpha particle emission into ^{218}Po, initiating a series of short half life decays of what are called "radon daughters" (or "radon progeny"). Succeeding decays and their principal particle emissions are (half lives in parentheses) ^{218}Po (3.05 minutes), 6.0 MeV alpha; ^{214}Pb (26.8 minutes), beta + 295 KeV and 352 KeV gammas; ^{214}Bi (19.9 minutes), beta + 609 KeV gamma; ^{214}Po (160 microseconds), 7.69 MeV alpha. Measuring techniques depend on detecting these emitted particles. There is a vast literature on these techniques for measuring concentrations of radon gas and its radioactive progeny, but we consider here only the most commonly used methods.

Health effects depend on working level and unattached fraction - these are defined and discussed in Appendix A - but the latter quantity is quite difficult to measure. Since radon progeny ordinarily attach to dust particles, the working level in a house can be greatly reduced by removing dust from the air through inadvertent circumstances (e.g. large areas of sticky surfaces) or by using electrostatic precipitators, ion generators, or other such devices. But the paucity of dust particles then results in fewer loci to which newly formed radon progeny atoms can attach and hence a higher unattached fraction. This is important because inhaled radon progeny attached to dust particles have only about 1-2% probability of sticking to the bronchial surfaces, whereas an unattached progeny atom has about a 50% probability and hence has a much greater health impact. Without measuring unattached fraction, working level is therefore not a reliable indicator of health impacts. The concentration of radon gas, however, is not dependent on dust level. Roughly, the effects of dust on working level and unattached fraction cancel each other, leaving health effects dependent, to a large extent, only on radon gas concentration; it is therefore what is ordinarily measured.

The situation is different in mines because dust levels are always so high that unattached fractions are negligible, leaving health effects dependent only on working level; it is therefore what is normally measured.

16.2 SCINTILLATION CELLS

The simplest method for measuring the instantaneous concentration of radon in air -- called "grab sampling" -- is by use of a scintillation cell (often referred to as a

"Lucas Cell"), a cylindrical chamber with a transparent window on one end and all other internal surfaces painted with ZnS(Ag) scintillator; the window is optically coupled to a photomultiplier tube with output fed into an amplifier and pulse height analyzer for electronic sorting of events by pulse height. The cell is evacuated and then the air to be studied is allowed to enter and fill the cell. The alpha and, less importantly the beta and gamma, particles from decay of radon and its progeny strike the scintillator lining the walls and the resulting flashes of light are detected by the photomultiplier tube and counted if their pulse heights are appropriate. With a well designed cell, the pulse height spectrum has a deep minimum between the low pulse height noise and the true pulses. The detection threshold should be set well above this minimum to minimize noise. The sensitivity of the cell must be periodically calibrated with a sample of known radon concentration obtained from a radon chamber. Since the radon progeny are largely removed as the air enters the cell, counting must be delayed for about 90 minutes for equilibrium between radon and its progeny to be established.

In some situations, the scintillation cell is treated as a portable unit; it is evacuated, carried to the site where the air is to be tested, and the valve is opened allowing the air to enter. Alternatively, the air can be collected in a gas sampling bag by use of a small battery operated pump, and later transferred into the evacuated scintillation cell. This allows the scintillation cell to remain in position in the laboratory, avoiding problems of coupling the scintillation cell to the photomultiplier and maintaining light tightness. If the sites to be tested are remote, several samples to be counted in a single cell can be collected in sampling bags before returning to the laboratory. Of course, the cell must be thoroughly evacuated between each measurement and the 90 minute waiting period must be observed.

In our laboratory, cells 15 inches long and 4 inches in diameter have been produced and used without difficulty, but commercially available cells are generally much smaller and hence less sensitive. With a large volume cell and sufficient counting times, radon levels as low as 0.37 Bq/m^3 (0.01 pCi/L) can be reliably determined by this method. (The USEPA limit is 4 pCi/L)

Scintillation cells are also useful for continuous monitoring by flowing air through the cell and printing out count rates. Of course such a system does not respond to very rapid changes in radon concentrations, but typical responses to such changes have a time constant well under one hour.

16.3 CHARCOAL ADSORPTION COLLECTORS

The most frequently used method for measuring concentration of radon gas in homes is adsorption in charcoal. In its simplest form, a container of charcoal is exposed to the air for a few days, sealed, and sent to a laboratory where the amount of adsorbed radon is measured. The quantity of radon adsorbed from surrounding air approaches equilibrium with a time constant of about 18 hours, so this method determines the average radon level in air during roughly the last 18 hours of exposure. A longer time average can be achieved by having the air enter and leave the charcoal container through a diffusion barrier -e.g. a relatively small hole covered with a filter paper -- which typically extends the time constant to about 85 hours. The laboratory measurement of captured radon can be made either by gamma ray counting of the characteristic gamma rays of the radon progeny (295,352, and 609 MeV) with a well shielded (several inches of lead on all sides) sodium iodide scintillation crystal (or Ge(Li) detector) feeding into a multichannel pulse height analyzer, or by transferring the radon into a liquid scintillation fluid which can then be counted with commercially available sample changers. In large scale routine programs, radon levels as low as $11 Bq/m^3$ (0.3 pCi/L) can bc rcliably measured with these methods.

Radon levels in a house can fluctuate substantially with time. Even with averaging over 18 hours, variations by a factor of 5 have been observed and variations of about 30% occur several times each month, with no recognizable explanation. Variations with shorter averaging times are much larger, and some houses have regular day-night or morning-evening variations. Since measurements with an open containcr arc largely controlled by the last 18 hours of exposure, they are susceptible to errors of that magnitude. This uncertainty is substantially reduced by use of the diffusion barrier. Elaborate studies have found that a single measurement with a diffusion barrier collector has a standard deviation of 15% from the true one month average; for an open container, the standard deviation is 30%. Another advantage of the diffusion barrier is that it can be made insensitive to humidity by inclusion of a desiccant bag in the barrier. With an open containcr, humidity can heavily influence the adsorptivity of the charcoal and must be corrected for by measuring the weight gain during exposure which requires weighing the collector before and after exposure. A disadvantage of the diffusion barrier is that it reduces the count rate in the measurement, which requires about a four times longer counting time to compensate.

The radon and moisture can be baked out of a charcoal collector by heating it for several hours above 100 degrees-C. This can be done several times without affecting its efficiency for collecting radon. This, of course applies only where measurements are made with gamma ray counting; when solvent extraction into an organic liquid scintillator is used, the charcoal must be discarded

Studies have shown that the sensitivity of charcoal collectors to temperature variations are small, about 0.3 % per degree-C for diffusion barrier collectors and perhaps a few times larger for an open container. Exposures to organic vapors which compete with radon for adsorption sites in the charcoal can be a problem. Studies indicate that if the odor of the solvent is still clearly evident when the container is unsealed, then the radon adsorptivity was reduced by about 15%, but it would be highly unusual for people to live in a house with strong odors persisting for several days.

The radon collection efficiency differs for different types of charcoal. Even different batches of the same brand of charcoal can have varying efficiencies. In our tests, we found an average of about 5% variations between batches ordered at different times from the same supplier.

16.4 ALPHA TRACK DETECTORS

Devices are available that eliminate the problem of fluctuations with time by using very long exposures not limited by the 3.8 day half life of radon. The most common method is to use alpha track detectors, consisting of a plastic sheet, typically 1 cm x 1 cm, made of a special material, most frequently one called CR-39. Alpha particles from decay of radon and its progeny leave damaged atoms along their paths which are etched out in a caustic solution to make the paths visible under a microscope. These visible paths, actually better described as "pits", are then counted to determine the concentration of radon in the surrounding air. The sample of air being studied is determined by the container which typically has the CR-39 sheet on the bottom and a hole covered by a filter on top to allow air to move in and out This method typically requires exposures of at least one month to obtain sensitivity to radon levels below 37 Bq/m^3 (1.0 pCi/L). Exposures of several months to one year are usual.

The alpha track method is subject to many potential problems and is not suitable for an amateurish approach. The plastic sheet must be stored prior to use at cool temperatures and handled so as to minimize exposure to air prior to the intended exposure (all air contains some radon). Leaching conditions must be carefully controlled, including temperature and time. Distinguishing between true tracks and background imperfections is not always easy. Counting the number of tracks per unit area was originally done by humans whose fatigue and boredom were often matters of concern. In recent years, automatic scanners were developed but these have added difficulty in recognizing the difference between true tracks and background - humans are aided in this by focusing in and out of the surface to get a depth perception of the track, and they are also more capable of pattern recognition. Pattern recognition is especially important because all tracks are not the same, depending on the energy of the alpha particle and its angle of entry, both of which are subject to very wide variations; alpha particles can originate from anywhere in a sizable volume of air losing varying amounts of energy before striking the detector. Alpha track detectors have the advantage of being completely insensitive to humidity and temperature.

16.5 ELECTRET DETECTORS

The third most common method for measuring time-averaged radon gas concentration is with an electric field set up by an electret, the electrical analogue of a permanent magnet. These devices are called "e-perms". Alpha and beta particles from radon and its progeny create ions in the air which are accelerated by the force from the electric field. This drains energy from the electric field, which shows up as a lowering of the voltage across the electret. By measuring this voltage before and after exposure, the number of ionizing events is determined which is calibrated to give the radon concentration. Depending on the geometry of the electret, these devices can integrate radon exposure over time periods varying from hours to months. They can be used for several measurements, with the final voltage for one measurement being the initial voltage for the succeeding measurement, until the voltage gets too low (its energy has been drained away), at which point the electret must be replaced, or recharged.

The sensitive air volume is determined by the areas of two parallel plates and the distance between them. Ions created by passage of alpha or beta particles through this volume are accelerated toward the plates and thus draw energy from the electric field. Before the exposure is started, the plates are very close together so there is essentially no air, and therefore no ions created, between them. To start the exposure, the plates are pulled apart, setting up the sensitive volume of air.

Manufacture of the electrets required very extensive development and the process is patented and proprietary. The e-perm method has advantages where measurements are made by professionals who can measure the voltage decrease on the spot in less than one minute, and then re-deploy the detector. The disadvantage is that the equipment is expensive. A charcoal collector or an alpha track detector cost less than one dollar and hence can be mailed to the householder who handles the exposure and mails it back to the laboratory -- if he fails to mail it back, the loss is negligible.

16.6 WORKING LEVEL MEASUREMENTS

Measurement of working level is most easily done by pumping air through a filter to collect the radon progeny. The radiations from the filter are then measured as a function of time by use of a scintillation detector, by alpha particle spectroscopy with detectors capable of separating the emissions from ^{218}Po and ^{214}Po, or by coincidence counting of alpha particles from one side of the filter and beta particles from the other side (the beta from ^{214}Bi is essentially in coincidence with the alpha particle from ^{214}Po). Simple and rapid methods are available for converting these measurements into working level determinations.

The quantity of air that has been pumped through the filter can be measured with a flow meter (like a rotameter), or by directing the exiting air into an inverted vessel initially filled with water and noting the volume of water displaced. A simple scintillation method of detection is to prepare a thin foil with ZnS(Ag) painted on one side, placing that side on the window of a photomultiplier tube, and placing the filter with its collected radon progeny on the back side. (The "paint" is easily prepared by mixing some ZnS(Ag) powder with Duco cement thinned with acetone to form a slurry.) The alpha particles passing through the thin foil generate scintillations which are then counted with the photomultiplier using suitable pulse height discrimination to eliminate background. By following the decay over time and fitting the data to a sum of the half lives involved, or by simply measuring the count rate after a few different time lapses (Kusnetz method), the working level may be calculated.

16.7 RADON IN WATER

The simplest method for measuring radon concentrations in water is by solvent extraction of the radon from the water into a liquid scintillator followed by counting of the latter in a commercial sample changing apparatus.

16.8 CALIBRATION CHAMBER

All of the measuring techniques discussed above require calibration which is carried out in a radon chamber. This consists of a closed box with air containing radon flowing through; there must also be facilities for taking devices into and out of the box, for monitoring the radon level in the box as with a flow-through scintillation cell, and for viewing the inside of the box through a window. The radon is generated by a radium source, with air pumped past it to produce the radon-laden air. There are also generally provisions for controlling and/or measuring the temperature and humidity.

16.9 CONCLUSION

There are four basic methods for detecting and measuring airborne radon and its progeny:

- Grab Sampling: Capturing an air sample, usually with a Lucas scintillation cell, and then analyzing it in the laboratory. This gives a "snapshot in time" of airborne radon concentration, and thus is not representative of average conditions.

- Charcoal Canisters: This is the most commonly used method because it is simple, acceptably accurate, and inexpensive. Charcoal in a canister is exposed to the air for several days, during which time airborne radon is absorbed by the charcoal. At the end of the sampling period the canister is sealed, and the radiation for the radon progeny is measured in the laboratory. The mean radon concentration is then determined from these data.

- Alpha Track Detectors: These are used to obtain average radon levels over long time periods – up to one year. A piece of calibrated plastic is exposed to the air, and the alphas from the airborne radon and its progeny produce microscopic damage to the plastic. The damage, which is proportional to the total radon exposure, is analyzed in the laboratory. The resulting data are used to infer the mean radon concentration during the exposure time.

- Electret: This device is somewhat analogous to an electroscope. An electric field is established in a chamber by charging two parallel plates to some voltage. The radon and its progeny in the air that fills the chamber ionize the air, and the ions are attracted to the plates, neutralize the charge, and thus decrease the voltage. The decrease in voltage is measured in the air that passed through the chamber. The result of the measurement gives the average radon concentration during the measurement time, which can be as long as several months.

16.10 APPENDIX A: RADON PROGENY AND THE WORKING LEVEL UNIT

The 3-min half-life of ^{218}Po and the 27 and 20-min half-lives of ^{214}Pb and ^{214}Bi are sufficient to allow a considerable separation of these from radon gas and from one another, especially in view of their rather different physical and chemical properties. In typical air sampling situations, the activity ratio to ^{222}Rn is close to 1.0 for ^{218}Po (because of its short half-life), 0.6 for ^{214}Pb, and 0.4 for ^{214}Bi and ^{214}Po. These ratios are subject to very wide variations depending on rates of air movement and exchange, and the availability of deposition surfaces. In extreme cases, the ^{218}Po/^{214}Pb/^{214}Bi ratios have been found to be as low as 1/0 05/0.02, whereas in stagnant open air during an atmospheric temperature inversion they approach 1/1.0/1.0. Since they have relatively high ventilation rates, the ratios in mines are generally somewhat lower than in open air; 1/0 5/0.25 is perhaps typical.

When radon progeny atoms are originally formed with the emission of an energetic alpha or beta particle, orbital electrons are knocked loose so the atoms are electrically charged. In this condition, they readily attach to surfaces, including dust particles in the air. All but a small fraction, called the "unattached fraction", become attached to particles large enough to leave their behavior uninfluenced by the electric charge. The unattached fraction is an important determinant of the probability for radon progeny to stick to surfaces of the bronchi when inhaled, and hence it heavily influences radiation dose and consequent health impacts.

For dosimetry and health effect purposes, the radon progeny are of much more importance than radon itself, so it would not seem to be adequate for these purposes to specify concentrations of radon gas in Bq/m^3 or pCi/L. The most accurate procedure would be to specify the concentrations of each of its progeny, but this would be both difficult to measure and cumbersome to use. As a compromise, the working level (WL) unit, developed for use in mines (originally, the maximum allowable level was 1.0 WL), is widely used. The WL is defined as any combination of radon progeny in I liter of air that will result in the ultimate emission of 1.3×10^5 MeV of alpha particle energy, which is numerically equal to the alpha particle energy released by the progeny in equilibrium with 3700 Bq/m^3 (100 pCi/L) of radon gas.

To show this, it may be noted that 100 pCi is equivalent to 3.7 decays per sec which is the initial decay rate for each member of the series in equilibrium. The number of atoms of each progeny is then 3.7 times the half-life in seconds divided by In 2; this is 980 for ^{218}Po, 8600 for ^{214}Pb, 6300 for ^{214}Bi, and <<I for ^{214}Po. The latter three are destined to decay with the emission of 7.68 MeV of alpha energy each (the energy of the ^{214}Po alpha), so their energy releases are 6.6×10^4, 4.8×10^4, and 0 MeV respectively; while the ^{214}Po atoms are destined to decay with emission of both their own alpha emission, which is 6.00 MeV and the 7.68 MeV alpha to give a total alpha energy release of 1.3×10^4 MeV. When all of these energy releases are added, the sum is 1.3×10^5 MeV. If all of the progeny products were in equilibrium, a radon concentration of 3700 Bq/m^3 (100 pCi/liter) would thus give I WL. However, in nearly all situations, the progeny are present in much less than the equilibrium concentration, so I WL is typically equivalent to something like 200 pCi/L of radon gas. The WL is often the quantity of interest, for if a particle deposits in the lung, all of its eventual alpha particle decay energy will be released at that point. Note that radon gas itself does not deposit in the lung, so its concentration is not directly relevant. The ratio of WL to radon gas concentration in units of 3700 Bq/m^3 (100 pCi/liter) is called the equilibrium factor, or the F-value.

The unit of integrated exposure is the working level month (WLM) which is the exposure at a level of I WL for 170 hr (a working month in a mine).

Chapter 17

EVERYTHING BUT THE COUNTING STATISTICS: PRACTICAL CONSIDERATIONS IN INSTRUMENTATION AND ITS SELECTION AND USE

Ronald L. Kathren

17.1 INTRODUCTION

17.1.1 Caveat Emptor

Proper instrumentation is essential to obtaining credible radiation measurement, the selection and use of suitable instrumentation is fraught with difficulties and pitfalls. Although a wide choice of instruments is available, published instrument specifications are quite often incomplete or misleading, often making the choice of a proper instrument a somewhat 'iffy' proposition, so much so that this chapter might have better been entitled *"Caveat Emptor,"* the Latin phrase used in law that translates to "let the buyer beware" and summarizing the rule that purchaser, or more broadly the user of an instrument system, must examine, test, evaluate and judge and for himself. Thus, the intent of this chapter is to provide awareness of and some practical insights and advice on factors to consider with respect to the purchase and use of radiation measuring instrumentation. In practice, there is but a single basic question confronting the person making or using radiation measurements:

How much confidence do I have in (or, how reliable is) my measurement? Implicit in this question is the idea that measurement -- quantification as opposed to detection -- of radiation, unlike measurement of the everyday physical quantities such as length, volume, time and speed, is not necessarily a simple and straightforward procedure. Indeed, the basic question gives rise to four subquestions which in themselves delineate the nature of the radiation measurement problem:

1. What type(s) of radiation(s) is (are) present?
2. Which one(s) do I wish to measure?
3. What levels do I wish (or need) to measure?
4. Will the presence of other radiations or other conditions interfere with my measurement?

It is the answers to these four derivative subquestions that largely determine (or should determine) the choice of instrumentation.

17.1.2 On Speaking the Same Language

Over the years, a degree of casualness has developed in the day to day usage of scientific and technical terms used by health physicists. The language of health physics, although quite precisely defined in the literature, may have different meaning to different practitioners of the profession, particularly in everyday spoken usage. Thus it is important to briefly and nonrigorously redefine, or perhaps more accurately, characterize, certain words and concepts basic to measurement of radiation, providing not only a refresher but ensuring a common basis that will hopefully minimize confusion and misunderstanding.

Consider, for example, the terms *radiation* and *radioactivity*. Radioactivity (or, alternatively, radioactive decay, or more correctly activity [see Chapter 2]) is a process by which unstable nuclides undergo spontaneous disintegration, and resulting in the emission of a particle and/or photons -- i.e. radiation. Although the previous sentence is not a rigorous definition, it does to be sure, illustrate quite clearly in plain language the difference between radioactivity and radiation. Looking a little more closely, and again without excessive rigor, radioactivity is a stochastic process governed by Gaussian statistics and is typically expressed in terms of transformations per time interval -- i.e. disintegrations per second or becquerel (Bq). The strength of a radioactive source can thus be expressed in terms of the quantity activity, which is expressed in units of Bq, or multiples or submultiples thereof. The term activity is typically used in a rather general way as a surrogate, as it were, to describe or characterize the disintegration rate; thus an activity measurement is one that provides a measure or indication of the rate of radiation (i.e. particle) emission, and is expressed in terms directly relatable to the decay rate.

Similarly, the term *radiological* will be used to describe the measurement of radiation in terms of energy deposition (i.e. dose) or *exposure*[*] in the special sense. Measurements made in terms of the various dose related quantities -- e.g. absorbed dose, equivalent dose, and *exposure* -- are thus radiological measurements. The term *radiation*, when used in conjunction with measurement, includes both radiological and activity measurements.

A few terms still remain to be defined before moving on to other topic; *calibration* of an instrument refers to determining or comparing the response of the instrument relative to that of a known reference standard [i.e. one relatable to a reference

---

[*] The practice introduced in Chapter 2 of italicizing the word *exposure* to delineate *exposure* in the special sense will be continued throughout this chapter.

laboratory such as the National Institute of Standards and Technology (NIST)] and should not be confused with the term *adjust,* which refers to making some alteration to the response of an instrument, such as changing a potentiometer setting, perhaps in an attempt to improve the agreement of the instrument reading with that of the standard or, in other and less precise terms, improve the calibration. Similarly the term calibration should not be confused with *check,* a rather broad term that typically refers to verifying the operability of the instrument, usually under field rather than laboratory conditions and with the aid of a radioactive source, although not necessarily a reference or so-called standard source.

17.2 MEASURING ACTIVITY AND ITS SURROGATES

17.2.1 Geometry and Geometry Factor

In activity measurements, two terms that are often incorrectly used interchangeably or confused are *efficiency* and *geometry*. Geometry, or, more precisely *geometry factor* refers to a physical parameter, namely the fraction of the total solid angle about the radiation source that is subtended by the detector. In other words, it is the theoretical upper limit of the fraction of particles emitted by a radioactive source that can enter the detector. A point source of radioactivity is a theoretical construct of a source of infinitely small size that emits its particles equally in all directions, or through 4π steradian. If such a source were placed in the center of a spherical detector, and hence completely surrounded by it, all particles or photons emitted by the source would of necessity be emitted right into the detector. In this admittedly theoretical case, the geometry factor would be 1, a situation given the special name of 4π geometry. Note what is perhaps the obvious, viz. that 4π geometry is 100% geometry, and that the geometry factor is expressed as a fraction while a percentage is used to characterize the "geometry" of the detector. Similarly, if a detector were placed atop a source such that all the particles or radiations emitted in the upward direction could enter the detector, whilst all the particles emitted in the downward direction could not, this would be 50% or 2π geometry and the geometry factor would be 0.5.

17.2.2 Efficiency

Efficiency, as contrasted with geometry, refers to the fraction of particles or radiation emitted by a source that is actually detected or recorded by the instrument, or by the detector, and is essentially the sum total of all factors which influence the ability of the system to register a count. What is usually meant or referred to by the term efficiency is the overall system efficiency, which is simply the fraction or percentage of counts are registered by the system for each particle emitted by the source. Thus, the system efficiency takes into account all the various factors that determine whether a count or ionizing event will be recorded (Table 17.1).

A related concept is detector efficiency, which refers to the likelihood or probability that a particle incident upon the detector will produce and transmit a signal or response to the remainder of the instrument system. For many instruments, system efficiency and detector efficiency are the same but in some cases the two may be quite different, for the system may be unable to count all the events or pulses produced in the detector. Ideally, a detector with 4π geometry should have 100% efficiency, but this is not necessarily the case. In theory, in 4π geometry, all the particles or radiations would be captured or interact within the detector. However, a number of mechanisms may alter the theoretical response. If the radius of the detector is smaller than the range of the particles being measured, a fraction might escape the detector without interacting or might be absorbed in the walls of the detector rather than in the sensitive volume of the detector. Or, interactions could occur but not get counted because of dead time or an inability of the counting system to resolve (i.e. separate) a specific particle interaction from other particles interacting at approximately the same time. And, since it is a practical impossibility to make a point source, there may be attenuation of some of the particles within the source itself. Thus the efficiency may well be and likely is less than the geometry.

Table 17.1 Factors Determining Counting Instrument Response

Geometry

Absorption/attenuation prior to the particle entering the detector

Absorption/attenuation in detector wall/window

Absorption/attenuation in the air or gap between the detector and source

Source self absorption/self attenuation

Probability of interaction within the sensitive volume of the detector

Probability of detecting the interaction

Backscatter

Ability of the system to count ionizing events in the detector

In many instances, manufacturers may choose to state geometry, particularly with the source close to the detector, not only because it is an easily determined, referenceable and reproducible physical parameter, but also because it typically has a larger numerical value than efficiency which could imply or suggest better or more desirable performance. In past years, less than scrupulous vendors have used such oxymorons as "geometric efficiency" in lieu of the proper term geometry or geometry factor, thereby misleading the unwary or uninformed. (*Caveat emptor*!) Manufacturers or vendors of counting apparatus may specify the geometry factor only at one specific location such as the top shelf in a multishelf shielded counting pig, thus putting their best foot (or highest value) forward, even though the particular location may be impractical to use. *Caveat emptor*, once again.

Efficiency (or more correctly the efficiency factor) is, of course, the fraction of disintegrations that is actually recorded by the counting apparatus, and is determined by many factors (Table 17.1), not the least of which is geometry. Detector wall and window thickness, applied voltage, fill gas and gas pressure in the case of gas filled detectors, and energy and type of radiation are major determinants of whether an ionizing event will be produced in the detector, and hence, of efficiency. In particular, the energy and type of radiation are of considerable import although not always considered. To illustrate the importance of energy and type of radiation, for a given specific geometry, a detector might have an efficiency of several per cent for the beta particles emitted by a nuclide such as ^{60}Co, but an efficiency an order of magnitude lower for the gamma rays from the same nuclide. This is true for many Geiger-Mueller tubes used in portable survey instruments. Similarly, the efficiency for the beta particles could be several times greater than the efficiency for beta particles of lower average energy.

Activity or source strength is quite properly reported in terms of disintegrations per unit time (usually disintegrations per minute [dpm, d/m, DPM] or second [dps, d/s, DPS]), quantities which are directly relatable to activity in units of Bq. Instruments, however, typically do not provide direct measurement of activity and thus are frequently calibrated in terms of *counts* rather than *disintegrations* per unit time. Thus counts per minute [cpm, c/m, CPM, ct/min etc.] and counts per second [cps] are frequently used on survey or count meter readouts. Occasionally, impulses per unit time (usually seconds or minutes) or events per unit time are the units of choice. CPM and related quantities must not be confused with activity; they can be converted to activity if the efficiency of the system for the particular geometry and nuclide(s) is known. Counts per second convert directly to activity in Bq by the simply dividing by the efficiency; for example, an instrument with an efficiency of 0.3 (which might be expressed in terms of a percentage, or 30% in this example) has a count rate of 150 counts per second; the activity is 150/0.3 = 450 Bq. For other time bases such as counts per minute, dividing by the efficiency converts to disintegrations *for the same time base.* In other words, dividing counts per minute (or per hour, or what have you) by the efficiency will yield disintegrations per minute (or per hour, or what have you)which in turn can be converted to activity in units of Bq or Ci, for those who prefer the traditional system of units.

17.3 SOURCES OF ERROR AND UNCERTAINTY

17.3.1 General Considerations

Exclusive of counting statistics, sources of error and uncertainty in radiation measurement can be divided into five broad and perhaps somewhat arbitrary overlapping categories:

1. Electromechanical.
2. Radiological.
3. Human factors.
4. Laboratory and sampling.
5. Statements of accuracy.

Each category contains a number of specific individual factors, with by far the largest included under the electromechanical category. The radiological category is next in number of individual factors, and is likely to be far more esoteric as well as likely to produce larger uncertainties or error.

17.3.2 Accuracy, Uncertainty and Precision

The term *accuracy,* refers to how close the measurement is the true or correct value, which in practice can never be known with certainty. To get around this problem, which is to some extent a semantic one, the term *conventionally true value* (CTV) has come into use; the CTV is a value that is agreed upon as being correct, and may in fact be, although this cannot be known with absolute certainty. However, use of the CTV provides a practical reference point without violating the sanctity, as to were, of the preciseness of scientific concept of accuracy. In any event, *accuracy* should not be confused with *precision*, which is a measure of, and basically synonymous with, reproducibility. The difference between precision and accuracy is illustrated quite simply by a gedanken experiment in which a sack of sugar truly weighing exactly 10 kg is weighed on the same scale 20 times. Each time the weight given by the scale is the same: 15 kg. The precision (or reproducibility) of the measurement is perfect; there is no deviation whatsoever and the result of the weighing is always 15 kg. The accuracy of the scale, however, is not so good, since it always yields a measurement that is 5 kg greater than what we know to be the true weight. In this simplistic example, the scale has a *bias* in the upward direction of 5 kg, or 50% at a true weight of 10 kg. In other words, the measurement has perfect precision but an error of 50%.

Note, however, that it is not possible to determine from the data given what the bias or measurement error will be for a sack of sugar with a true weight of 20 kg. The scale in question may have a constant 5 kg bias -- i.e. every weighing shows an increase of 5 kg over what it should be. In this case, the measurement would be 25 kg and the measurement error would be 25%. However, another possibility is that the scale consistently registers 1.5 kg for each kg of true weight. In this case, the measured value would be 30 kg and the measurement error would be 10 kg, and hence 50% of the actual weight. The response of the scale, however, may be very different than any of the possibilities thus far proposed and is impossible to determine from the information available. In an effort to determine the response characteristics of the scale, suppose another measurement had been made, this time with two sacks of sugar, which weighed exactly 20 kg, and the scale again read 5 kg high, or 25 kg. This would be a hint, perhaps, of a constant systematic error of 5 kg higher than the true weight. Suppose, however, we weighed the two sacks and got a result of 30 kg, indicating that the scale seemed to always be 50% high. Or perhaps we weighed the two sacks, and got the correct answer, viz. 20 kg. There are whole range of reasonable possible explanations. Maybe with three sacks, the result would drop to 25 kg, suggestive of a nonlinearity problem. The moral of all this is the caution not to overinterpret or draw conclusions from very limited data, and to note that errors can take many forms.

Since in the real world the true value is typically unknown, accuracy or deviation from the truth cannot absolutely be specified. Instead, *uncertainty* is what is normally specified. Uncertainty is simply the estimate of the total or combined effect of all factors that may affect the measurement (or response or instrument reading) and is usually expressed as a percentage or as an estimated standard deviation of the measured value. Uncertainty refers to the likely range of values that would include the true value, with the center of the range being the best estimate based on the measurement.

What is required, recommended or specified by expert bodies with respect to uncertainty is by and large reflective of what is reasonably achievable with the current state of the art. In its Report No. 57 published in 1978, the National Council on Radiation Protection and Measurements (NCRP) squarely addressed the question, noted that a factor of 2 is acceptable at $\leq$ one-fourth of the MPD; $\pm$ 30% at MPD levels, and $\pm$ 20% at levels above the MPD. This sliding scale quite correctly recognized that uncertainty was not constant over the entire range of measurement, an important factor to consider in the selection and use of an instrument. In a subsequent report issued more than a decade later, the NCRP reviewed its earlier recommendations concluding that the levels cited in the earlier report may be difficult to achieve, but in essence let them stand (NCRP 1991). Other standards setting and regulatory bodies have specified acceptable uncertainties giving similar values to those put forth by NCRP, including the Nuclear Regulatory Commission, the International Electrotechnical Commission (IEC) and the American National Standards Institute (ANSI). However, it should be noted that the uncertainties specified may be highly specific to a particular application. Frequently uncertainty is cited as percentage with the clear implication that it is a percentage of the instrument reading when in fact it may be a percentage of the full scale or decade reading; this often the case in manufacturer's literature. Thus an instrument with an uncertainty of $\pm$2% at full scale would, by extension, have an uncertainty of $\pm$20% at one tenth of full scale. Once again, *caveat emptor.*

17.4 ELECTROMECHANICAL EFFECTS

Many electromechanical factors have the potential, in some cases quite subtle, for affecting the response of an instrument and thus adversely impacting the accuracy of a measurement. The most likely of these to affect response include the following:

A. Power supply variations
B. Temperature
C. Pressure
D. Humidity
E. Tracking error
F. Response time
G. Geotropism
H. Non-ionizing radiations
I. Extracameral response
J. Noise and vibration

The effects produced by most of the ten factors in the above listing are likely to be small, but some, and in particular humidity, tracking error, and response time, may result in rather large measurement errors or limitations on instrument usage and thus merit special consideration.

17.4.1 Power supply variations

Instruments with so-called regulated power supplies drawing their power from the mains are often thought to be immune from variations in voltage as power from the mains, particularly in the United States and Canada, is generally quite reliable. Nonetheless, it may be quite variable. Indeed, one might well ask this question first: What is the standard alternating current voltage in the United States? For many years, 110 volt two phase ac was standard; then 120, 125 and now, perhaps, 130 volts, with various other values in between. (If you have any doubts as to what the standard is, simply inspect the voltages printed on standard incandescent light bulbs of various manufacture or sometimes even the same manufacturer!) Also, three phase 220, 230, or even 240 volt ac is given as a standard, particularly outside of North America.

But in truth, there is considerable variability in the so-called 'standard' ac voltage at the mains, and although the power supplies of many instruments are equipped to handle a wide range of voltages, voltage can nonetheless affect the response. In particular, this is true of older ac powered instruments, or instruments with unregulated or poorly regulated internal power supplies. Many instruments exhibit line voltage dependence, which in some cases can result in considerable variation in instrument response. It is important to note that the mains voltage from even a single outlet may vary considerably, depending on circuit loading within the building or line loading or system demand on the utility providing the power. Even such mundane activities as the automatic or routine startup of an air conditioner, a resistance heater, or an electric motor may cause line voltage drops of several volts. Conversely, turning off other power consuming sources may cause a rise in line voltage. Voltage provided by the power company may not be constant. It may be lower during periods of peak usage, and may surge when new generating capability is put on line. It may also vary depending on distance from the substation, and which substations are in use at a given time.

Startup or shut down of power consuming equipment may produce a slight shift in line voltage, typically without significant effect. However, switching equipment on or off may produce a wide instantaneous voltage swing in either the negative or positive direction, known as a transient or spike, which can adversely affect or alter instrument response, even in instruments equipped with transient suppression circuitry or components. This is true of both ac and battery powered (dc) instruments. Transients can do strange things to instruments, including dropping a portion of an accumulated measurement, or adding a disproportionate amount to a measurement. They may also make an instrument insensitive for a period of time after the transient, and can even temporarily or permanently disrupt or damage electronic components within the circuitry. Large transients in mains power supplies can result from lightning storms, and can cause serious adverse effects, including inducing changes in computer default parameters which can sometime go undetected for long periods, or even rendering an instrument system completely inoperable. Potential problems from transients and voltage changes can by and large be controlled in ac powered instruments through the use of surge protectors and constant voltage transformers, and good practice would suggest the use of such devices. No such simple and effective mechanism exists for dc powered instruments; in these, it is important to evaluate the effect of transients induced by range switching.

Line frequency variations are usually quite small and likely to produce insignificant, if any, changes in instrument response. Thus, line frequency variations are typically not a problem. Standard American line frequency is 60 Hz, and is usually maintained within ± 0.5 Hz or better. However, standard line frequency in Europe and South America is 50 Hz, and an instrument designed to operate at 50 Hz may not operate properly or in accordance with published specifications at the standard American frequency of 60 Hz. This is particularly true if an instrument contains oscillators or other components whose operating frequency is derived from the line frequency.

Instruments operating on direct current are ordinarily powered by batteries. Battery voltage, which may decrease after long use, or at low temperatures, is an important parameter, and should be evaluated with the instrument under continuous load, and preferably under conditions of maximum power drain (e.g. speakers turned on for portable survey instruments thus equipped). Low temperature tests are also recommended to ensure that the instrument, which may be rated and indeed operate at temperatures below zero, in fact does so and for the appropriate length of time with the battery pack in place. Some battery powered instruments may include horns or warning devices which draw a considerable amount of current, and these may be rendered inoperable at low temperatures because of reduced battery output.

Under more normal conditions, batteries should provide a more or less constant source of power, and thus have no significant or discernible impact on proper instrument function. However, use under load may drop the battery output voltage, and although the instrument may still appear to be functioning normally, at some critical voltage, instrument response characteristics may change dramatically. Such changes may not be readily detectable during the measurement, and the errors so introduced are unfortunately likely to result in underestimates rather than overestimates of dose with potential serious consequences. Fortunately, such errors are usually relatively small.

17.4.2 Temperature

Temperature effects on electronic instruments can be subtle as well as pronounced. Observed effects may be attributable not only to those on the power supply or battery, but also to temperature dependence of various specific components. The overall effect of temperature on instrument response may thus be variable, and not conveniently or correctly expressed in terms of a simple temperature coefficient. Nonetheless, a temperature coefficient that gives the change in response per degree change in temperature is a common way of expressing temperature dependence. A typical temperature coefficient might thus be specified as + 0.2%/°F, which indicates a two-tenths per cent increase in response per degree F in the ambient temperature. The wording on this specification, however guileless in intent, may be inadvertently deceptive. First glance would suggest that any temperature induced change would be minimal. However, if, for example, a portable instrument calibrated indoors at 72°F is taken outdoors in winter, a temperature differential of 50 degrees and possibly even more could result, lowering the reading by 10% or even more. Moreover, the statement does not include a specification of over what range the purported temperature coefficient applies. A properly stated temperature coefficient would include a phrase such as "over the range 40 to 110 °F" as a qualifier. While this bounds the applicability of the stated temperature coefficient in terms of a range of temperature, it provides no clue as to what happens outside the stated range. In some cases, even a temperatures a few degrees above or below the stated upper or lower limits, respectively, temperature effects might be very large, even to the point of total instrument inoperability.

A better specification of temperature dependence is terms of an envelope or range, perhaps best stated as a percentage, over a specified temperature range. Thus, temperature dependence might be specified as "+10, -15 % from 30 to 120 °F; no response below 0°F" gives a far more informative picture of temperature dependence. However, even such a specification needs to be interpreted and applied with care, for a major and frequently unrecognized pitfall in radiological measurements occurs when an instrument is cycled through various temperatures as would be the case when the instrument is carried in a warm (or cooled) vehicle and brought outdoors into a much cooler (or warmer) ambient environment. In going from a warm location to a cold one, condensation may occur within the instrument that could render the measurement temporarily or even permanently unusable or inaccurate; cyclical temperature changes can result in erratic readings which may or may not be obvious to the instrument user. The cause may never be detected as the malfunction cannot be reproduced in the laboratory.

Still better would be express temperature dependence in the form of a plot of response as a function of temperature. Such a plot should be based on actual measurement and should specify whether it applies to the complete instrument system, the detector, or specific portions of the system. Even so, care should be taken in interpreting such plots, for a sometimes used practice is to evaluate the electronic package of an instrument by feeding an electronically generated pulse into the system, and then evaluating the readout under different conditions. This, of course, will not test the response of the detector, or cables between the detector and electronic package, but nonetheless is a practice that is widely used.

As a final caveat, it should be borne in mind that passive integrating dosimeters such as film badges and TLD's are not necessarily free of temperature effects. Some photographic emulsions are quite temperature dependent, and may fog at elevated work temperatures (e.g. >110 °F) particularly if humidity is low. If humidity is high, elevated temperature may accelerate the latent image fading phenomenon. Similarly, low temperature traps in TLD materials can be emptied at high ambient temperatures, such as may be encountered in a closed automobile on a hot day, resulting in possibly large underestimates of dose, measurement errors which are all but undetectable.

17.4.3 Atmospheric Pressure.

In general, effects attributable to changes in air pressure are small, although in extraordinary cases, changes in pressure have been known to rupture windows of gas filled and other detectors. Changes in air pressure usually affect unpressurized ion chambers, which are sometimes known in the vernacular as "air breathers". It is important to be aware that measurements made at high altitudes (e.g. Los Alamos, Denver, in an airplane) with instruments calibrated at sea level, or vice versa, will not be correct. Correction for changes in temperature and pressure can be easily made with the following equation

$$F = \left(\frac{273.2 + T_2}{273.2 + T_1}\right)\left(\frac{P_1}{P_2}\right) \tag{17.1}$$

in which F gives the correction factor, T_1 and P_1 are the temperature and atmospheric pressure during calibration and T_2 and P_2 are the temperature and pressure during the measurement. In the above equation, temperature must be in units of degrees Celsius, while the pressure can be in any units so long as both pressures are in the same units. Small effects of temperature and pressure may be significant in radiological standardization work involving unpressurized ionization chambers used as transfer or laboratory standards.

17.4.4 Humidity

Of all the electromechanical factors identified above, humidity can cause large errors in measurement, but is quite often ignored. Perhaps best known are the effects of high relative humidity on photographic emulsions, which may produce fading of the latent image and hence produce large underestimates of dose. Humidity effects on photographic emulsions are interrelated with temperature in a complex fashion, and should thus be considered together, not only for film badges but for other instruments as well. Recall that relative humidity is a function of air temperature for a given absolute humidity.

Some detectors, notably NaI and certain TLD materials, are hygroscopic, and can be ruined by excessive humidity. For this reason, NaI detectors are typically "canned" or hermetically sealed, but pinhole leaks in the can ultimately result in disastrous consequences preceded by a lengthy series of erroneous measurements. Condensation has already been briefly mentioned in connection with temperature. As is the case with temperature, cycling can also introduce error in measurements through changes in the characteristics of electronic components, as well as effects on the detector, which in extreme situations can result in arcing or intermittent grounding with resultant loss of signal.

To protect against humidity effects, small bags of silica gel or other moisture absorbers (desiccants) are sometimes placed inside the cases of instrumentation containing potentially sensitive electronic components. The effectiveness of this procedure is questionable, and in any case, the moisture absorbing materials may quickly become saturated or in equilibrium with the surrounding environment and hence will require frequent replacement if they are to be of any real value.

17.4.5 Tracking Error

Tracking error, also and not quite correctly termed *linearity* refers to the ability of an instrument such as a portable survey meter or rate meter to correctly follow changes in radiation level. Numerous factors may affect tracking, and resultant errors can be quite large. Not the least of these is the human factor, i.e. the ability to correctly and properly read the scale. Analogue readouts may be linear, logarithmic or quasi-logarithmic; logarithmic readouts, because of compression at the upper end of the decade and non-linear interpolation, are particularly prone to readout errors. Digital readouts with rapidly changing values are also prone to erroneous readout, particularly if the decimal point is moving (or, for that matter, has moved), or if the display is not consistent with the standard symbols.

Tracking errors may be particularly pronounced at the extremes of the range. This is of particular significance at the high end of the range (and thus of most concern from a dose measurement standpoint) where the indicated response is likely to be low rather than high. Ranges may not properly overlap, and differences in readings made near the high end of one range and the low end of another may be unacceptably large. For example, suppose an instrument has a reported tracking uncertainty of $\pm$ 5 per cent of full scale, a apparently reasonable and quite acceptable situation. A reading of 0.95 -- 95% of full scale -- is obtained on the 1X range. But on the 10X range, the instrument reads 12% of full scale, or 1.2. The instrument is performing within its claimed range of specifications but the reading obtained at the low end of the 10X range is fully 25% greater than the reading obtained on the 1X range, posing potential problems with respect to exposure limitations and stay times. Although in the example cited a happy medium can be obtained by averaging the two readings, a technique which sometimes works, interrange differences and associated errors can be large and in operational situations can readily compromise ALARA.

Switching transients may introduce significant error into readings from instruments with long time constants. An upscale value introduced into an analogue instrument by a switching transient may take several seconds to resolve or return downscale to the correct reading if the time constant is relatively long as is true of some ion chamber survey meters. To rectify this problem, some meters are equipped with a reset button, a feature which may prove desirable. Digital readout autoranging instruments are far less susceptible to tracking errors than are analogue readouts with manual range changes. However, autoranging digital readouts have been known to jam when in high fields.

17.4.6 Response Time

Over the years, the term *response time* has undergone several definitions. The current definition, as put forth by ANSI, is "the time interval required for the instrument reading to change from 10% to 90% of the final reading (or vice versa) following a step change in the radiation field" (ANSI 1997). Note that this definition includes both the downscale as well as the upscale response. Upscale and downscale response times may be quite different, a factor that needs to be borne in mind when making measurements in variable fields.

Response time is determined by and related to, but should not be confused with, the *time constant* which is the product of the resistance and the capacitance in an electronic instrument and is generally given as the time required for the instrument to come to $1 - e^{-1}$ or 63% of the final reading. Thus, two time constants will provide only about 86% of the final reading. The time constant is relatively easily determined from design or measured electronic factors, and because of this sound basis and the fact that it is typically smaller than the response time, is often quoted by vendors, manufacturers and designers. In instruments with a number of ranges, higher resistance is used to obtain more sensitive ranges. Thus the time constant (and hence response time) values typically are greater on the lowest ranges, increasing by the interrange scaling factors. These are usually 10, so an instrument may well have a 100 fold greater time constant or response time on the 1X range than on the 100X range.

While short time constants are usually desirable, they can result in excessive meter fluctuations or bobble. Hence, variable and externally adjustable time constants can be designed into pulse counters to provide a damped or more stable readout. However, there are certain inherent limitations of detectors, such as resolution time or dead time in Geiger-Mueller counters which cannot be overcome even with modern fast electronics. The limiting factor may thus may well be the detector, and knowledge of its response characteristics may not be as readily obtainable as those of the associated electronics. In particular, various detectors or instruments may not have the ability to resolve pulsed sources such as some accelerators and some types of x-ray generators. Integrating ion chambers are notoriously unable to accurately monitor pulsed beams because of recombination losses, and may give an apparently plausible result that is as much as an order of magnitude low. Even so-called passive detectors and dosimeters, including photographic film, are not free from such effects, although they are generally adequate for most health physics related pulsed source measurements.

17.4.7 Geotropism

Geotropism, or position dependence, refers to changes in an instrument or meter reading as a function of orientation with respect to the plane of the earth. A pocket ionization chamber may read differently as it is held to the eye and rotated through 360°. This effect, which is largely gravitational, may be quite pronounced in some situations, although is usually relatively small -- on the order of a few per cent change -- and is not limited to pocket ionization chambers but affects various other types of instruments, including portable survey meters, as well.

17.4.8 Non-ionizing Radiations

Non-ionizing radiations include the entire electromagnetic spectrum below x-rays and may adversely affect measurements with a number of different devices. Perhaps most obvious -- indeed classic -- are the effects of light leaks on photographic film dosimeters, photomultiplier tubes and scintillators of various kinds. Light can also affect some TLD's. Electronics, particularly if inadequately shielded, can respond to or otherwise be influenced by low energy portions of the electromagnetic spectrum, more specifically radio frequency and microwave radiations, sometimes resulting in altered measurement accuracy in ways so subtle that the error will never be detected even though it could be relatively large. Similarly, magnetic fields may produce effects on instrument electronics, detectors, or meter readouts. These effects too can be quite subtle, but nonetheless may be significant and may surprise the unwary making measurements in the vicinity of an accelerator with an associated high magnetic field, particularly when the magnetic field is intermittent. Instruments to be used in magnetic fields should thus be equipped with mu metal or other ferromagnetic (magnetic absorbent) shields; similarly, RF cages made of copper or other metallic mesh can mitigate the effects of radiofrequency and microwave radiations.

17.4.9 Extracameral Effects

Extracameral effects refer to the effects of the radiation field on portions of the instrument other than the detector. They are by and large trivial, although in some cases can be significant. It is possible that an instrument with a remote detector head -- i.e. one in which the electronics and readout are outside the field -- may give a different response from one in which the electronics are in the field. Where the entire instrument is in the ionizing radiation field, the effect may be attributable to a radiation response on the part of the electronics or to interference or disruption of electronic signals by ionization within the electronics package or even the components themselves. Cables used to make the connection between a remote detector and associated electronics are notorious for their response to high radiation fields, producing signals in some cases which exceed those of the detector. Increased resistance, capacitance, line losses and other factors in cabling used for remote detectors may also have an effect, and instruments calibrated *in situ* may have quite a different response from those in which the detector and electronics are separately calibrated under laboratory conditions.

17.4.10 Noise and Vibration

The final electromechanical effect to be considered is that from noise -- more correctly sound pressure -- and vibration. Sound pressure is essentially air pressure, and for all practical purposes, is normally insignificant insofar as instrument response is concerned. However, there are rare instances in which a resonance occurs in a specific component leading to instrument malfunction or erroneous measurement. Vibration effects may manifest themselves as intermittent effects from loosened components, or total measurement capability failure, but can also produce more subtle changes in instrument response.

17.5 RADIOLOGICAL EFFECTS

17.5.1 Energy Dependence

Energy dependence refers the response of the detector with respect to different energies of the same radiation. Any detector exposed to ionizing radiation will measure the dose it receives, and that dose is determined by the energy absorption probability for the exposing radiation. For photons, the energy absorption probability or coefficient is strongly a function of the atomic number of the absorber and energy of the incident radiation. Thus, for any given photon radiation, the dose to the detector may not be the same as the dose to tissue. Basic to the statement of energy dependence is the answer to the question, "relative to what?" The specification of energy dependence must give a clear and unambiguous reference point. Normally, the response (dose or energy absorbed) of a detector is specified relative to air or soft tissue, but even with such a specification, there are numerous pitfalls. In virtually all cases, detectors used for radiological measurements differ from tissue in their response to radiation, and yet it is the tissue dose that usually is of most concern. Energy dependence of photon detectors is largely dictated by their effective atomic number (Z), which may differ significantly from that of tissue. Since energy absorption via the photoelectric effect predominates at photon energies below a few hundred and certainly 100 keV, and is approximately proportional to Z^5, photon dose measurements may be significantly influenced by the construction materials of the detector. For materials with Z higher than soft tissue, even with fairly low Z such as aluminum, the effect can be quite significant, typically peaking in the 30-50 keV energy region. At lower energies, absorption in the wall of the detector becomes significant, reducing the overresponse relative to tissue. Energy dependence is typically most pronounced at photon energies below about 100 keV, but is also important even with high energy photons such as those associated with the decay of ^{60}Co or ^{137}Cs, for scattered photons may comprise as considerable fraction of the photon field that is being measured, particularly with broad beam conditions that are normally encountered in the field.

So-called air equivalent and tissue equivalent detectors are designed to respond much like air or tissue to a wide range of photon energies. But an important and oft overlooked caveat relates to what happens with photons with energies above or below the effective energy range of these specially designed detectors. At very high photon energies, such as those associated with ^{16}N decay in reactor containments, or around high energy (20+ MeV) photon generating accelerators used in therapy, a properly functioning detector calibrated at ^{60}Co energies is likely to underrespond relative to tissue, perhaps by as much as 30-40%, because the wall is insufficiently thick to produce charged particle equilibrium. This is true for passive integrating dosimeters such as TLD and pocket ionization chambers as well. The solution is to add additional buildup material outside the detector, but this, of course, changes the characteristics of the response to lower energy photons, and may reduce it significantly in the energy region of the annihilation radiation associated with these high photons. A similar problem is encountered at lower energies; addition of absorbers to 'flatten' the photon energy response may produce a lower energy cutoff that is unacceptable or introduces significant undermeasurement of dose. Simply stated, there is no single detector that covers the complete photon energy range of interest in health physics, although many detectors can provided a suitable range for most measurement purposes.

Energy dependence is not limited to photons, but to betas and neutrons as well. With these radiations, the health physicist is more likely to be flying blind as data regarding beta and neutron energy response are far fewer and less likely to be available or reported in manufacturers' literature. Suffice to say that the potential effects on measurement accuracy are likely to be large and much more likely to be unquantifiable than is the case with photon radiations. Neutron instruments are particularly susceptible to the effects of energy dependence, and conversion from measurement of fluence rate to equivalent dose is a process fraught with error. Even instruments designed and marketed as dose equivalent meters exhibit an often pronounced energy dependence, particularly for low energy neutrons. Thus, in a situation in which there is a neutron spectrum rich in thermal or scattered neutrons, special calibration may be indicated.

17.5.2 Angular dependence

Were a detector to be a perfect sphere, with completely isotropic response, angular dependence would not be a source of error in radiological measurements. However, detector geometries are anything but spherical, and the incident radiation may be attenuated or absorbed scattered by differences in wall thickness, sensitive volume depth, connectors, and attached electronics packages. Wall materials in the case of detectors with a window, such as an end window cylindrical ionization chamber and thickness may vary depending on the angle of incidence. Personal dosimeters such as film and TLD badges, because of their planar geometry, also exhibit angular dependence, and edge on exposures are likely to produce rather different results than the same field measured at a normal angle of incidence. Angular dependence, at least for photons, is likely to be more pronounced at energies below a few hundred keV, and at some energies and angles can result in errors exceeding an order of magnitude. Thus, energy and angular dependence are intimately related, and in a complex manner. Errors introduced by angular dependence can be very large or very small, depending not only on energy but also on the specific characteristics of the field. *Caveat Emptor!*

17.5.3 Response to unwanted ionizing radiations

Unwanted radiations, both ionizing and nonionizing, i.e. radiations other than those the detector is designed to detect -- may in fact have an effect which is usually manifested as an overresponse. Measurements in mixed fields are thus fraught the potential for error. Even BF_3 and other neutron detectors, which many believe are insensitive to photons, will in fact respond to photons if the levels are sufficiently great to produce pulse pileup, an effect that often is associated with pulsed accelerators. In such cases, the gamma pulse arrives first and triggers an instrument pulse; the neutrons which arrive later may or may not be appropriately detected depending on the field intensity and the recovery characteristics of the system. Similarly, photon measuring instruments may respond to high energy betas if the wall thickness is insufficiently great; a wall thickness of about 1000 mg-cm^{-2}, roughly equivalent to a thickness of 1 cm of plastic, is required to fully attenuate betas from commonly encountered beta emitters. A wall with this thickness may well interfere with or significantly reduce the response of the detector to low energy photons. Many beta-photon survey meters of the cutie pie type are constructed with wall thicknesses of 440 mg-cm^{-2} as a sort of compromise value; this thickness is usually sufficient to attenuate most betas while at the same time allowing for flat energy response down to 50 keV or so.

Measurement of photons in mixed neutron/photon fields needs to take into consideration the possible effects of neutrons on the photon measuring instrument. Such effects can range from direct ionization in the detector (usually a small problem) to the more subtle effects produced by short lived activation products in the materials of the detectors. NaI detectors may be particularly prone to such effects, as may various gas filled detectors with stainless steel walls or metallic electrodes.

17.5.4 Relevance of calibration

Ideally, an instrument should be calibrated in the field, or at least in a laboratory situation that duplicates the exact energy and angular characteristics of the field to be quantified. Clearly, this may be impossible and also non-relatable to so-called 'standard' conditions of the primary or secondary calibration laboratory. The ideal calibration considers not only the radiological aspects of the exposing field, but also other factors that are likely to affect the resultant measurements, including environmental factors such as temperature, pressure and humidity, and perhaps even dose rate. Wherever possible, the energy spectrum to be encountered in the field should be used for calibration purposes. Alternatively, a combined energy/angular dependence study of the detector should be carried out, and suitable interpolated or averaged correction factors developed.

Calibrations carried out under so-called "good" geometry -- i.e. narrow beam conditions -- are not ordinarily comparable with broad beam conditions encountered in the field and simply may fail to adequately measure the effects of scattered radiation, which under some circumstances can be appreciable.

17.6 HUMAN FACTORS

To some extent, human factors have been discussed briefly above under the section entitled "Tracking Error". Obvious errors attributable to human error include misreading or bias in misreading a scale or meter (many people will tend to round off to even numbers -- e.g. 2, 4, etc., or to quartiles or quintiles, for example); more subtle are reading errors introduced by such factors as distortion from the meter cover glass and parallax. Such errors are likely to be small, although accidental misreading, particularly of the range or multiplier, can obviously result in a large error. Larger errors can be introduced by failure to wait a sufficient time for the instrument to respond or to recover, a particularly important consideration with digital readout survey instruments, which may fail to respond when scanned across a small point source. Analogue instruments are not immune to this problem, and scanning speeds when making contamination surveys should be sufficiently slow to permit the instrument to respond suitably.

17.7 SOURCES OF ERROR IN ENVIRONMENTAL SAMPLING

Measurement in the laboratory is no guarantee of eliminating or minimizing error. Consider, for example, sample collection and replication. Perhaps the first question might be: How representative is the sample? Widely differing results may be obtained depending on whether a grab or continuous sample is collected, and whether the sampling location may be depleted of the radionuclide(s) being sought as may be the case in which the top lays of soil are continually remove from the same sampling spot. Inherent in meaningful air samples are such factors as the particle size distribution of the aerosol and the collection efficiency of the sampler as a function of particle size. Isokinetic sampling conditions are essential in stack sampling and in duct sampling as samples obtained under anisokinetic or turbulent conditions may not be suitably representative. While such considerations as plateout of radioactivity in sampling lines and the necessity for flow monitoring or constant volume or displacement pumps are often stressed in textbooks, in practice such considerations may be improperly implemented, if implemented at all. Although it seems implausible, such obvious factors as holes in filter media, leaks in the filter holder that allow the sampled air to bypass the filter or filters loaded to the point that air passage through them is a virtual impossibility are all too frequent source of error in measurement of ambient air radioactivity concentrations.

Gaseous activity presents special problems with respect to measurement error. Radioiodine or noble gases may adsorb onto particulate collectors, not only producing erroneously high measurement values, but compounding the problem by offgassing in the counting chamber and contaminating it. Flow through gas monitors -- i.e. tritium sniffers and real time radon monitors -- may be measuring more than just the gaseous activity if the air is not filtered prior to entry into the chamber to remove particulate activity. And, although it should go without saying, it nonetheless should not hurt to mention that most flow through gas monitors are current measuring devices whose readout is directly proportional to energy deposition in the chamber. Obviously an alpha particle with an energy of several MeV will deposit a much greater amount of energy, and hence produce significantly more current per particle than, say, a beta particle from tritium decay, so appropriate calibration is a must.

Improper or inadequate handling of samples after collection may also inadvertently introduce measurement errors, some of which are quite subtle in nature. Volatile fractions such as radioiodine or noble gases on filters may offgas, and offgassing in turn may be enhanced by high temperature or retarded by high humidity. Particles in solution or suspended in liquids may agglomerate, and radioactivity of any type may plateout or be adsorbed onto the walls of the laboratory glassware. Chemical yield measurements and the use of carriers and tracers have their benefits with respect to minimization of errors, and if properly used are unlikely to introduce significant additional error or measurement uncertainty. Precipitation may remove a portion of what is being measured, and ion exchange columns are subject to wide variations in removal of radioactive ions. Factors such as temperature, tightness of column, resin size should be carefully controlled; even a few degrees difference in the temperature of the solution passing through the ion exchange resin can result in errors of greater than 20 per cent in the measured result.

17.8 NON-STATISTICAL ERRORS IN COUNTING

Counting error or uncertainties may arise from other than counting statistics, and incredible as it may seem, the so-called counting statistics errors may be small to the point of triviality in relation to other errors of counting. Environmental effects -- temperature, atmospheric pressure, relative humidity -- generally do not result in significant errors, although some instruments or components are notoriously temperature dependent, notoriously the photomultiplier tube and certain solid state devices.

Power supply stability is vital to accuracy in measurement; even small changes in the applied high voltage may have large and dramatic effects on counting results. Surprisingly large errors -- certainly greater than a few per cent -- can be introduced by even small changes in detector voltage; high voltage may in some cases be produced by stepped rather than continuously variable transformers, and any defects or errors attributable to voltage could well be introduced or intensified by a stepped transformer.

A potentially large source of error relates to counting efficiency. Clearly, the radioactivity, geometry, and to the extent possible source strength of the calibration source must be identical or at least relatable to the sample being analyzed. But astonishingly, the actual sample counting geometry may be different from that of the calibration source. Samples are normally also counted on this same shelf, but if a particularly weak sample is obtained, it might be moved up a shelf or two without due regard being paid to the change in geometry factor. The reverse is also true; samples may be moved away from the detector, or partially shielded, again at the cost of a change in the geometry factor, which, if not properly taken into account, might be overlooked and thus become a significant source of error.

Many factors, not properly or adequately controlled, may introduce small errors; these include backscatter, self-absorption or attenuation and detector edge effects. Resolving time losses and corrections are not limited to Geiger-Mueller counters but may affect other detectors and counting systems as well. Others, perhaps more subtle, include decay of the nuclide during counting, and the need to correct for decay back to the time of sample collection. To do this implies that one has an accurate record of the time of collection. Timer errors, while not common and usually not large, may occur, and particularly so with variations in line frequency, frequency mismatches, or voltage changes. Spectroscopy introduces an entirely new realm of pitfalls whose discussion is beyond the scope of this presentation.

17.9 ACCURACY STATEMENTS AND SPECIFICATIONS

Like most of the potential pitfalls already discussed, an entire tome could be (and indeed some have been) written on this particular topic area and the brief following discussion will focus on what the vendors' literature provides. Clearly, the vendor wishes to put his instrument in the best possible light, but at the same time does not wish to provide out and out false statements. Specifications given by various vendors are regrettably thus often not comparable to one another; for example, a vendor might pick a specific energy -- the one that presents the best energy dependence or smallest lower level of detection L_c -- as the reference point. The competitor, also putting his best foot forward, picks a different point; how can these be properly compared?

In an example from real life, one vendor published in the specifications for a high level monitor the capability of measurement of 10^8 rad/h, referenced to air. Of course the vendor could not test the detector response at this level, since available existing sources could only provide at most a level of 5×10^6 rad/h, or 5% of the indicated full scale range of the detector, very much lower than this outrageous claim of 10^8 rad/h. However, the vendor could calculate what current the detector would be expected to produce from such a dose rate. And since the ammeter provided with the instrument was of sufficient range, the claim was made, notwithstanding the fact that such as level may not be possible to achieve, even in a Chernobyl situation, and even if it were, would likely very quickly melt the detector!

17.10 SOME WORDS IN CONCLUSION

All is not totally bleak, for over the past two decades a number of instrument evaluation programs have been developed and implemented, and consensus performance standards published by ANSI. Knowing that an instrument meets the published ANSI performance standards can go a long way towards answering the basic questions posed in Section 17.1.1; indeed, the standards themselves provide an indication of what to expect in the way of instrument performance capabilities. However, it is unlikely that the questions posed by the wary instrumentation purchaser and user will be completely answered, so *caveat emptor* should remain the watchword; with careful examination of manufacturers and vendor claims and specifications, the asking of knowledge based questions and insistence upon answers is the means to avoid unanticipated errors and perhaps prevent purchase of a pig in a poke.

17.11 REFERENCES

American National Standards Institute (ANSI). Performance specifications for health physics instrumentation -- portable instrumentation for use in extreme environments. New York, NY: American National Standards Institute. ANSI N42.17C-1989; 1994.

American National Standards Institute (ANSI). Performance specifications for health physics instrumentation -- occupational airborne radioactivity monitoring instrumentation. New York, NY: American National Standards Institute. ANSI N42.17B-1989; 1994

American National Standards Institute (ANSI). Personnel dosimetry -- criteria for performance. New York, NY: American National Standards Institute, ANSI N13.11; 1993.

American National Standards Institute (ANSI). Standard performance criteria for active personnel radiation monitors. New York, NY: American National Standards Institute. ANSI N42.20-1995; 1995.

American National Standards Institute (ANSI). American National Standard Radiation Protection Instrumentation Test and Calibration, Portable Survey Instruments. New York, NY: Institute of Electrical and Electronics Engineers; ANSI N323A-1997; 1997.

Becker, P. H. B.; Matta ,L. E. S. C.; Moreira, A. J. C. Guidance for selecting nuclear instrumentation derived from experience in the Goiania accident. Health Phys. 60:77-80; 1991.

Eisenhauer, C. M.; Schwartz, R. B. Analysis of measurements with personnel dosimeters and portable instruments for determining neutron dose equivalent at nuclear power plants. Washington, DC: U.S. Nuclear Regulatory Commission; NUREG/CR-3400; 1993.

Eisenhower, E. H. Measurement quality assurance. Health Phys. 55:207-213;1988.

Eisenhower, E; Welch, L; Wiblin, C. Performance of portable radiation survey instruments. Washington, DC: U.S. Nuclear Regulatory Commission; NUREG/CR-6062; 1993.

European Communities. Intercomparison of environmental gamma dose rate meters: A comprehensive study of calibration methods and field measurements, Washington, DC: European Community Information Service; 1989.

Liu, J. C.; Mao, S.; McCall, R. C.; Donahue, R. The effect of static magnetic fields on the response of radiation survey instruments. Health Phys. 64:59-63; 1993.

National Council on Radiation Protection and Measurements (NCRP). Instrumentation and Monitoring Methods for Radiation Protection; Bethesda, MD: National Council on Radiation Protection and Measurements; NCRP Report No 57; 1978.

National Council on Radiation Protection and Measurements (NCRP). A Handbook of Radioactivity Measurement Procedures, 2nd Edition, Bethesda, MD:National Council on radiation Protection and Measurements, NCRP Report No 58; 1985.

National Council on Radiation Protection and Measurements (NCRP). Calibration of Survey Instruments Used in Radiation Protection for the Assessment of Ionizing Radiation Fields and Radioactive Surface Contamination; Bethesda, MD: National Council on Radiation Protection and Measurements; NCRP Report No 112; 1991.

Swinth, K. L. Performance expectations for electronic dosimeters. In: Higginbotham, J., ed. Applications of new technology: external dosimetry. Madison, WI: Medical Physics Publishing; 1996: 151-172.

Swinth, K. L. and J. L. Kenoyer. Evaluation of health physics instrument performance. IEEE Trans. Nucl. Sci. NS-32:923-933;1985.

Swith, K. L.; McDonald, J. C.; Sisk, D. R.; Thompson, I. M. G.; Piper, R. K. Performance testing of electronic personnel dosimeters. Washington, DC: U.S. Nuclear Regulatory Commission; NUREG/CR-6354; 1995.

Swinth, K. L.; Roberson, P. R.; MacLellan, J. A. Improving health physics measurements by performance testing. Health Phys. 55:197-205; 1988.

Chapter 18

INSTRUMENTATION FOR THE MEASUREMENT OF LOW FREQUENCY ELECTROMAGNETIC FIELDS (0-3 kHz) AND LASER OPTICAL RADIATION

John A. Leonowich

18.1 INTRODUCTION

Extremely low frequency (ELF) fields are a type of nonionizing electromagnetic energy in the form of electric and magnetic fields. For this chapter, we will consider ELF as extending from direct current (dc) fields into the radio-frequency part of the spectrum, up to 3 kHz. ELF electromagnetic energy comes primarily from the generation, transmission, distribution and use of electric current.

Laser sources exist with outputs from the infrared to the UV. The eye is generally the most sensitive organ and the one that must be protected. However, ultraviolet and high power lasers may also be hazardous to the skin. In addition, high-powered lasers may become fire hazards. The eye is transparent to radiation from 400 to 1,400 nm and is termed the ocular hazard region since the radiation incident on the cornea is focused on the retina.

This chapter will review the instrumentation available to the radiation safety professional to make measurements in these portions of the electromagnetic spectrum.

18.2 INSTRUMENTS TO MEASURE ELF FIELDS (0 – 3 kHz)

Unlike RFR, ELF fields are incapable of heating tissue. Standards (ACGIH 2000) are based on the neurological effects of induced currents. The exposure metrics are the electric field strength **E** in V m^{-1} and the magnetic flux density **B** in tesla. In recent years, most research has centered on the effects of magnetic fields on humans, so that the bulk of the instruments commercially available measure the magnetic flux density **B**. Because the human body is relatively transparent to ELF **B** fields, measurements made in air are a good indicator of exposure within the body.

The primary method of measuring ELF electric fields is to place two small, conductive plates, which are connected by a wire, in the volume filled by the field and measure the current between the plates. This is called a displacement sensor, or a free-body E-field meter. The lines of electric force are incident upon one plate, thereby inducing a current. The wire carries the current to the other plate, which develops a voltage and an electric field. A meter is installed in the wire between the plates to measure the current flow induced by the voltage. The field strength is equal to the measured voltage divided by the distance between the two plates, E = V/d.

ELF magnetic fields are usually measured with loops or coils. The oscillating magnetic field will induce a voltage in the loop, which is measured and can be related to the current flowing in the loop. Loops are highly directional. The current induced in a loop reaches a peak value when the loop opening is perpendicular to the flux lines of a magnetic field and is zero when the opening is parallel to the direction of the field vector, as predicted by the Biot-Savart Law.

Another method of magnetic field measurement relies on the Hall effect. An object in a magnetic field will develop a voltage in a direction perpendicular to the magnetic field that can be measured. Hall effect sensors are less sensitive than loops, but can be used for field surveys where fields in the millitesla range are expected, and are useful in measuring static magnetic fields.

Just as in RFR instruments, loops are often placed together in mutually orthogonal arrays in order to produce an isotropic response to the field. An isotropic response is independent of sensor orientation and field polarization, i.e., it is non-directional. An example of an ELF magnetic-field survey instruments using orthogonal detector array is shown in Fig. 18.1.

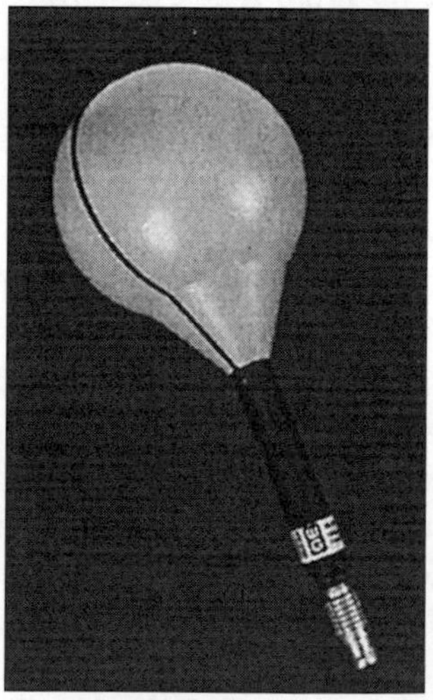

Fig. 18.1 - Orthogonal ELF Magnetic Field Probe (Courtesy NARDA Microwave, Inc.)

In addition to field instruments, there exists a number of small **B** field densitometers which can be worn by an individual to quantify his time averaged exposure during a workday (Hitchcock 1995).

18.3 LASER INSTRUMENTATION

The watt (W), the fundamental unit of optical power, is defined as a rate of energy of one joule (J) per second. Optical power is a function of both the number of photons and the wavelength. Each photon carries an energy that is described by Planck's equation:

$$Q = hc / \lambda \quad (18.1)$$

where Q is the photon energy (joules), h is Planck's constant (6.623×10^{-34} J s), c is the speed of light (2.998×10^{8} m s^{-1}), and λ is the wavelength of radiation (meters). All laser measurement units are spectral, spatial, or temporal distributions of optical energy. Laser MPEs are given in terms of radiant exposure in J cm^{-2} or irradiance in W cm^{-2}.

Lasers are classified by the American National Standards Institute (ANSI 2000) in numbered classes where the higher the number the greater is the potential hazard.

Class 1 - denotes exempt lasers or laser systems that cannot under normal operating conditions produce a hazard. This includes lasers that are completely enclosed and interlocked or so low in power than no hazard exists. No warning label is required.

Class 2 - denotes low power visible lasers or laser systems which because of the normal human aversion responses (< 0.25s), do not normally present a hazard but may present some potential for hazard if viewed directly for extended periods of time (like many conventional light sources).

Class 3a - denotes lasers or laser systems that normally would not produce a hazard if viewed for only momentary period with the unaided eye. They may present a hazard if viewed using collecting optics. A sign with the words "Caution - Laser Radiation – Do not stare into beam or view directly with optical instruments" should be clearly visible. Eye protection should be worn.

Class 3b - denotes lasers or laser systems that can produce a hazard if viewed directly. This includes viewing of reflections from various smooth reflective surfaces. Diffuse reflections are not hazardous. A sign which reads "Danger-Laser Radiation" is used.

Class 4 - denotes lasers or laser systems that can produce a hazard not only from direct or specular reflections but also from diffuse reflections. In addition, such lasers may produce fire hazards and skin hazards. A sign that reads "Danger –Laser Radiation - avoid eye or skin exposure to direct or scattered radiation" shall be clearly visible. Lasers in this class are capable of producing serious eye injury. These lasers should be enclosed if possible, and operated as Class I lasers. If this in not possible, eye protection is required and access must be controlled when the laser is in operation. The skin should also be protected. The laser should be operated remotely if possible. Highly reflective surfaces must be removed or painted. Only qualified and trained personnel should operate the laser.

Because the ANSI Z136.1 laser standard, as well as the Federal Laser Products Performance Standard (FDA/CDRH 2000) categorizes lasers, routine measurement of laser systems is not normally necessary. Z136.1 recommends that measurements only be made if the laser has not been categorized, or if it is used in a situation where the uncertainties of propagation may influence exposure. An example of such a situation would be outdoors use of a Class 3b or 4 laser.

In general, the amount of power emitted by a laser is measured using a power meter and a detector tuned to the wavelength of the laser. The types of detectors fall into three main categories: silicon photodiodes, thermopiles, and pyroelectric infrared.

Planar diffusion type silicon photodiodes are perhaps the most versatile and reliable sensors available, particularly for visible and ultraviolet wavelengths. The P-layer material at the light sensitive surface and the N material at the substrate form a P-N junction which operates as a photoelectric converter, generating a current that is proportional to the incident laser light. Silicon cells operate linearly over a ten-decade dynamic range, and remain true to their original calibration longer than any other type of sensor. For this reason, they are used as transfer standards at NIST. Silicon photodiodes are best used in the short-circuit mode, with zero input impedance into an op-amp. The sensitivity of a light-sensitive circuit is limited by dark current and Johnson (thermal) noise. The practical limit of sensitivity occurs for an irradiance that produces a photocurrent equal to the dark current.

The thermopile is a heat sensitive device that measures radiated heat. The sensor is usually sealed in a vacuum to prevent heat transfer except by radiation. A thermopile consists of a number of thermocouple junctions in series that convert energy into a voltage using the Peltier effect. The Peltier Effect was discovered in 1822, and refers to the reversible heating, or cooling which occurs at a contact when current flows from one connector to another. The reversible heat quantities are small and hence somewhat hard to measure. For this reason, liquid crystals are ideal for indicating temperature change as it increases or decreases. Thermopiles are convenient sensor for measuring the infrared, because they offer adequate sensitivity and a flat spectral response in a small package. Thermopiles suffer from temperature drift, since the reference portion of the detector is constantly absorbing heat. The best method of operating a thermal detector is by chopping incident radiation, so that the modulated reading zeros drift out.

Pyroelectric Infrared (PIR) Detectors converts the changes of incoming infrared light to electric signals. The pyroelectric materials are characterized by having spontaneous electric polarization, which is altered by temperature changes as infrared light shines the elements. By choosing appropriate receiving electrodes, they serve a wide range of infrared wavelengths.

Excimer lasers operate in the ultraviolet portion of the electromagnetic spectrum (100 – 400 nm). Because of their relatively short wavelength, these types of lasers can produce large amounts of scattered radiation. Although this scattered radiation is no longer coherent, it is still potentially a hazard. It is therefore important to always make ultraviolet measurements around such systems. Handheld units that are calibrated to the ACGIH UV action spectrum (ACGIH 2000) are now available. These units read out in effective irradiance, based on exposure to pure 270-nanometer UV light. This type of instrumentation makes simplifies compliance with the ACGIH limits for ultraviolet radiation. This is probably the most common type of measurement made around lasers today.

18.4 REFERENCES

American Conference of Governmental Industrial Hygienists. Threshold limit values and for chemical substances and physical agents. Cincinnati, OH: ACGIH, 2000.

American National Standards Institute. American national standard for safe use of lasers, ANSI Z136.1 –2000. Orlando, FL: Laser Institute of America, 2000.

Food and Drug Administration/Center for Devices and Radiological Health. Federal laser produce performance standard (FLPPS), 21 CFR 1040. 2000.

Hitchcock, R.T.; McMahan, S.G.; Miller, G.C. Extremely low frequency (ELF) electric and magnetic fields. Fairfax, VA: American Industrial Hygiene Association, 1995.

Chapter 19

RADIATION INSTRUMENTATION USED IN DECOMMISSIONING

Shane Brightwell

19.1 INTRODUCTION

The past few years have seen many changes in the regulations governing decommissioning at nuclear facilities. The most recent rule that went into effect in August 1997 regarding nuclear facilities, other than uranium and thorium recovery facilities, requires that the facility, separate building, or outdoor areas be released for unrestricted use if the residual dose equivalent from all sources to an average member of the public (more specifically, the critical dose recipient group) does not exceed 0.25 millisievrts (25 millirem) in any one year. This is a very small value with respect to background radiation dose values in the United States that average 3 mSv y^{-1} and are as high as 6 mSv y^{-1} in Denver, Colorado.

As you can see, the cleanup criterion is only about 5-10% of the background value. This poses some challenges when trying to detect residual radiation, especially when using gross measurement methods or trying to quantify residual radionuclides that are also present in the background component. In some cases, it is even difficult to quantify residual radionuclides that are not present in background simply due to inevitable measurement interferences. Worse yet, significant variabilities in background radiation values can cause serious problems, especially in land areas and buildings with a wide assortment of construction materials.

Fortunately, there is a statement in the decommissioning rule that allows licensees to avoid taking unreasonable measures to quantify residual radiation where background interference is extensive. The rule specifically states that a site "...will be considered acceptable for unrestricted use if the residual radioactivity that is distinguishable from background radiation does not exceed..." 0.25 mSv y^{-1}. This implies that a licensee must use reasonable methods to quantify the dose equivalent contribution from residual radioactivity values in excess of background radiation. It does not, however, mean that a licensee can drag a GM pancake probe over the floor or swing a Model 19 one-inch sodium iodide over the ground and call it good simply because there were no clicks or the needle never jumped.

The Nuclear Regulatory Commission (NRC) has put forth considerable effort to provide the nuclear industry with regulatory guidance that presents what, in their opinion, are some acceptable methods for demonstrating compliance with the new decommissioning rule. These guidance documents (described in the next section) address most major issues that arise with the new rule, including survey methods using standard industry radiation detection instrumentation. The majority of this chapter will be devoted to discussing such instrumentation, including instrument selection, calibration, detection limits, quality control, and operation.

19.2 DECOMMISSIONING REGULATIONS, REGULATORY GUIDANCE

The following are descriptions of regulations governing decommissioning and associated regulatory guidance.

Title 10 Code of Federal Regulations Part 20 (10 CFR 20) Subpart E (NRC 1997): This regulation establishes the dose-based unrestricted release criteria for license termination and decommissioning at nuclear facilities other than uranium and thorium recovery facilities. It basically states that a facility will be considered for unrestricted release if the level of residual radioactivity distinguishable from background will not result in a total effective dose equivalent (TEDE) to an average member of the critical group in excess of 0.25 mSv (25 mrem) in any one year, and that such residual radioactivity levels have been reduced to levels that are considered ALARA.

10 CFR 30.36 (NRC 1997)*:* This regulation is often referred to as the "timeliness rule." It basically establishes time limits for reporting and decommissioning following the cessation of principal licensed activities at a facility, separate building, or outdoor area. It is essentially intended to prevent a licensee from ceasing principal activities and isolating a facility, separate building, or outdoor area, then leaving it unattended for extended periods of time. This regulation also establishes some conditions under which a decommissioning plan is required (or not required) to be approved by the regulatory authority.

10 CFR 40 Appendix A, Criterion 6(6) (NRC 1999)*:* This regulation establishes cleanup criteria for uranium and thorium recovery facilities. It basically states that the residual levels of $^{226}Ra/^{228}Ra$ resulting from the uranium/thorium recovery process in soils at a licensed or disposal site, when averaged over 100 m^2, cannot exceed 0.19 Bq g^{-1} (5 pCi g^{-1}) in the top 15 cm of soil, or 0.56 Bq g^{-1} (15 pCi g^{-1}) in subsequent 15 cm layers. ALARA is an implicit requirement in this regulation.

Demonstrating Compliance with the Radiological Criteria for License Termination (Draft Regulatory Guide 4006) (NRC 1998)*:* This regulatory guide provides guidance for implementing the cleanup criteria in 10 CFR 20 Subpart E. It only covers the release of buildings and soils, not equipment or materials. It provides detailed guidance, based on regulatory position for decommissioning issues such as dose modeling, final status surveys, and ALARA evaluations.

Federal Register Notice Volume 64, Number 234, Pages 68395 - 68396 (December 7, 1999) (NRC 1999)*:* This notice published screening activity concentrations for selected radionuclides in soils. The values were derived from the NRC's DandD code with default parameter values. It implies that if a licensee can adequately demonstrate that the site meets the applicable screening value concentrations in this list (or as determined by DandD using conservative default parameter values), then the site is a candidate for unrestricted release under all conditions, assuming that ALARA criteria have been met.

Guidance on using Decision Methods for Dose Assessment to Comply with Radiological Criteria for License Termination (NUREG-1549 DRAFT) (NRC 1998)*:* This NUREG is intended to be used to assess applicable dose modeling scenarios and site-specific parameters for determining the derived concentration guideline levels (DCGLs). DCGLs are activity values (activity concentration for soils, activity per unit area for building surfaces) corresponding to the dose-based cleanup criterion (0.25 mSv y^{-1}).

Multi-agency Radiation Survey and Site Investigation Manual (MARSSIM, NUREG-1575) (EPA 1997)*:* This manual has become a standard guide for performing surveys to demonstrate compliance with the cleanup criteria in 10 CFR 20 Subpart E. MARSSIM provides guidance on major issues such as survey development, instrument selection, calibration, and operation. MARSSIM also provides a look into what is called "Scenario B" in which the investigation is geared toward determining if residual radioactivity levels are distinguishable from background levels. Scenario B is detailed in *NUREG-1505 - A Nonparametric Statistical Methodology for the Design and Analysis of Final Status Decommissioning Surveys* (NRC 1998).

Minimum Detectable Concentrations with Typical Radiation Survey Instruments for Various Contaminants and Field Conditions (NUREG-1507) (NRC 1998)*:* This document is instrumental to radiation surveys where the pursued radiation/contamination limits are in the range of background, as is the case with the decommissioning criteria. NUREG-1507 addresses many factors affecting the quality of radiological surveys, such as instrument selection, detection limits, and detection efficiencies.

19.3 SURVEY INSTRUMENTATION

In the realm of decommissioning, direct measurements using portable radiation detection instruments have long been used and accepted as a standard method of demonstrating compliance with release criteria. However, the new cleanup standard has essentially raised the performance bar, which compels the licensee to spend a little more time considering what instrumentation to use.

The instruments described below are those I have used and evaluated in the course of decommissioning surveys. This list is by no means comprehensive. I hope and expect to expand my experience to include other quality instrumentation commonly used in decommissioning as the opportunities arise.

19.3.1 Gross Alpha and Beta Contamination Measurements

Residual radioactivity in buildings undergoing decommissioning is usually present in the form of surface contamination. Depending on the type of licensed activities that took place in the facility, there might be both alpha- and beta-emitting radionuclides present. Additionally, some of these alpha- or beta-emitters might also emit characteristic gammas.

If the licensee has decided to use surface contamination measurements to quantify the level of residual contamination in a facility, there are several standard industry detectors to choose from that measure gross alpha and beta components. Two types of detectors commonly used for gross alpha and beta measurements are gas flow proportional counters and dual phosphor scintillators.

19.3.1.1 Gas Flow Proportional Counter (GFPC). This detector consists of a sealed gas chamber (active volume) with a thin mylar window covering the measurement surface to allow alpha and beta particles to penetrate to the active volume. Once a radioactive particle enters the active volume, it interacts with the purge gas, resulting in a number of ion pairs proportional to the incident energy of the particle. The most common purge gas recommended and used in GFPCs is called P-10 (90% argon, 10% methane). A high-voltage bias applied between an internal wire (anode) in the active volume and the chamber walls (cathode) causes the ion pairs to be collected within a very short period of time. The result of this charge collection is an electronic "pulse" that is proportional in size to the number of ion pairs, and thus, to the incident energy of the incoming particle.

GFPCs come in a variety of models and window sizes. Common models and sizes include

- Ludlum Model 43-68 hand-held, 126-cm² window GFPC that can be used in any orientation for surveying floors, walls, and ceilings (Fig. 19.1).
- Ludlum Model 239-1F Floor Monitor, which incorporates a Model 43-37 wide-area, 544-cm² window GFPC mounted on a cart used for surveying floors (Fig. 19.2).

As would be expected, there are trade-offs to using one detector over another. The larger the window area, the more surface area a surveyor can cover in a given period of time. However, a larger window area also means a higher background count for both alpha and beta components. This can cause some problems for a surveyor who is attempting to locate elevated contamination areas that are considerably less than 500 cm^2 in size (commonly called "hot spots"), because the larger window detector tends to average the hot spot over the entire 544 cm^2 area of the detector window. Conversely, the smaller window detector would average the same hot spot over a smaller window area, in which case the impact on overall count rate would be more substantial, resulting in an increased probability that the hot spot would be identified.

Fig. 19.1 - Ludlum Model 2360 Dual Alpha/Beta Channel Data Logger and Model 43-68 GFPC.

Fig. 19.2 - Ludlum Model 239-1F Floor Monitor, incorporating a Model 2360 Dual Alpha/Beta Channel Data Logger and Model 43-37, 544 cm² GFPC.

On the other hand, in a situation where the surface contamination is relatively uniform and spread over larger surface areas, the larger window area detector would be preferable. Not only could the surveyor cover considerably more floor area in a given time, he/she could take advantage of the upright operating mechanism of the larger detector, as opposed to using the smaller hand-held detector while on hands and knees.

19.3.1.2 Dual Phosphor Scintillator. This detector incorporates two scintillating materials - one for detecting alphas, another for detecting betas - into a single detector window. A thin silver-activated zinc sulfide (ZnS(Ag)) crystal, used for detecting alphas, is placed directly between the mylar window and a plastic scintillator crystal, which used for detecting betas. Essentially all alphas that penetrate the mylar are fully absorbed in the ZnS crystal, each resulting in a pulse of visible light that is proportional in intensity to the incident energy of the particle. Most betas that penetrate the mylar also penetrate the thin ZnS crystal and are absorbed in the plastic scintillator, each resulting in a pulse of visible light proportional in intensity to the incident beta energy.

The light pulses resulting from both particles are directed (typically through a light pipe) to the photocathode window of a photomultiplier tube (PMT), where they are converted into electron signals and amplified by many orders of magnitude, respectively, before being transmitted to the instrument electronics. The instrument is typically set up with different operating windows so that the alpha and beta signals can be registered independently. One type of dual phosphor scintillator is a Ludlum Model 43-1-1, 75 cm^2 Detector (Fig. 19.3).

There are some advantages to selecting the dual phosphor scintillator over the GFPC for performing surveys. The main advantage is that the scintillator does not require any type of gas purge. This means that the surveyor can start using it almost immediately, without having to charge it with gas or drag around a gas tank. On the other hand, the scintillator gets expensive if a large window area is desired, considerably more expensive than the GFPC. Additionally, the scintillator is more susceptible to mechanical shock, and the replacement of a scintillator crystal is more difficult to perform in-house than the replacement of a platinum anode wire in a GFPC. A breach of the mylar also has a potentially greater impact on getting the scintillator back in operation. Unless it is repaired in-house under darkened conditions, a scintillator exposed to residual light during repair has to set and 'decay' (delayed phosphorescence) for up to 24 hours before use; whereas, a repaired GFPC is only down for the duration of a required gas purge.

19.3.1.3 Electronic Instrument There are numerous instruments available to operate both types of detectors described above. An optimum instrument should register both alpha and beta radiation independently and simultaneously.

Typical alpha energies are on the order of four or more megaelectron volts (MeV), while spectral average beta energies range from about 0.15-2.5 MeV. Since pulse size is proportional to particle energy, it is possible to set up and calibrate an instrument to a GFPC such that alpha and beta components can be measured independently and simultaneously. In this way, a surveyor can measure/monitor both alpha and beta components at the same time, which can reduce survey times considerably.

One example of an instrument that can simultaneously distinguish and register both alpha and beta components is the Ludlum Model 2360 Scaler/Ratemeter

(Figs. 19.1, 19.2, 19.3). The Model 2360 is programmable and has data logging capabilities. It can be programmed and controlled using either a dedicated hand-held terminal (keypad) or using a computer with accompanying download software installed. Ratemeter and scaler alarm setpoints can be set to assist the surveyor in identifying locations that might exceed known limits.

Fig. 19.3 - Ludlum Model 2360 Dual Alpha/Beta Channel Data Logger and Model 43-1-1 Dual Phosphor Scintillator.

One important factor in simultaneous, dual channel measurements of both alpha and beta radiation is cross talk between channels. The two channels are typically set up so that the larger alpha signal is registered in the upper channel and the smaller beta signal falls in the lower channel. It is possible for an alpha particle to lose enough energy in the mylar, resulting in a reduced signal that falls in the beta channel. Fortunately, under normal background count rate conditions, this does not have a major effect on beta count rate. On the other hand, high alpha count rates might also incorrectly register high beta count rates, which can complicate the quantification of true beta contamination levels. For this reason, it is important to use a mylar density thickness that is low enough to be conducive to simultaneous, dual channel measurements.

Additionally, cross talk can occur in the other direction, where beta signals are registered in the alpha channel. This usually occurs as a result of a significant shift in operating high-voltage. Such a shift can occur as a result of physical adjustment (increase) of the high-voltage setting above the optimum value for the set channels. Another shift, which is common in GFPCs, occurs when the detector and associated electronics are calibrated at a low elevation, then operated at a high elevation. The optimum high-voltage for dual channel operation of a GFPC characteristically decreases as elevation increases (described in a later section).

The consequence of beta signals being registered as alpha signals can be significant. A typical ambient alpha background count rate is no more than 1-2 cpm per 100 cm^2. Therfore any additional contribution to the alpha count rate due to cross talk can incorrectly imply elevated surface activity.

19.3.2 External Gamma Measurements

There are numerous options for performing gamma radiation measurements in the field, ranging from a hand-held sodium iodide probe calibrated with a portable instrument for measuring exposure rate, to an in situ gamma spectroscopy system calibrated to identify and quantify each gamma-emitting radionuclide present above detectable levels. There are advantages to each option, depending on the needs/conditions of the survey, as described below.

19.3.2.1 Thallium-activated Sodium Iodide Detectors (NaI(Tl) or NaI). NaI detectors have long been used for measuring environmental levels of external gamma emissions. Common sizes of portable survey NaI detectors are 2.5 cm x 2.5 cm (1 inch x 1 inch), 5 cm x 5 cm (2 inch x 2 inch), and 7.5 cm x 7.5 cm (3 inch x 3 inch) cylindrical crystals.

When gammas enter and interact in the NaI crystal, the result is a pulse of visible light proportional in intensity to the incident energy. The crystal is optically coupled to the photocathode window of a PMT, where the light pulse is converted into an electron signal that is amplified by many orders of magnitude before being transmitted to the instrument electronics.

If the detector is calibrated with the instrument for exposure rate measurement, the amplification of the signal through the PMT is such that all detector output signals are essentially the same size, regardless of incident gamma energy. If the detector and instrument are calibrated for single channel analysis (SCA) operation, the output signal from the PMT is proportional to the incident gamma energy.

The larger the NaI crystal is, the more sensitive it is to low levels of gamma radiation. The NaI detector has a varying response to incident gamma energy, as will be described in the section on calibrations.

19.3.2.2 High-pressure Ionization Chamber (HPIC). The HPIC is another detector commonly used for environmental-level gamma measurements. It consists of a 10-inch diameter stainless steel spherical chamber (cathode) containing an active volume of argon gas at a pressure of 25 atmoshperes and a 2-inch diameter stainless steel anode.

Gammas that pass through the HPIC interact with both the steel wall and the argon gas to generate ion pairs in the argon gas. The quantity of ion pairs produced is a function of the number of incident gammas (gamma flux), their associated energies, and the angle of incidence. A high-voltage bias across the chamber results in the ion pairs being continuously collected, generating an electric current proportional to the gamma flux and incident energies.

In the associated electronic instrumentation, the electric current is converted to useful readout of air-equivalent exposure rate units of microGray per hour (microRoentgen per hour or $\mu R\ hr^{-1}$). This means that the response of the HPIC will be proportional to the ionization produced in air by the gamma flux. The HPIC energy response is relatively flat in comparison to that of the NaI, as will be described in the section on calibrations.

19.3.2.3 Gamma Spectroscopy (Gamma-spec). Gamma-spec has been successfully used for decades for environmental measurements to reliably identify and quantify gamma-emitting radionuclides. It has gained more recognition and merit in recent years by regulatory authorities that have actually experienced its application in decommissioning surveys first-hand.

Two common detectors used for *in situ* gamma-spec are NaIs and high-purity germanium (HPGe) detectors. Both have their advantages, but it seems that the HPGe has become more popular, especially for measurements where a multitude of gamma-emitters is expected, due to its higher energy resolution. The HPGe can distinguish between gamma energies separated by only a few KeV; whereas, the NaI in this application would likely identify the same two gamma energies as a single gamma, thereby possibly masking the identification of one of the contributing gamma-emitting radionuclides.

The instrument that is the key to gamma-spec is the multichannel analyzer (MCA). The MCA can store energy-specific information in dedicated channels. Most MCAs are set up to maintain 8192 such channels. The result is a histogram of the information collected from the gamma energies incident to the detector.

This chapter does not go into detail, beyond this section, on the application of gamma spectroscopy in decommissioning surveys.

19.4 INSTRUMENT CALIBRATION

An integral part of getting a survey instrument ready to go into the field is assuring that it will accurately determine what the surveyor has set out to determine. In other words, the surveyor has to be confident that the instrument readout accurately reflects the level of radiation/contamination present at the location of interest. The consequences of using instrumentation that gives inaccurate readouts could include:

- Falsely concluding that a contaminated area is free of residual contamination in excess of the established limits, in which case future occupants of the area are at risk of receiving radiation doses that exceed the established criteria. This scenario does not go over well with regulators.

- Falsely concluding that a clean area is contaminated in excess of the limits, in which case unnecessary remediation could take place. This scenario can be expensive and time consuming, and extends the occupational health risks associated with remediation.

The most common method of minimizing the likelihood of occurrence of either of these two scenarios is to assure that the instrumentation is properly calibrated. "Calibration" is a somewhat general term that typically includes:

- Electronic calibration to assure that the electronics of the instrument are properly converting the detector output into the correct readout/recorded data;

- High-voltage calibration to assure that the detector bias is set in the proper operating range, and

- Detection efficiency calibration to assure that the instrument output is accurately converted into the desired radiation measurement units (Bq 100 cm^{-2}, $\mu Gy\ hr^{-1}$, etc.)

19.4.1 Electronic Calibration

Electronic calibrations are those performed on the instrumentation to assure that the electronics are in good order and also to establish certain operating parameters. One common practice is to have the instruments electronically calibrated by the manufacturer or other qualified instrument calibration facility on a routine (typically annual) basis. In fact, many licensees have this requirement written into their license and/or operating procedures, which leaves little flexibility.

However, it is possible for a properly trained and experienced person to perform electronic calibrations to satisfy the intent of the routine calibration requirements. Most manufacturers provide instrument-specific and detector-specific calibration instruction manuals with newly purchased equipment. These manuals can also usually be obtained upon request. With the right equipment (Fig. 19.4), a properly trained person should be able to calibrate, maintain, and troubleshoot instrumentation in the field. This can be very important during the throes of decommissioning surveys where down time for sending instruments back-and-forth to the manufacturer is prohibitively expensive and time consuming.

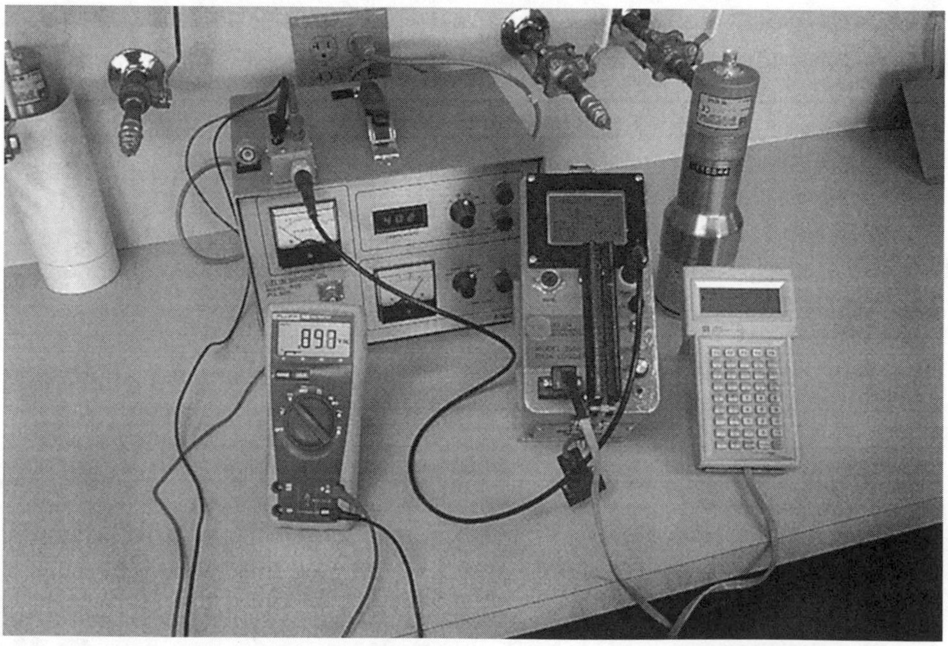

Fig. 19.4 - Electronic calibration equipment.

The following sections describe some of the electronic calibrations typically performed on counters.

19.4.1.1 Pulse Calibration. The pulse calibration verifies that the output count rate of the instrument is consistent with the input count rate from the detector. The pulse calibration is normally performed by replacing the detector signal with a simulated signal from a pulser that can be adjusted to an exact pulse rate over several orders of magnitude.

19.4.1.2 High-voltage Check. This check assures that the high-voltage reading on the instrument is the same as the actual high-voltage being applied to the detector. This check can be performed using essentially any DC voltage meter. Most detectors operate in the range of 1000 VDC, so if a high impedance voltmeter ($\geq$ 1000 megohm input resistance) is not used, a voltage divider (typically $\div$1000) should be placed in-line between the instrument output and the voltmeter. The consequence of not doing this could be substantial voltmeter damage.

19.4.1.3 Threshold and Window Setting (Pulse Height Discrimination). This calibration establishes the operating band in which input signals from the detector must fall in order to be registered on the instrument readout. The threshold setting, also known as the input sensitivity, establishes the minimum pulse height (usually in units of millivolts or mV) below which a pulse input from the detector will not register. It is important not to set this value too low; otherwise, electronic noise (inherent in the electronics of almost any instrumentation) can be inadvertently registered as true radioactivity measurement. Setting the threshold too high carries a consequence of actually discriminating against the lower end of the real radioactivity measurement signal. Typical threshold values are 4 mV for GFPCs, 10 mV for NaIs, and 35-40 mV for phosphor scintillators.

Window settings are usually a function of intended use. For example, a NaI calibrated with an instrument for exposure rate measurement will not typically have an upper window setting because the nature of the instrumentation setup is that all pulses are amplified to essentially the same high value (i.e., pulse height is not a function of incident energy). Therefore, an upper window setting might incorrectly discriminate against part of the real radioactivity measurement signal. If a NaI is calibrated with an instrument for single channel analysis (SCA mode), the threshold and window settings are very important. The primary purpose of operating in SCA mode is to filter out all input signals from the detector except those corresponding to the gamma energy of interest. If the threshold and window are incorrectly set, then the readout of the instrument will not reflect what the surveyor assumes is the correct response to the radiation of interest.

19.4.2 High-voltage Calibration

The standard operating parameter of most any survey instrument is the detector high-voltage. The high-voltage typically sets up an electronic field in an area where charge collection is a factor, such as in the active volume of a GFPC. In other applications, such as the PMT for scintillating detectors, the high voltage is a factor in signal amplification.

Each type of detector has at least one optimum high voltage setting. Some detectors operate on a plateau voltage, where inadvertent shifts in voltage have limited effect on detector response. Other detectors operate in a proportional region, where shifts in voltage have a proportional effect on signal size for a given incident particle energy.

The plateau high-voltage setting is usually established using a check source and a high-voltage ramp routine. The high-voltage is initially set to a value expected to be lower than the optimum operating voltage, the detector is exposed to the source, and a measurement is taken. This process is repeated, maintaining constant detector-source geometry, and increasing the voltage in 25-50 volt increments until the readings are observed to level off then increase, indicating that the upper end of the plateau has been reached. The calibrator's best judgment is then used to select an optimum high-voltage setting in the middle of the plateau.

When it comes to performing the high-voltage calibration for GFPCs, there is a caveat. The optimum operating high-voltage for a GFPC is sensitive to changes in elevation. For example, a Ludlum Model 43-68 was calibrated with a Model 2360 for dual alpha/beta channel operation. This calibration, including the operating high-voltage determination, was performed by the manufacturer at an elevation of about 650 meters (Sweetwater, TX). When the instrumentation was set up for operation at the site, at an elevation of about 1800 meters (Boulder, CO), the alpha background was observed to be unusually high, and the beta background was somewhat lower than expected. As a result, the operator decided to repeat the high-voltage calibration. The results of the re-calibration, along with the manufacturer's data and "as-found" data, are summarized below.

Parameter	Manufacturer's Setting	On-site "As-found" Value	On-site Post-calibration
High-voltage (VDC)	1625	1625	1475
Alpha BKG (cpm)	2	156	2
Beta BKG (cpm)	204	153	233

It appeared from the above data that the change in elevation affected the beta window settings, shifting some of the beta counts up into the alpha channel. This was confirmed by observing, under "as-found" on-site conditions, an overwhelming count rate response by the alpha channel on the instrument to a beta source (thousands of cpm).

When the same source was measured after onsite calibration, the alpha channel only registered an expected 2 cpm.

19.4.3 Detection Efficiency Calibration

Once all electronic and high-voltage calibrations have been completed, the detector should be ready for detection efficiency calibration. This is the calibration that determines what the output of the instrument actually means in useful terms and units.

19.4.3.1 Surface Activity Detection Efficiency. Detection efficiency (ε_d), also known as total efficiency, is the product of the instrument efficiency (ε_i) and the contamination source surface efficiency (ε_s). The value of ε_d is used to convert the count rate measured by the detector into units corresponding to surface activity, as shown in Eqn. 19.1:

$$A_s = \frac{R_{s+b} - R_b}{\varepsilon_d W} \qquad (19.1)$$

where

A_s	≡	surface activity (dpm per 100 cm^2)
R_{s+b}	≡	gross surface count rate including background (cpm)
R_b	≡	background surface count rate (cpm)
ε_d	≡	detection efficiency = ε_I x ε_s [cpm per (dpm per 100 cm^2)]
W	≡	detector window surface area (cm^2).

If the instrument is calibrated with a source that exhibits characteristics similar those of the measurement surface (i.e., radiation type and energy, source-detector geometry, self-absorption, and backscatter), then there may be no need to evaluate the ε_i and ε_s components independently. However, calibration sources are only available in a limited number of configurations, and rarely do they meet the above criteria. Therefore, it can be important to evaluate these two efficiency components separately. One method of doing this is to determine the ε_i component using the most appropriate calibration source, then estimate the ε_s component using a related reference value. NUREG-1507 (NRC 1998) is a good reference for estimating ε_s values. NUREG-1507 has a detailed collection of ε_s values categorized by detector, radionuclide, surface material type, and surface material self-absorption properties.

Another important factor in determining the ε_d value has to do with window surface area. Most mylar detector windows incorporate a protective screen. Manufacturers often report the total surface area of the window as well as the 'active' window area (the surface area of the window minus the surface area of the screen). A common practice by personnel in the field has been to use the reported active window area in Eqn. 19.1 to calculate surface activity. If the calibration is performed with the protective screen in place, and the surface area of the calibration source is several times larger than the screen pitch, then the screen's effect on ε_i is accounted for and the total window area should be used in calculating surface activity.

Sometimes a licensee just wants to send off a detector and instrument to the calibration facility and receive them back ready-to-use. The calibration facility might be asked to perform the efficiency calibration in addition to all other services. Most calibration facilities have standard efficiency calibration protocols in-place for doing this, and they will do it upon request without hesitation. But it is important that the calibrator know enough about the intended survey parameters to perform an appropriate calibration; otherwise, the information could be useless to the survey from a quantitative standpoint.

The following are some additional considerations regarding the ε_d of detectors used for surface activity measurements:

- The mylar density thickness can have a minor effect on beta efficiency and a significant effect on alpha efficiency.
- If a calibration source is used that has a small active surface area as compared to the detector window area, the source should be used to take multiple measurements systematically over the entire window area in order to establish the average ε_i value.
- When selecting among available scintillator detectors, be aware that the orientation of the light pipe has a measurable effect on ε_i. Characteristically, a straight light pipe oriented at 0° (such as that of the detector shown in Fig. 19.3) has the highest value; as the magnitude of the angle increases, the ε_i value tends to decrease.

19.4.3.2 External Gamma Detection Efficiency. Determining the detection efficiency of an external gamma measurement detector, such as the NaI, can be a bit more involved than for the surface contamination measurement detector. For instance, surface contamination detectors detect 2π emissions outward from the surface directly under the window, with the area of influence being essentially the same as the window area. An unshielded NaI is typically used for measuring exposure rate in three-dimensional (4π) space, with a sphere of influence on the order of meters in diameter.

Another characteristic that can complicate the use of NaIs for performing quantitative exposure rate measurements is the fact that the NaI detector response is energy dependent (Fig. 19.5). This means that if a NaI detector has been calibrated for exposure rate against a ^{137}Cs source by the manufacturer and then used to measure exposure rate at a site that is contaminated with a higher energy gamma-emitter, such as ^{60}Co, the detector will tend to underestimate the exposure rate by about 30%. Conversely, if the same detector were used to measure exposure rate at a site contaminated with a lower energy gamma-emitter, such as ^{133}Ba, the result would be an overestimate of exposure rate by a factor of about 3.5.

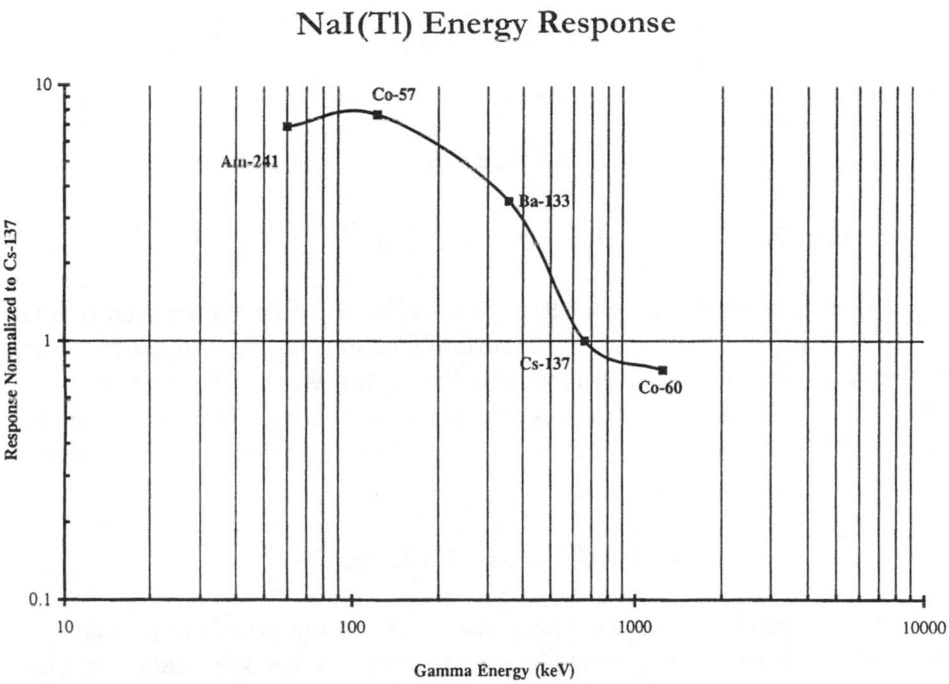

Fig. 19.5 - Energy Response of a NaI(Tl) detector, normalized to ^{137}Cs.

One effective method of using NaIs to accurately quantify exposure rate values without contending with the differences between calibration and on-site gamma energies is to perform a site-specific secondary calibration using a HPIC. Since the HPIC has a relatively flat energy response (Fig. 19.6), it will report the actual exposure rate as a function of incident energy (described in an earlier section).

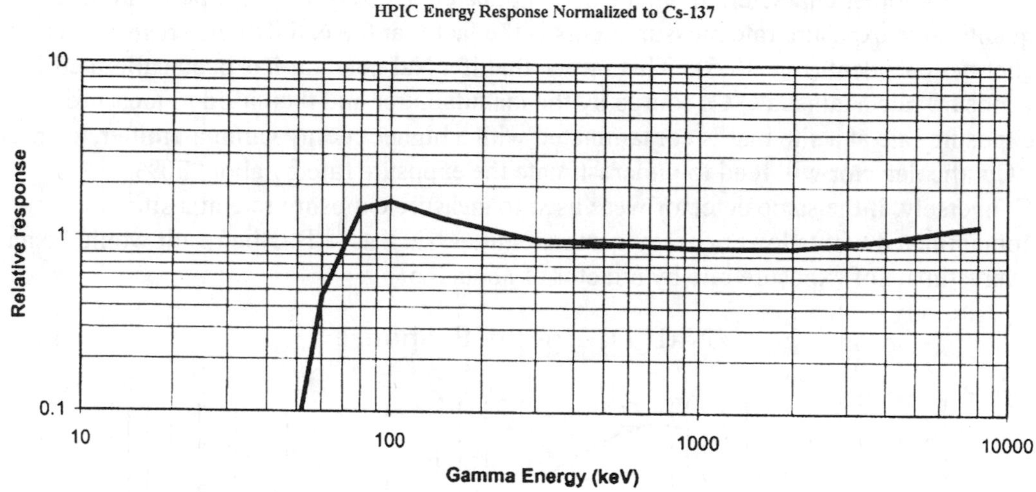

Fig. 19.6 - Energy Response of a HPIC, normalized to ^{137}Cs.

One straightforward approach to a site-specific secondary calibration is to take sequential HPIC and NaI measurements at several locations on-site that exhibit a range of exposure rate values from background up to the highest value of interest. A correlation can then be developed between the HPIC and NaI readings, which can then be applied to the site survey data taken with the NaI to determine the actual exposure rate values.

19.5 DETECTION LIMITS

Decommissioning surveys are performed to determine whether the residual contamination levels of a facility or outdoor area meet or exceed the established cleanup criteria. The cleanup criteria (in terms of activity concentrations) are known as derived concentration guideline levels (DCGLs). DCGLs are typically radionuclide-specific, but can also be derived in terms of gross alpha or beta surface activity if enough information is known.

The DCGL is a very important component of the survey, because it determines survey parameters (such as count times and scan rates) based on the detection capabilities. In fact, a DCGL that is relatively small in comparison to the associated background value may preclude some instrumentation with relatively high detection limits from being used in surveys.

The activity detection limit of a detector is also known as the minimum detectable concentration (MDC) or minimum detectable surface activity (MDSA), both terms of which are often used interchangeably. For simplicity, the term "MDC" is used in this text. MDC is basically the minimum activity per unit volume (soils) or surface area (buildings), determined *a priori* (prior to surveys), that a detector should be capable of detecting with 95% confidence. The MDC is dependent on instrument characteristics such as background and ε_i, as well as survey/field conditions such as ε_s.

Eqn. 19.2 can be used for determining the *a priori* MDC where the background and survey count times are not the same:

$$MDC = \frac{3 + 3.29\sqrt{R_b T_{s+b}(1 + \frac{T_{s+b}}{T_b})}}{\varepsilon_i \varepsilon_s T_{s+b}} \qquad (19.2)$$

where

R_b ≡ background count rate (cpm)

T_{s+b} ≡ survey count time (minutes)

T_b = background count time (minutes).

If the background and survey count times are the same (T), then the above equation reduces to

$$MDC = \frac{3 + 4.65\sqrt{C_b}}{\varepsilon_i \varepsilon_s T} \qquad (19.3)$$

where

C_b ≡ background counts for given time T.

As you can see, MDC is dependent on the $\sqrt{C_b}$ term, which is also known as the theoretical standard deviation of a single background count. If the variability in background counts is expected to be lower than that predicted by the value of $\sqrt{C_b}$, a good way to decrease the MDC value is to determine the actual standard deviation, s_b (Eqn. 19.4), by taking multiple background counts, and replacing the $\sqrt{C_b}$ term in Eqn. 19.3 with s_b.

$$s_b = \left(\frac{\sum \left(C_b - \overline{C_b} \right)^2}{n-1} \right)^{0.5} \tag{19.4}$$

where

$\overline{C_b}$ ≡ arithmetic average of **n** background count values.

19.6 QUALITY CONTROL

Even after all calibrations and survey parameters have been determined, it is important to assure that survey instruments are operating within expected parameters. The following describes methods that can be used to assure that counting instruments are operating properly with respect to precision and accuracy both before and during surveys.

19.6.1 Precision

A good method for determining whether a counter is operating as it should with regard to precision is to perform a *Reliability Factor* (RF) test (Schleien, Terpilak 1984). This test involves taking multiple source counts (usually a minimum of ten consecutive counts) in a constant source-to-detector geometry, and comparing the actual standard deviation (s_s) to the theoretical standard deviation of the average source counts.

The following are steps for performing the RF test:

1. Place the detector in a constant geometry over the source.
2. Initiate and record a scaler count; the source strength and count time should be such that a minimum of 10,000 counts are collected. Repeat this step until a minimum of ten (10) observations are recorded.
3. Calculate the average source count = $\overline{C_s} = \dfrac{\Sigma C_s}{n}$ (19.5)
4. Calculate the average source count theoretical standard deviation $= \sigma_s = \sqrt{\overline{C_s}}$ (19.6)

5. Calculate the average source count standard deviation

$$= s_s = \left(\frac{\sum \left(C_s - \overline{C_s} \right)^2}{n-1} \right)^{0.5} \quad (19.7)$$

6. Calculate the reliability factor = $RF = \dfrac{s_s}{\sigma_s}$ (19.8)

7. Plot RF on a *Statistical Limits of Counter Reliability* graph (Schleien, Terpilak 1984). If RF plots between the P = 0.10 and 0.02 lines or between the P = 0.90 and 0.98 lines, repeat this procedure. If RF plots above the P = 0.02 line or below the P = 0.98 line, the instrument should be removed from service pending further operability assessment.

NOTE: This test should be performed using GROSS counts in an area where background is not expected to change during the duration of the test.

This test has proven to be effective in identifying instrumentation malfunctions and improper operating settings. On one occasion when a detector consistently failed RF high (RF above P = 0.02), it was returned to the manufacturer and found to have a failing PMT. On another occasion when a detector consistently failed low (RF below P = 0.98), it was discovered that the input sensitivity was set too low and electronic noise was interfering, consequently resulting in an artificially tight set of source count data.

19.6.2 Accuracy

There are many ways of determining whether a detector is giving accurate results. One way is to measure a source of known activity and compare the measurement results to the theoretical (calculated) results. The following is an example of a procedure for determining the accuracy of a NaI detector, calibrated for exposure rate measurements against ^{137}Cs, to yield measurement results based on known source and calibration parameters.

1. Secure the probe in a low scatter geometry.
2. Collect and record an integrated background count (C_b).
3. Place the ^{137}Cs source a given distance (d) from the face of the detector. The source-to-detector distance should be a minimum of three times (preferably no more than five times) the maximum dimension of the detector or source (the purpose of this distance range is to reduce the volume effects while still

obtaining adequate source counts within a given time). For example, if a 7.5 cm x 7.5 cm NaI crystal is being tested using a point source, the distance between the source and the center of the crystal should be at least 30 cm (12 inches), based on three times the crystal diagonal dimension of about 10 cm (4 inches).

4. Collect and record an integrated gross source count (C_{s+b}). The count time should be sufficient to yield a C_{s+b} value that is at least 5000 (preferably 10,000) counts above background to optimize between a reasonable count time and acceptable counting statistics.
5. Calculate the net source count = $C_s = C_{s+b} - C_b$ (19.9)
6. Use C_s along with the applicable detector calibration information to calculate the measured exposure rate (XR_m).
7. Calculate the theoretical exposure rate (XR_t) value at the given distance for the ^{137}Cs source:

$$XR_t = \frac{A\Gamma}{d^2} = \_\_\_\_\_\_\_\_ \mu R/hr, \tag{19.10}$$

where

A ≡ ^{137}Cs source activity (μCi)

Γ ≡ ^{137}Cs specific gamma-ray constant = 3300 μR cm^2 hr^{-1} μCi^{-1} (Schleien, Terpilak 1984)

8. Compare the XR_m value to the XR_t value. If the two results differ by a value that is unacceptable to the measurement application (± 10% is commonly used) and the disrepancy is determined to be instrumentation-related, the instrument should be scheduled for repair and recalibration before it is returned to the field.

19.6.3 Bias

It is important to assure that detectors are operating within an acceptable range of accuracy on a day-to-day basis in the field. One common method is to perform routine source checks. Source checks usually involve taking measurements on a check source in a fixed geometry for a set count duration, then comparing the results to specified acceptable limits. If the detector passes, it can continue to be used. If it fails, it should be placed out-of-service and its operability should be evaluated before it is returned to service.

There are two types of source checks commonly used in the field. One is the simple method where the average source count is determined early and an acceptable operating band is established around that average (such as ± 10%). This method can readily detect if the operating characteristics of the detector shift significantly from the initial conditions. However, it does not allow for inherent/inevitable shifts such as those inherent in NaIs, especially if the detector is used and its source checks are evaluated over many months without reestablishing the average and acceptable operating band.

In the event that a detector will be used for many months, a better option for evaluating source checks might be the development of control charts. Control charts can usually be developed after a minimum of thirty data points have been collected under routine conditions. If the daily data are then added real-time, the average will be reestablished on a daily basis. The acceptable operating bands for control charts are usually ± 1 standard deviation (1SD) and ± 2 standard deviations (2SD). This application actually establishes detector-specific acceptable operating bands and does account for long-term changes in operating characteristics.

The following is a list of recommendations for performing source checks.

- Background Radiation Levels - Establish a location for performing source checks where the background levels and variability are sufficiently low. Evaluate the net (gross minus background) source check values unless there are reasons for including background variabilities, in which case, evaluate the gross values.
- Frequency - At least one source check should be performed each operating day. It is recommended that as many as three be performed each day: one before the day's start, one at lunch, and one at the end of the day. This provides data "sandwiching" from a quality control (data validation) standpoint.
- Simple Method – Use the average of the first 5-10 source check measurement values. These values should not be collected all at once; they should be collected under the same routine conditions as subsequent values so that the true day-to-day operating characteristics of the detector are factored in.
- Control Charts – Use at least the first 30 measurement values to generate the initial chart. These values should not be collected all at once; they should be collected under the same routine conditions as subsequent values so that the true day-to-day operating characteristics of the detector are factored in. Additionally, keep the control chart updated routinely, weekly if not daily. If a source check measurement value exceeds ±1SD, repeat the measurement until it falls within ±1SD. If the measurement value exceeds ±2SD, the next measurement should fall within ±1SD if the detector is to be considered for

field use; otherwise it should be placed out-of-service pending further operability assessment. If seven consecutive measurements all fall either above or below the mean, even if they all fall between the mean and either 1 or 2 SD from the mean, the instrument should be removed from service and thoroughly tested. The probability of obtaining seven consecutive counts either above or below the mean is less than 0.01.

19.7 OPERATION

The theoretical operation of all standard radiation detection instrumentation is well established and studied for the basic understanding of how the instruments are supposed to function. However, reality always comes into play when theory is put to practice.

In the interest of time and space, I have decided not to go into the details of performing decommissioning surveys in this text. There are numerous references (MARSSIM, for example) that provide detailed instructions regarding decommissioning surveys. But what are sometimes lacking in such documents are the small things that can actually make a difference. This section contains descriptions of some of these small things.

19.7.1 GFPC

The following are operational items that should be considered when performing surface contamination surveys using GFPCs.

- Purge Gas - The purge gas commonly used for GFPCs (P-10) is considered a "specialty gas" by many suppliers. It is mixed/prepared in batches as needed. Although supply companies have quality control measures in place, it is possible to receive a supply of gas that affects the operation of the GFPC differently from previous supplies. Therefore, a good practice is to test each new supply of gas with known operating parameters, such as the performance of a series of efficiency and/or source checks. If the operating characteristics of the GFPC with the new gas supply vary significantly from previous characteristics, it might be a good idea to try to obtain a bottle from the old batch; otherwise, it may be necessary to perform a full re-calibration of all applicable detectors, starting with optimum operating high-voltage, and including efficiency.

- Static Purge vs. Continuous Purge - Some GFPCs, such as the Model 43-68, are promoted as being able to operate for extended periods of time on a single static gas purge, obviating the need to drag around a gas supply bottle. However, a good practice is to operate the GFPC in continuous purge mode, with a steady gas supply at all times, even though it means dragging around the bottle. It should be okay to disconnect from the bottle for short periods of time, so long as care is taken during disconnection (simultaneous disconnection of dual gas connectors is vital in maintaining purge). If it is decided that a GFPC will be used with a static gas purge for extended periods of time, a good practice is to increase the frequency of source checks to every 1-2 hours, and recharge the detector at the first reliable sign of reduction in counts.

- Gas Flow Rate - When operating a GFPC in continuous purge mode, try to maintain a steady flow rate. The Models 43-68 and 43-37 recommend maintaining gas flow rate at 30-50 $cm^3\ min^{-1}$. Keep in mind that the gas pressure in the active volume chamber of the detector is not directly proportional to flow rate (i.e., a factor of two increase in flow rate does not double the pressure), so detection efficiency is not so adversely affected by small changes in the gas flow rate. It is okay if the flow rate deviates slightly from the set value, as it will inevitably tend to do. However, the flow rate should be routinely checked and adjusted to maintain it within a reasonable operating band.

- Mylar Window - The mylar window is fragile. Most surveyed surfaces have some type of debris, and the mylar is always in jeopardy of being breached. A small mylar breach can likely be identified by routine monitoring of gas flow (the detector supply flow rate may increase, and the return will decrease or drop to zero). If a mylar breach occurs that exceeds the makeup capacity of the continuous purge system, it might also be recognized by a steady decrease in beta channel count rate, dropping continuously from hundreds of cpm to zero cpm in a matter of minutes.

- High-voltage - The optimum high-voltage setting of a GFPC is sensitive to changes in elevation. If the GFPC is to be used at a location that differs in elevation from the calibration facility by more than 150 meters (about 500 feet), the optimum high-voltage setting should be reevaluated on-site.

19.7.2 Dual Phosphor Scintillator

The following are operational items that should be considered when performing surface contamination surveys using dual phosphor scintillators.

- Mylar Window - The mylar window is fragile. Most surveyed surfaces have some type of debris, and the mylar is always in jeopardy of being breached. A small mylar breach on a dual phosphor scintillator can likely be identified by a significant increase (or saturation) of the audible signal and a zero reading on the meter. If a mylar breach is suspected, expose the window to alternating dark and light conditions and monitor the audible and visual indications.

- Scintillator Decay - If a dual phosphor scintillator is repaired in an environment where the scintillator material is exposed to any ambient light, the repaired detector should be allowed to sit idle for at least 24 hours before it is returned to service. This idle time allows for adequate light decay (delayed phosphorescence) of the scintillator material.

19.7.3 NaI

The following are operational items that should be considered when performing surveys using NaIs.

- Energy Dependence - The NaI is energy dependent, as was discussed in an earlier section. This means that if the instrumentation is calibrated for exposure rate measurements at one energy, it will closely estimate the true residual air ionization exposure rate if measurements are performed in a field consisting predominantly of gammas at the same energy. If the same instrumentation is used in fields of higher (or lower) gamma energies, the measured exposure rate value will be underestimated (or overestimated) by a predictable factor, as shown in Fig. 19.5. In this case, steps should be taken to correct the measurement results for energy dependence of the instrumentation.

- SCA Mode - Some instruments, Such as the Ludlum Model 2350-1 Data Logger, allow for SCA mode operation using a NaI. This mode allows the instrumentation to register only gamma energies that fall within a specified energy range by performing an equivalent "gain" adjustment by way of high-voltage adjustment in the proportional region below the high-voltage plateau. However, if such instrumentation is to be operated in SCA mode, the "gain" (high-voltage) adjustment has to be performed routinely (usually daily) due to inevitable drifts in operating characteristics on the proportional region of the operating high-voltage curve.

- Shielding Effects - NaIs are typically calibrated in a diffuse or uniform collimated gamma field. Therefore, the calibration usually applies to unshielded, unperturbed gamma fields. If the detector is to be used under survey conditions involving shielding (collimators, surface contact with soil or other surfaces, etc.), special consideration should be given to the effects that these conditions could have on the measurement readout of the instrument. Even body shielding (holding the detector close to the body while surveying) can have a noticeable effect on measurement results.

19.7.4 General

The following are general operational items that should be considered when using survey instrumentation.

- Instrument Connections - Instruments and detectors are designed such that when the instrument is turned on, the bias is applied to the detector in a manner that will not damage components. This protection typically does not extend to the scenario where the detector is connected to an instrument that is already on. A good practice is to turn the instrument off and make connections (detectors, keypads, download cables, etc.) before the power is applied.

- Changing High-voltage - Some instruments are equipped to operate more than one detector at different operating high-voltages by way of a toggle selector switch or digital command. Although the switch or detector settings can be changed instantaneously, the actual high-voltage changes (ramps) relatively slowly, especially when going from a higher to a lower operating high-voltage. A good practice is to wait a minimum of five seconds for every 100 volts that the operating high voltage changes to allow stabilization. Better yet, turn the instrument off before changing the detector and selector switch if possible.

- Detector Cables - One thing that is often overlooked during surveys is the continued stress placed on detector cables. A damaged cable can contribute to spurious, erratic count rate that can go essentially unnoticed with detectors operating at high count rates. Even a few spurious counts contributed by the cable could adversely affect detector performance, especially where low count rate alpha contamination measurements are concerned. Furthermore, a damaged cable usually only interferes when it is being cycled, so the problem may not be recognized during static measurements (such as source checks) when the cable is stationary. A good practice is to perform routine cable condition checks by disconnecting the detector and rigorously flexing the cable (taking care not to touch the end connector), while simultaneously monitoring audible and visual indications. Any cable that shows signs of wear or damage should be replaced before returning the instrumentation to service.

- Detector Contamination - Another issue that is sometimes overlooked is the potential for detector contamination. It makes sense that a detector used for surface contamination surveys could collect some residual (or even natural background) contamination. A good practice is to perform routine cleaning of detector contact surfaces and windows using forced air or dry towels. If more aggressive decontamination methods are needed, mild soap and water can be used. NOTE: It is very important to turn off the instrument before performing any decontamination.

19.8 SUMMARY

Radiation detection instruments are necesary and integral components of radiological decommissioning and site assessment processes. With the new decommissioning rules establishing cleanup criteria that are in the range of background radiation levels, close attention must be paid to the selection and operational parameters of instrumentation. The consequences of poor instrument selection or setup could include releasing a site that exceeds the release criteria, or performing unnecessary/expensive remediation of a site that meets the release criteria.

19.9 REFERENCES

U.S. Environmental Protection Agency. Multi-agency radiation survey and site investigation manual (MARSSIM). Washington, D.C: EPA; NUREG-1575, EPA 402-R-97-016; December 1997.

Schleien, B; Terpilak, M.S. The health physics and radiological health handbook. Second Printing. Maryland: Nucleon Lectern Series; 1984.

U.S. Nuclear Regulatory Commission. Title 10 Code of Federal Regulations Part 20, Subpart E. Radiological criteria for license termination. Washington, D.C: NRC; 62 FR 39088; July 1997.

U.S. Nuclear Regulatory Commission. Title 10 Code of Federal Regulations Part 30.36. Expiration and termination of licenses and decommissioning of sites and separate buildings or outdoor areas. Washington, D.C: NRC; 62 FR 39090; July 1997.

U.S. Nuclear Regulatory Commission. Minimum detectable concentrations with typical radiation survey instruments for various contaminants and field conditions. Washington, D.C: NRC; NUREG-1507; 1998.

U.S. Nuclear Regulatory Commission. Demonstrating compliance with the radiological criteria for license termination. Washington, D.C: NRC; Draft Regulatory Guide 4006; August 1998.

U.S. Nuclear Regulatory Commission. Supplemental information on the implementation of the final rule on radiological criteria for license termination. Washington, D.C: NRC; 64 FR 68395; December 1999.

U.S. Nuclear Regulatory Commission. A nonparametric statistical methodology for the design and analysis of final status decommissioning surveys. Washington, D.C: NRC; NUREG-1505; 1998.

U.S. Nuclear Regulatory Commission. Title 10 Code of Federal Regulations Part 40, Appendix A, Criterion 6(6). Criteria relating to the operation of uranium mills and the disposition of tailings or wastes produced by the extraction or concentration of source material from ores processed primarily for their source material content. Washington, D.C: NRC; 64 FR 17506; April 1999.

19.10 ACKNOWLEDGMENT

I want to thank Shepherd Miller, Incorporated (SMI) of Fort Collins, Colorado for their contributions and continued support of my work on this manuscript. Most of my experience with radiation detection instrumentation used for decommissioning was gained while I was employed by SMI. Additionally, SMI offered their equipment, references, and guidance throughout the development of this manuscript. For all that SMI has done, I am truly grateful.

Chapter 20

DECOMMISSIONING SURVEYS WITH AUTOMATED EQUIPMENT

Joseph J. Shonka

20.1 INTRODUCTION

At the onset of the Protestant Revolution, seizure of lands held by the Catholic Church in England drove the need to develop instrumentation and methods for land surveys. There was a need to accurately divide large parcels of church-owned property as spoils for the noble class in England. One can trace the origins of many of the civil engineering instruments and methods in use today for surveys to that era. The first widely used chain (a common distance measuring device) was developed by Aaron Rathborne, who went on to write the classic text "The Surveyor in Foure Books" published in 1616. His guidance from 400 years ago is somewhat more succinct than the Multiagency Radiation Survey and Site Investigation Manual (MARSSIM), and still applies today (U.S. NRC 1997):

> *"Before we enter the fields to survaie, I hold it necessarie wee provide us of fitting furniture for the purpose, lest by our neglect therein, those by whom wee have imployment receyve no lesse losse and preiudice, then our selves shame and reproch: Wherefore let us first consider what instruments are most usuall, and then of those, what most fit for our present purpose."*

(Quotation courtesy of Jane Insley, Environmental Sciences Curator, National Museum of Science and Industry, London, England)

Before we enter the fields to perform a radiation survey, we must consider the selection of equipment that is most fit for our purposes. In order to do this, it is useful to decompose the functional elements required for the survey. One such decomposition is to consider that five elements are required: a sensor for radiation, a sensor for location, a means to readout and record the sensor(s), a platform on which to integrate all of these components, and the data management system used for mathematical treatment and reporting of the data.

All radiation surveys have integrated these five elements. A successful survey is one that integrates these elements while keeping appropriate balance among cost, performance and schedule. A Radiological Controls Technician (RCT) who uses orienteering (or a fixed grid laid out with a measuring tape) to locate his instrument in a room and performs a survey using a pancake Geiger-Mueller counter with a count rate instrument, and logs the data with a pencil and paper represents one extreme of the continuum of choices. A robotic vehicle that autonomously performs a radiation survey and uplinks the data into an automated data management system represents another extreme. Either can perform a successful survey if the balance among cost, performance and schedule is well considered.

20.2 RADIATION MEASUREMENT

The basic physics of radiation detectors has been understood for many decades, and is well described in many textbooks. Detectors based on gaseous, liquid and solid phase media are commonly available in a wide range of materials. Detection principles vary from collection of light from scintillation to collection of ionization in the media. With sufficient gain, gas phase detectors can exploit media gain via the Townsend discharge in the gas. Scintillation detectors exploit gain in photomultiplier tubes and avalanche photodiodes. A comparatively large industry of vendors provides an impressive number of options for a systems designer.

The nuclear instrumentation industry has provided integration of the radiation sensors into measurement systems, which have seen widespread commercial use. Such systems include hand-held radiation survey instruments as well as fixed object counting applications such as conveyor systems or portal monitors. The industry has not developed integrated systems, for the most part, for large-scale surveys in support of Decontamination and Decommissioning (D&D). This is likely due to the declining U.S. nuclear market and the small size of the potential D&D marketplace. In this void, most users are faced with three alternatives: applying commercially available systems to their needs, contracting for services from companies that have proprietary integrated systems, or integrating systems for their own internal use. The growing computer power (and ease of use) available to users is lowering the threshold for integration of systems directly by users.

The handling and treatment of data sets from automated systems is different in scale from traditional methods, and the individual considering automating instrumentation should devote the preponderance of effort in consideration of data management, rather than the automation alone. Modern computerized radiation detectors are capable of generating one million (or more) measurements per hour. Additionally, without computerization most instruments (other than those in use in a laboratory) have been used in a mode in which the data is not preserved for a posteriori analysis. Such use intrinsically limits the capability to assert a detection limit to the Currie formulation (Currie 1968).

The MARSSIM Manual and supporting documentation have comparisons and guidance for selection of instrumentation (U.S. NRC 1997). This guidance compares the sensors on the basis of the individual detector alone. Users often improperly limit themselves by considering the use of only one detector at a time. As detection systems become computerized, the user is capable of providing significant enhancements to the system by considering combinations of sensors. Sensor combinations can be used to measure a variable component (either in space or time) present in background in order to subtract it, thus providing a measurement without the influence of the added component.

For example, in March 2000, surveys for free release of surplus office trailers were required at the Haddam Neck Nuclear Plant. The trailers were located in highly variable background radiation fields due to the presence of the Steam Generators, which had been removed for shipment to a waste site. The survey was performed with two computerized radiation detectors, one of which was beta sensitive, and the other identical detector incorporated a low atomic number window that stopped all beta radiation. Subtraction of the two readings from the identical detectors at the same location removed the spatial variability of the background gamma radiation from the desired surface contamination reading*. Use of two or more automated beta detectors with increasing filtration would permit, in principle, application of a Feather analysis to both the average beta reading in a survey unit and to any localized hot spots.

The automation of radiation detection equipment offers many opportunities for novel system integration and novel measurements that better fit the need of complex surveys encountered in decommissioning.

20.3 LOCATION MEASUREMENT

Establishing the location of a measurement using automated means requires that a system designer consider how the survey will be performed. Surveys can either follow a pattern, or can be taken in an unpatterned sequence.

Traditionally, most users have employed simple orienteering and tape measures to establish reference grids for surveys. Electronic versions of the tape measure are readily available with survey wheels (which can measure distance using a precision wheel encoder), as well as laser and acoustic electronic tape measures. These systems can be used to reduce the effort required to grid a survey unit. If the survey is performed in a regular pattern (e.g. back and forth rows with both the count rate and distance traveled data logged), then automating with these systems permits the location to be recorded with no need to separately place a grid in the area.

---

* Dubiel, R. W. Private Communication, March, 2000.

Significant effort is spent in establishing a grid and documenting the locations of measurements with reference to the grid. Systems are categorized as "high tech" when an automated location measurement method is applied to radiation surveys. These automated systems are often in widespread use in civil engineering. They are in widespread use because they significantly lower the cost of civil surveys.

Any automated position system will have limitations and even may suffer degradation or loss of data from time to time. Successful deployment requires that the user consider how data will be treated when accurate positional data are lost. Some systems assign the data lacking positional information by placing them with an assumption that the data were spaced evenly between the last two valid fixes of the positioning system. Greater reliability can be obtained by using two or more methods to establish location; with the redundancy in information providing a greater assurance that unambiguous location data is always available. A wheel encoder offers a low cost way to provide measurement of distance, and the output of the device is a pulse stream, which is similar to that obtained from most radiation counters. Thus, automation of a wheel encoder provides a basic system for automating the location of a reading within a survey block.

20.3.1 Total station

The surveyor's theodolite has been radically transformed in recent years through robotics and automation. The instrument is now called a total station, and uses lasers and wheel encoders to automate the theodolite functions. In recent years, autotracking and robotic total stations have become available. Following setup, these stations are capable of providing highly accurate location measurement. The first user of the Total Station may have been Argonne National Laboratory, who used a Leitz Model SET 3 for surveys for the State of Illinois (ANL 1988). Robotic Total stations are available for $25,000, or can be rented for $50 to $100 per day. Total stations can locate a point within the line of sight of the instrument within centimeters. The auto-tracking total station is used for location for the Laser Assisted Ranging and Data System (LARADS), which uses a Topcon Autotracking Total Station in conjunction with an Eberline Instrument Corporation's Model E600 data logging monitor (see http://em-52.em.doe.gov/pubs/itsrs/itsr1946.pdf).

20.3.2 GPS

In recent years, growing numbers of users have begun incorporating Global Positioning System (GPS) into survey systems. Use of a differential correction improves the resolution from 20 meters to better than 2 meters for an instantaneous reading.

Greater accuracy can be obtained by remaining stationary and averaging the readings for a longer time Post-processing of GPS data offers improvement to precision centimeter levels.

20.3.3 Sensor Readout

Computer based readout of detectors has become widespread, with most vendors offering computer compatible electronics. The choice of readout device often permits choices for the data that would be presented to the operator, and the data that are logged for subsequent analysis. These choices are driven by the methods and procedures that will be used. As will be discussed later in data analysis, there is a substantial difference between a conventional survey that scans (but does not record the observation unless a reading in excess of a detection limit is observed), and a scan and record operation, in which all of the data are logged. Logging all of the data permits a posterior analysis of the data, which can improve the understanding of how impacted an area is, often by up to an order of magnitude.

20.3.4 Platform

Three examples of platforms will be shown in the following section, a motorized cart for beta surveys, a land vehicle for outdoor gamma surveys, and an automation of a technician using conventional instrumentation. In general, the job requirements serve to drive the platform choices. The system integrator should seek to obtain a solution that is independent of platform.

20.4 SYSTEM INTEGRATION

Given the choices to be made in radiation detection and location sensors, computers to read out the sensors and platforms for deployment, the preferred implementation is one, which optimizes the selection for a given survey. The U.S. Department of Energy has tested numerous technologies, and provides reports on the test results (see http://em-52.em.doe.gov/itsrall.html#s) Examples of systems are as follows:

20.4.1 SCM

The Shonka Position-Sensitive Radiation Monitor System (see http://em-52.em.doe.gov/pubs/itsrs/itsr1942.pdf) consists of a cart-mounted surface contamination monitor (SCM) and a survey information management system (SIMS) (see Figure 20.1). The SCM consists of a position-sensitive proportional counter that acts as the equivalent of hundreds of individual detectors aligned in a 6-foot-long detector row. Using a low-cost wheel encoder, the detector scans over the surface in a series of 6-foot-wide strips. The wheel encoder is calibrated such that when a survey starts, the system automatically records the radiological and location information.

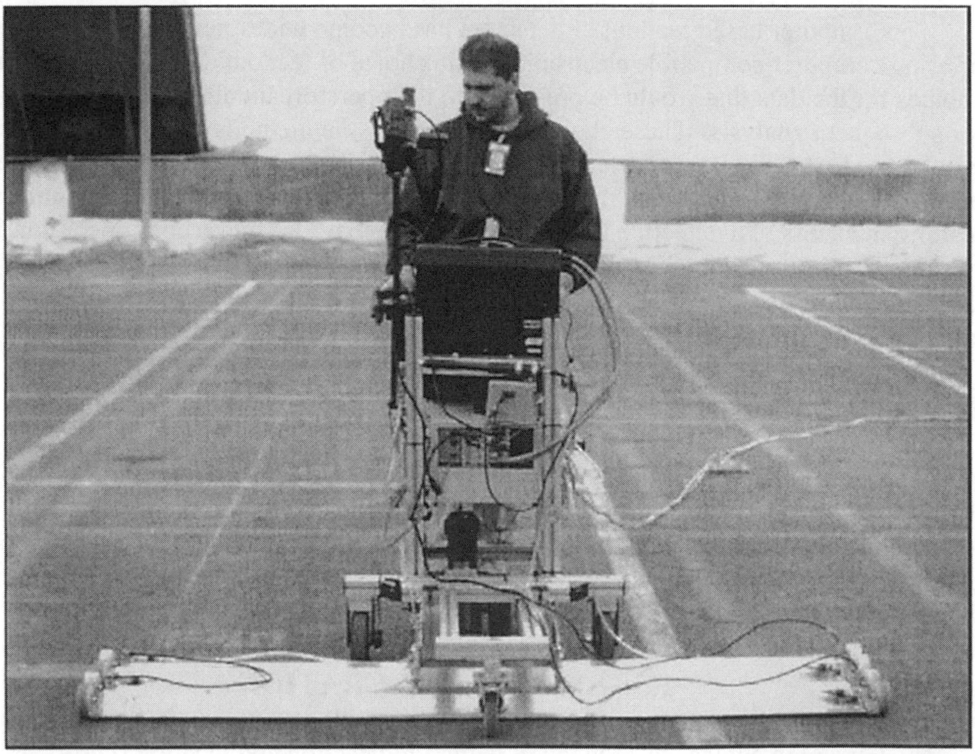

Fig. 20.1 - The Surface Contamination Monitor (SCM).

20.4.2 SMCM

With Small Business Innovative Research (SBIR) funding from the Nuclear Regulatory Commission (NRC), SRA has developed the Subsurface Multispectral Contamination Monitor (SMCM), which provides a scanning in situ measurement method.

The SMCM method refines a gamma ray scan by recording the entire spectrum rather than a gross count or from a single channel analyzer. If the spectra are further analyzed, this method limits or avoids the need for investigations, since the data that would be collected in a fixed spectrometer measurement are already available.

The location of the scanning spectrometer system is established by differential global positioning system (GPS) measurements that are recorded simultaneously with the gamma spectra. The GPS data are accurate to about one meter. The data can be compiled for areas as small as 1-meter square. The detectors are placed at a height where the individual detectors have a largely uniform response over the assigned area, which is called a pixel. When adjacent detectors are used to interpolate a response between the detectors (often called sum zones in process monitors such as laundry monitors), the overall capability to respond to a point source anywhere in the measured field is not compromised. Although the detector is responding to hundreds of square meters, the data are assigned to areas ranging from 1 to 16 square meters. Successive measurements actually overlap the area measured to a considerable extent. A four-detector system is capable of surveying more than an acre per hour when the detectors are placed at 2-meter height with the data assigned to a 4-meter pixel. The height fixes the image resolution of the data. The survey speed is then chosen to meet the requirements given by the DCGLs.

20.4.3 Platform

Fig 20.2 shows a 4-detector version of the SMCM deployed on an all terrain utility vehicle for surveys of moderately rough terrain. The motorcycle in the foreground was used in support of operations.

20.4.4 Radiation Sensor (Detector)

The SMCM scanning measurements use five-inch diameter by two-inch thick (5X2) NaI(Tl) scintillation detectors. These detectors have 50 times the scintillator mass as a 1X1 (e.g. Ludlum Model 44-2) that might be used for monitoring background gamma radiation fields.

20.4.5 Data Analysis

SRA has developed, with NRC funding, the application of spectral components to the interpretation of gamma ray surveys. This analysis substantially improves the information gamma ray spectra collected using the SMCM.

Fig 20.3 is a 20 second gamma ray spectra taken with a 3X3 NaI(Tl) detector placed 0.5 meters above the ground adjacent to a sewage pond. The spectrometer was set to have an energy range of 0 to 3 MeV. The “X” axis is plotted in channels from 0 to 512, and the "Y" axis is in counts. The marker (vertical line in the spectra extending

Fig. 20.2 - The Subsurface Multispectral Contamination Monitor (SMCM).

above the data) is placed in channel 250, where a peak is evident, corresponding to ^{40}K in the clay soils. The data from the channel that the marker is set on is below the graph and shows a 4-count value (in 20 seconds), corresponding to 0.2 cps. This would not be considered an adequate spectrum for spectroscopy purposes. Fig 20.4 is another spectra in the same area with the cursor moved to channel 112, corresponding to ^{137}Cs. ^{137}Cs is present, and a peak is noticeable to a trained eye, but is not well defined. The 609 KeV peak from ^{214}Bi (a progeny of naturally present uranium) interferes with the ^{137}Cs to a variable degree. This interference is difficult to remove, and affects all similar field measurements.

A little more than 100 such spectra were taken at one-meter intervals around the sewage pond. A waterfall plot can be used to portray these spectra in one image. In this plot, each spectrum is plotted as a line of color dots, with the color of the dot scaled to the count rate. Successive spectra are plotted as additional rows of dots. Fig 20.5 shows the waterfall plot from 200 spectra taken as a NaI(Tl) detector was moved around the sewage pond. The "X" axis is energy from 0 to 3 MeV, and the "Y" axis corresponds to the spectra as the detector was moved around the sewage pond from the northwest (NW) corner in a clockwise direction. Because the spectra can all be seen at once, the ^{40}K, present at about 15 pCi/g, can be more clearly seen as a vertical blue line in the center of the image. The ^{137}Cs corresponding to the spectrum shown above, along with adjacent acquisitions, can be seen as a bright line that extends to the right halfway between SW and NW. The ability to see ^{137}Cs is impacted by the background from primordial radioactive materials, primarily potassium and uranium. Different color scales would also show lower level peaks extending as vertical lines in the image.

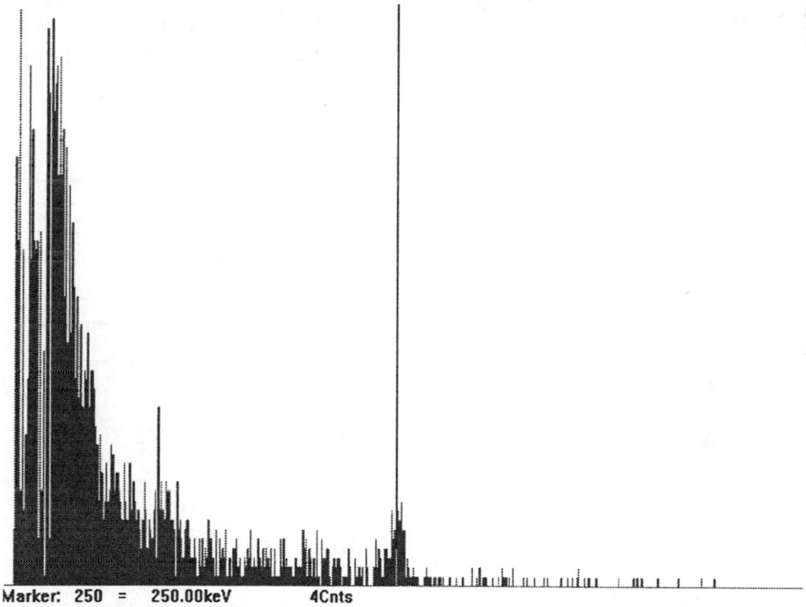

Fig. 20.3 - NaI(Tl) Spectrum without ^{137}Cs

When detectors are used in a manner where successive data are related to the prior data, smoothing can result in considerable reduction of the inherent statistical variability. In SRA's method, a Bayesian method is used to smooth the data. By way of explanation, in conventional gamma ray spectroscopy, the spectra are often "smoothed" by passing an averaging filter of a few channels through the data. This is extended by SRA in two more dimensions with post-survey processing of the data. The filter method does not pass a fixed filter (such as a fixed sliding 5 channel average), but rather the size of the filter is dynamically changed based on the local standard deviation of the data. By comparison, this might be thought of as a process for a single spectrum that would reduce the number of channels averaged when rapid changes in data occur such as in the vicinity of peaks. This method preserves the peak information while averaging background. This

Bayesian filter serves to greatly reduce the statistical uncertainty in the spectra since each spectrum uses the information in adjoining acquisitions to establish the mean behavior. Rather than spend minutes of counting time in one location (in order to get a good spectrum), seconds are spent for each spectrum, which are then averaged over an area consistent with the point response function for the same detector.

Fig 20.6 shows the data that have been smoothed, with the background component removed. The ^{137}Cs is now clearly visible. The ^{137}Cs was present in localized areas at levels close to the DCGLs suggested for this survey. This image shows that the spectral components method is a valuable adjunct to traditional in situ gamma ray spectroscopy. This provides significant improvement over simple single channel analyzers.

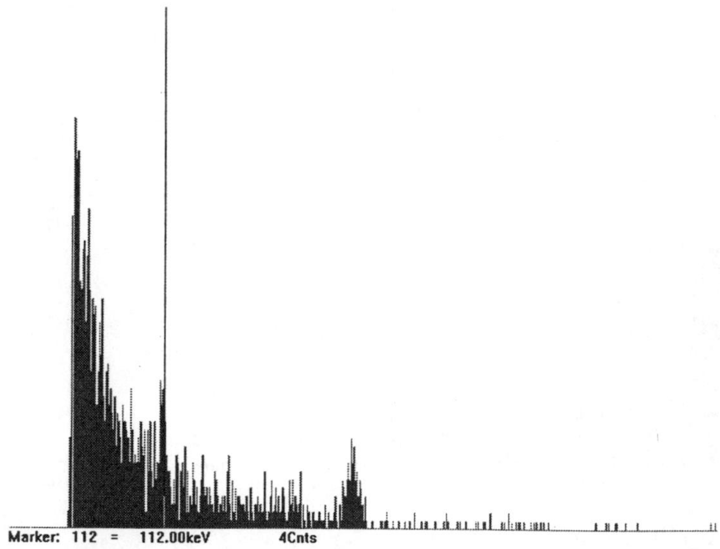

Fig. 20.4 - NaI(Tl) Spectrum with ^{137}Cs

Fig 20.7 presents the ^{137}Cs data as a map around the pond, with data adjoining the meter closest to the fence added. The image is scaled to provide more detail for lower contamination levels by adjusting the scale such that all data greater than about 5 pCi/g (gross) are shown as white. The data have not been adjusted for background to permit the viewer to more easily see the edges of the survey domain. (Black areas represent areas in which no measurements were taken, including inside the fence surrounding the pond). The figure shows that the contamination extends in an east-west direction from the midpoint of the fence surrounding the west side of the pond, and also shows lower level contamination beginning at a more southern point along the western fence. The fixed grid measurements established the background ^{137}Cs in this area as an average of 0.5 pCi/g. This is the level of ^{137}Cs from fallout in this area of the country. The red and black mottled areas with no evidence of elevated contamination are measurements at this level. Additional localized areas with elevated ^{137}Cs to the south can also be seen. The survey was extended until heavy woods prevented scanning 100% of the area. Nearly one thousand measurements were recorded to produce this image.

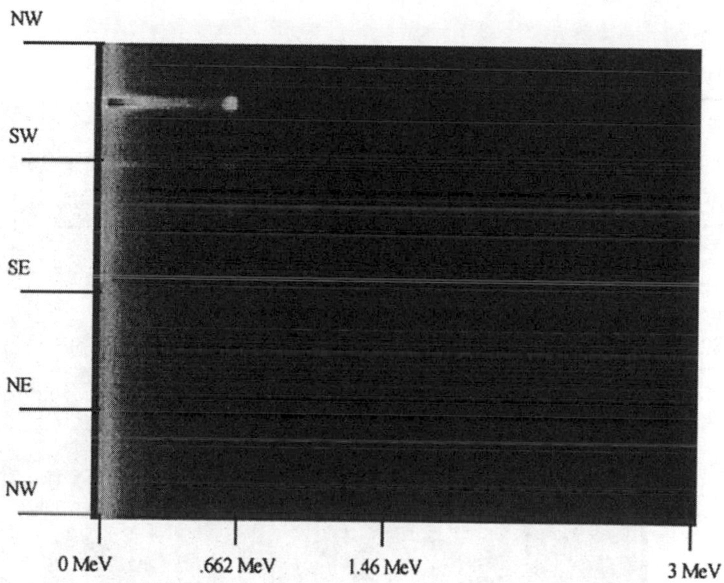

Fig. 20.5 - Waterfall Plot of Spectra

In Fig 20.7, the backgrounds around the pond are consistent with 0.25 to 0.50 pCi/g ^{137}Cs from fallout. In the SE to SW pass, as one proceeds up the vertical scale in the background corrected waterfall plot in Fig 20.6, a faint line just below the SW tick mark can be seen. The brightest objects in the image are the in the upper quarter of the figure, corresponding to the SW to NW traversal. Those artifacts correspond to localized contamination that ranged from tens of pCi/g to over 100 pCi/g. The areas were small, typically less than a meter wide, and were from various sources such as an overflow line for the sewage pond.

SRA's method could ultimately replace *in situ* spectroscopy. In the Environmental Measurements Laboratory (EML) method, a triangular grid of 5 meters is laid out and a measurement made at each grid point for 5 minutes or more. Calculations are made to define how large a source might have been present between measurements and not be seen. In the SRA method, the time is spent scanning all of the area, and Bayesian methods are employed to smooth the data, providing more detailed information while not sacrificing the detection limit of the system. Spectral components (developed originally in the remote sensing field for sonar) are used to separate the spectra from ^{40}K from that of ^{137}Cs.

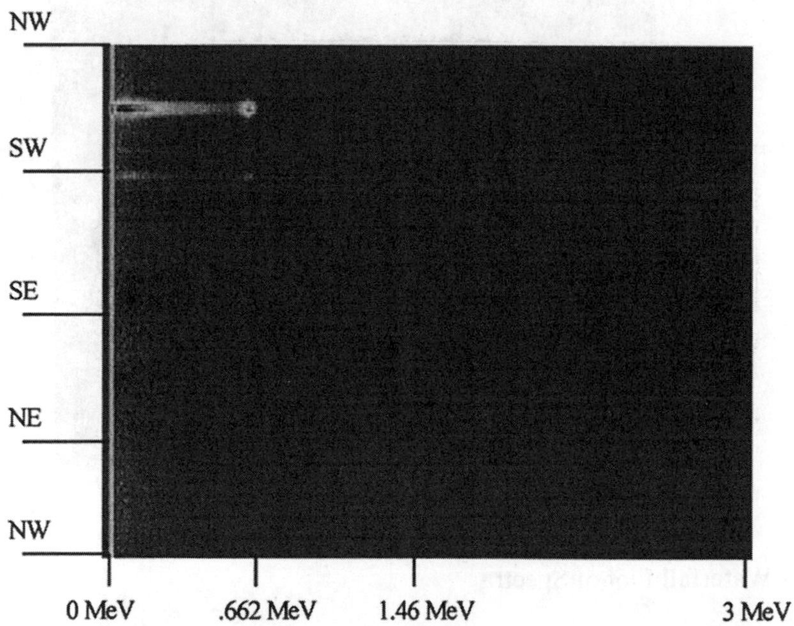

Fig. 20.6 - Waterfall Plot, Smoothed and Background Adjusted.

20.5 BACKGROUND DETERMINATION FROM SURVEY DATA

An estimate of background is needed to correct any gross measurement of contamination. Guidance for the determination of background has focused on finding an equivalent area of similar construction, age, and use that can be confirmed to have never been used in a manner that would result in radiological contamination. This process results in an inexact estimate of background. Various guidance documents have recommended this a priori method. As a practical issue, this has resulted in the traditional practice of estimating the surface contamination through sampling. A good deal of effort is spent in optimization to estimate the minimum number of samples required to meet some data quality objective. The cost for a sampling program of measurements scales with labor, which scale with the number of required measurements to meet the data quality objective. However, unless a similar area is found, unquantifiable uncertainty exists as to the exact value to use for background.

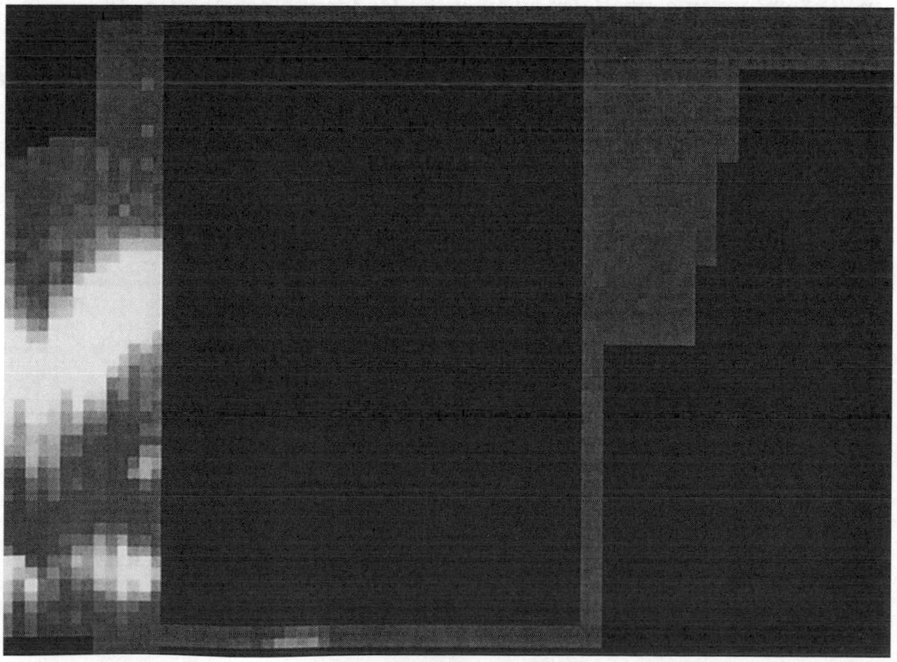

Fig. 20.7 - Pseudocolor Image of Pond Survey Color Trimmed to 5 pCi/g (gross).

New classes of automated instrumentation have changed that traditional paradigm. If the differential cost of a measurement is negligible, measurement of thecomplete surface can be made. When the complete surface is measured, one can establish background from the data itself (Shonka 2000).

The traditional approach to background determination is to define clean or reference area based on previous characterization data and historical site knowledge, then take samples within these representative areas prior to actual survey. Release criteria may dictate rigor of background determination (e.g. background may be able to be ignored if limits are high in comparison).

When using automated systems, an a posteriori (after the fact) statistical analysis based on actual data collected can be used. Background is established from unaffected (uncontaminated) surface areas within the area being surveyed. This can best be seen in the Cumulative Frequency Distribution (CFD) plots that are generated for each survey area. CFD's present count data as the probability of occurrence as a function of activity. CFDs are an extremely useful tool for visually investigating contamination data. They permit easy recognition of data that are not distributed in a normal fashion, as well as any statistical outliers that may be present in the data.

Fig 20.8 shows an asphalt driveway at the Haddam Neck Nuclear Plant, including a section that was replaced to repair a water pipe. This surface was 100% scanned in using the SCM. Fig. 20.9 shows an image plot of the same area. The area behind the fence on the left of the photo was not surveyed and appears black in the image. The asphalt section that was repaired appears brighter, with significantly more counts.

Fig. 20.8 - Photograph of varying aggravate asphalt materials at the Connecticut Yankee Haddam Neck Plant.

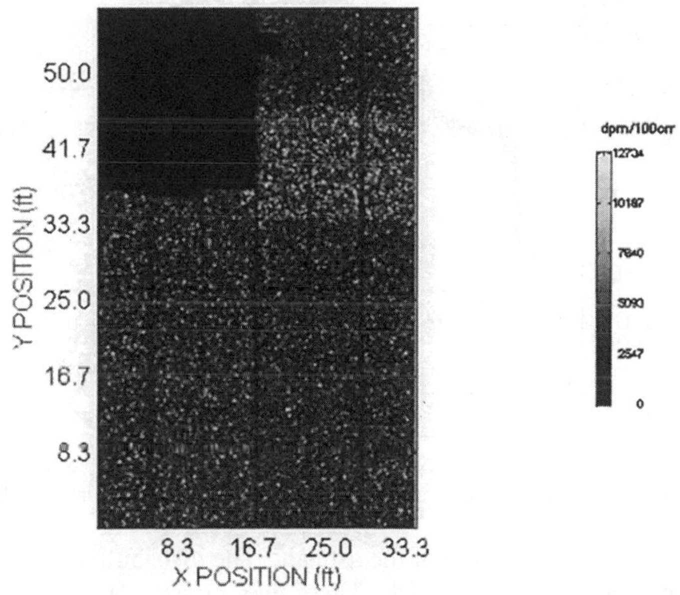

Fig. 20.9 - 2-dimensional image plot of same asphalt surfaces shown in Fig 20.8.

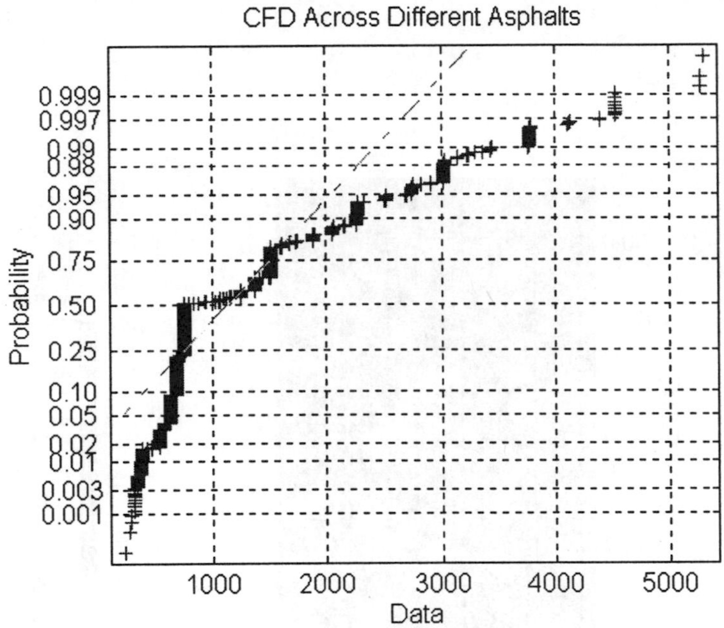

Fig. 20.10 - Cumulative frequency distribution plot for asphalt surface shown in Fig.20.8.

Fig. 20.10 provides a CFD of the asphalt. The distribution does not appear as a straight line, but rather as two segments. These two segments are correlated with the darker and lighter areas of the image. CFD's allow us to recognize that the elevated region (left) is likely an artifact of a different aggregate asphalt material and not surface contamination. The distributions are roughly a factor of two apart in their means. The means can be determined permitting the correct background to be applied to each type of asphalt in the survey unit.

20.6 REFERENCES

Argonne National Laboratory. Estimate of Volume of Radium Contaminated Soil on Five Sites in Ottawa, Illinois. Argonne National Laboratory; ANL/ESH/TS-89/100, September-October, 1988.

Currie, L. A. Limits for qualitative detection and quantitative determination. Analytical Chem. 40:586-693; 1968.

Shonka, J. J.; Weismann, J. J.; DeBord, D. M. Establishment of Background for Surface Contamination Measurement from the Area Surveyed. Paper presented at the Health Physics Society Meeting, Denver, CO, June, 2000.

U.S. Nuclear Regulatory Commission. Multiagency Radiation Survey and Site Investigation Manual (MARRSIM). Washington, D.C. U.S. Nuclear Regulatory Commission; NUREG-1575, 1997.

Chapter 21

THE GEORGIA INSTITUTE OF TECHNOLOGY RESEARCH REACTOR DECOMMISSIONING

R.D. Ice, N.E. Hertel, E. Jawdeh, R.S. Eby, S.G. Marske, and L.A. Lundberg

21.1 INTRODUCTION

21.1.1 Historical Perspective

In 1957 the Georgia Nuclear Advisory Commission recommended that a research and teaching reactor be built and operated. With a combination of private money and state financing, a nuclear reactor was built on the Georgia Institute of Technology campus in Atlanta, Georgia. The campus is located in downtown Atlanta. The reactor was built in the early 1960's at a total cost of $4.5M and was dedicated in January 1963 as a component part of the Neely Nuclear Research Center Facility. At that time, the facility was the largest single building project undertaken at Georgia Tech. The Georgia Tech Research Reactor first went critical on Dec. 31, 1964 under the auspices of a 30-year reactor license. The reactor operated at 1 MW (t) until 1974. In 1974, the license was amended and the reactor upgraded to operate at 5 MW (t). Based upon research and training needs, the reactor was operated until Nov. 1995. The total operation was about 40,000 MW-hr.

21.1.2 Description of the Reactor

The Georgia Tech Research Reactor (GTRR) was a 5 - MW thermal, heavy-water-moderated and -cooled reactor, fueled with plates of aluminum and highly enriched uranium alloy. The reactor core was approximately 2 feet in diameter and 2 feet high. When the reactor was fully loaded with fuel, it contained 19 fuel elements, spaced 6 inches apart. Each element contained 16 fuel plates per assembly; each plate was 0.05 inches thick, 23.5 inches long, and 2.85 inches wide. Each element also contained 11.7 g of 93% enriched uranium. The fuel elements were centrally located within a 6-foot aluminum reactor vessel that provided for a 2-foot thick D_20 reflector that completely surrounded the core. The reactor was controlled with four cadmium shim-safety blades and one cadmium-regulating rod. The reactor vessel was mounted on a steel support structure and suspended within a thick-walled graphite cup. The graphite provided an additional 2 feet of reflector, both radially and beneath the vessel. The core and reflector system was completely enclosed by the lead and concrete biological shield. The bioshield was housed inside a cylindrical containment building made of steel and concrete. The containment building was approximately 82 feet in diameter and 50 feet tall.

The reactor included a heat removal system, a D_20 purification system, and a shield cooling system, a D_2O storage system, a radiolytic gas recombiner system, and a ventilating system. The heat removal system was composed of a primary heavy-water system and a secondary light-water system. All components in contact with the D_2O were fabricated of stainless steel or aluminum.

Since the GTRR was intended for research applications it was designed with a variety of experimental beam ports that allowed for a wide range of investigations. Experiments that required high-intensity neutron or gamma-ray beams could be accommodated, as well as those that required a uniform thermal neutron flux throughout a large 2 feet by 2 feet sized beam. The reactor produced a thermal flux of greater than 10^{14} neutrons/cm^2/sec at a power of 5 MW. Irradiations of short duration and irradiations that required rapid sample recovery could also be accomplished with a pneumatic tube system. In addition, the reactor contained a biomedical irradiation facility, although no biomedical experiments were ever performed.

The reactor was housed in a containment building with three floors. The basement contained the heat exchangers, the ventilation equipment and two rooms with pneumatic tube access and space for experimental systems. The main floor provided space for the reactor bioshield and for experiments that used one of the available horizontal beams from the reactor. The control room was located on the third floor on the top of the biomedical irradiation facility. Access to the reactor was via an air lock connected to the adjoining laboratory building. In addition, there was an emergency airlock available that led to the outside and airtight sealed truck access door for moving in and out heavy and/or large equipment.

21.1.3 Decision To Decommission the Reactor

The decision by the administration to decommission the reactor involved numerous facets and personnel. No single factor led to the GTRR decommissioning. However, one factor catalyzed the decision process—some events of the 1996 Olympic games were going to be held on the Georgia Tech campus. Furthermore the campus was going to be transformed into the Olympic village, temporary home of over ten thousand athletes from over 160 countries. The high visibility of the Olympics led to increased questions about reactor security during the Olympics and increased public interest.

The reactor's 30-year operating license was due to expire in 1994. The reactor license renewal process also, heightened public interest as evidenced by the filing of a petition alleging 21 contentions with GTRR operations. As a result, there was an extensive review by the Nuclear Regulatory Commission (NRC) of the contentions. The

NRC concluded that there was no basis for 19 of the contentions but two of the contentions deserved additional evaluation and should be aired via pubic hearings. After the public hearing, the last two contentions were also dropped, but meanwhile, much publicity, mostly negative, clouded reactor operations. When the licensure process was finished, NRC decided to relicense the reactor for an additional five years.

Another factor effecting decommissioning was the status of the reactor. The reactor had been designed with an anticipated 30 years lifetime and the reactor had reached the design lifetime. It was now time to reevaluate systems and to upgrade areas within the reactor. Staff analysis indicated that to bring the reactor up to current standards for use would require a one-time funding of about two million dollars.

The reactor was used both as a research facility and a training facility. However, overall utilization of the reactor was less than 15% of available time. This low utilization factor can be attributed to the decreased enrollment of nuclear engineering students and the increased regulatory burden required of faculty.

Finally, the reactor had been initially fueled with highly enriched uranium (HEU), i.e. 93% uranium-235. As a matter of public policy, HEU fueled reactors were scheduled by the DOE for replacement with low enriched uranium (LEU), i.e. 20% ^{235}U. The GTRR was high on the DOE list for fuel replacement and new LEU fuel was fabricated. Thus the HEU fuel was scheduled for return to DOE. Operations staff had planned to make the fuel exchange in 1995. The opportunity was obvious, remove the HEU before the Olympic games and replace the fuel with LEU after the games, thus obviating an Olympic security problem.

The reactor was shut down by administrative decision November 17, 1995 and the HEU fuel removed after 90 days cool down. The Olympics came and went. On July 1, 1997 the administration announced the decommissioning of the reactor, despite relicensure and already fabricated fuel. The reactor had no fuel on site, thus the decommissioning factors outweighed the potential restarting factors.

21.1.4 Decommissioning Requirements

Georgia Tech's decommissioning decision involved removing all NRC-licensable materials from the site, termination of the license, and release of the site for unrestricted use. The site release guideline for the facility selected was NRC *Regulatory Guide 1.86* "Terminating of Operating License for Nuclear Reactors," June 1974, with unrestricted use guideline values of 5,000 dpm/100 cm^2 for total contamination and 1,000 dpm/100 cm^2 for removable contamination for the isotopes of interest. Georgia Tech selected the DECON alternative, which includes the removal of all fuel assemblies, source material, radioactive fission and corrosion products, and all other radioactive and contaminated materials that have activity levels above the unrestricted release values.

Currently, reactor decommissioning activities are no longer performed under *Regulatory Guide 1.86* and are guided by NRC's *Draft Guide DG-4006, NUREG-1575 Multi-Agency Radiation Survey and Site Investigation Manual (MARSSIM)*, and by the new *Code of Federal Regulations* 10 *CFR* 20 Subpart E "Decommissioning Criteria." These latter and more current guidelines use a dose-based rather than a radionuclide contamination based approach.

21.2 HISTORICAL SURVEY

During the 30 years of operations their had been many changes in staff and faculty positions at the facility. A consolidated evaluation of incidents, contaminations, spills and problems had never been produced. Thus, while not required, it was determined that a Historical Survey of the GTRR should be done.

A student, co-majoring in health physics and public policy, elected to do the historical survey as a MS dissertation project. The objective of the historical survey was to document the "corporate memory" from all available current and previous staff and faculty. A list of persons, who may have memory of incidents, etc. at GTRR, was developed by existing staff and given to the student for follow-up. The student then contacted the identified former staffer/faculty/student. She then made either an appointment to meet the individual and to tape his or her response to a series of predetermined questions or sent out an extensive questionnaire. The data generated were then compiled and analyzed.

The resultant thesis gave the current staff a good document that aided in identifying spill occurring areas and identifying the scope of radioactive materials for which decommissioners would have to be concerned. The dissertation also turned out to be a positive factor in decommissioning as it kept knowledgeable personnel informed of the reactor status and they in turn served to be a great source of information to the public. One negative observation was noted in the dissertation, i.e., petty internal politics dominated some of the questionnaires.

We concluded that the historical survey is an integral part of the decommissioning process. However, one needs to separate out those goals associated with decommissioning from internal selective memory by former employees.

21.3 CHARACTERIZATION SURVEY

Unlike the historical survey that relied on archived records and individual memories, the characterization survey was a detailed analytical survey of the nature and extent of radiological contamination in the facility to be decommissioned. Such a survey involved a detailed understanding of the reactor, sensitive instrumentation, knowledge of statistics and criteria expected for use of the data as part of the GTRR. The facts determined by the characterization study served as a basis for design of the decommissioning plan and also the basis from which the decommissioning cost estimate was made.

The Georgia Tech Office of Radiation Safety (ORS) is staffed on the basis of the number of authorized users of radioactivity and the number of radiation workers. The need for a nuclear reactor characterization survey was above and beyond the ORS current staff duties Georgia Tech therefor contracted for this characterization survey with a radiological engineering firm. The selection of the consultant contractor to do the survey was without the input from either the head of the nuclear engineering program or the radiation safety officer. Basically, when the decision was made to shut down and decommission the reactor, we were also told of the hiring of a radiological engineering contractor to aid us through the decommission process. While the intent was good—in reality the contractor relationship with GTRR and RSO personnel was non-synergistic (Ice, Jawdeh and Strydom).

The consultant characterization contractor did benefit Georgia Tech with the change of the operating license to a possession only license, by their suggested reorganization of the oversight committee (separating decommissioning activity from the campus radiation safety committee), and by their suggestion of an on site Executive Engineer during decommissioning. In contrast the consultant contractor did not understand our specific reactor and thus we basically had to write our own technical specifications for the license change, write our own decommissioning plan and supplement their characterization survey personnel. Their characterization report was found later to be inadequate, included significant errors and led to significant increase in decommissioning costs by the decommissioning contractor via change orders associated with discovery of radioactivity not identified in the characterization survey. Finally, the consultant contractor underestimated decommissioning costs by a factor of two! One of our basic lessons learned from this decommissioning project is that a good characterization survey is essential to future planning and cost estimating. For every dollar spent on characterization you save an estimated $10 of future decommissioning costs.

21.4 DECOMMISSIONING PLAN

21.4.1 Scope

The consultant contractor was hired to do the characterization survey of the facility and write the decommissioning plan (DP). NRC approved the decommissioning plan, after three iterations, on July 22, 1999, as an amendment to the existing possession only license (Amendment 14 to Facility Operating License No. R-97). Once submitted, NRC approval of the DP required approximately one year. The NRC approved the plan within weeks of the prescribed one-year allowance. One advantage to issuing the DP as an amendment to the license as opposed to a stand-alone decommissioning order was that minor changes to the DP can then be approved by the licensee (in this case, Georgia Tech) through an established 10 *CFR* 50.59, "Changes, tests or experiments" safety screening and evaluation process, as opposed to sending all changes back to the NRC for NRC action. Our Decommissioning Contractor (DC) has successfully used this 10 *CFR* 50.59 process to change: (1) the sequence of the decommissioning tasks; (2) the means of radioactive material control; (3) the method of plug storage vault removal; and (4) the reactor bioshield demolition approach. In addition, the process has been used to address the adoption of the current NRC Policy Issue for free release standards of tritium and ^{55}Fe.

21.4.2 General ALARA Plus Specific ALARA

21.4.2.1 ALARA - Worker Exposure Based on the Georgia Tech *Research Reactor Radiological Characterization Report* and past experience, the DP reported an estimated overall Total Effective Dose Equivalent (TEDE) budget of just below 8.0 person-rem for the project. At the start of each work activity, an as reasonably achievable (ALARA) plan is developed and an as low ALARA dose budget allocated. At four months into the decommissioning, the actual cumulative ALARA dose budget was approximately 1.75 person-rem. Yet the reportable (actual) cumulative TEDE (from Optically Simulated Luminescent [OSL] and Internal Dose [in-vitro analysis]) indicated a total of less than 1.0 person rem to date. Then a significant increase in dose occurred which correlated with the removal of the reactor vessel and activation products in the surrounding material. We fully expect to complete the project about 20% above the cumulative ALARA budget plan in the Decommissioning Plan.

21.4.2.2 Specific ALARA - Unique Georgia Tech Approach The Georgia Tech approach includes a unique aspect for decontaminating hot spots below the *Regulatory Guide 1.86* toward achieving the institute's ALARA goal of 20% of the regulatory guide values. As the contractor is decontaminating the facility, if hot spots are identified, the DC may propose a change order to the Executive Engineer to further reduce the contamination even though the release limits as defined in *Regulatory Guide 1.86* are met. The Executive Engineer and Georgia Tech have developed a procedure to assess the value of reduced dose compared to the cost of the decontamination effort to achieve the reduced dose. The requests may or may not be approved based on the overall dose reduction versus cost effectiveness assessment. As a result of this approach, Georgia Tech hopes to develop a cost versus dose reduction assessment, which may prove helpful to other parties interested in achieving specific ALARA goals in their decommissioning activities.

21.4.3 Decommissioning Funding

During the budgeting process, Georgia Tech took a unique approach to (ALARA), setting aside 10% of the decommissioning budget for specific ALARA. The money is used to implement existing Georgia Tech ALARA level policy as it relates to *Regulatory Guide 1.86* requirements for unrestricted release. The unique GT approach to ALARA involves specific State of Georgia funding that was set aside to decontaminate the facility below the *Regulatory Guide 1.86* and to meet the Georgia Tech ALARA goal of 20% of requirements. The approved budget was:

Decommissioning Contractor	$ 5.5 M
Contingency	0.4 M
ALARA Allowance	0.5 M
Other	0.1 M
Total	$ 6.5 M

21.5 SELECTION OF CONTRACTORS

In Georgia, the Georgia State Financing and Investment Commission (GSFIC) acts as the owner on construction jobs working for the using agency; in this case, Georgia Tech. Neither the GSFIC nor Georgia Tech had any previous experience decommissioning radioactive equipment, much less decommissioning a nuclear reactor. While waiting for approval of the DP, the GSFIC then hired, through a technical competitive procurement, CH2M HILL Inc. to serve as the Executive Engineer on the project, acting as the project management eyes and ears for the state and for Georgia Tech during the decommissioning project. CH2M HILL has extensive experience in decontamination and decommissioning major nuclear facilities and significantly assisted Georgia Tech in developing a model research reactor decommissioning.

After public notice of decommissioning bidder qualification, eleven contractors submitted their credentials for carrying out the GTRR decommissioning. Three open meetings were held giving tours of the proposed project site and answering all questions. A selection committee of the Georgia Tech Project Manager, Facility Director, Radiation Safety Officer, Executive Engineer and GFSIC representative then prequalified five potential Decontamination & Decommissioning (D & D) contractors and the low bidder IT Corporation subsequently won the job as DC. IT was awarded the contract on June 30, 1999.

21.6 DECOMMISSIONING OPERATIONS

21.6.1 Mobilization

One of the first tasks was a joint chartering session held among participating personnel from Georgia Tech, the State of Georgia, the Executive Engineer, and the DC. The purpose of the chartering session was to develop vision and mission statements for the project. The established vision of this project was "A facility (reactor, containment building and grounds) left in a condition that meets required safety codes and is suitable for conventional demolition/construction (i.e., unrestricted free release plus ALARA). The mission of the project team is to work together in an enjoyable teaming relationship to bring to fruition the vision while concurrently accomplishing the following:

- Remain a good neighbor to the surrounding communities
- Win the confidence of the local people and the State of Georgia
- Become a model for learning, as befits one of the premier locations in the country for the study of health physics
- Avoid negative impact to ongoing operations in the remainder of the facility

During the chartering session, critical success factors for the project were developed and barriers to achieving the goals were identified. Actions were taken to prevent or mitigate those barriers. Following the chartering session, the DC spent the first 60 days developing eight major project specific plans and procedures for executing the decommissioning contract, including policies and procedures for Health and Safety, Radiation Protection, Decommissioning Work plan, Quality Assurance, and the Waste Management Plan. These plans contained more than 70 individual procedures. These documents were submitted to the Executive Engineer and Radiation Safety Officer for subsequent approval by the Georgia Tech Technical Safety Review Committee (TSRC) prior to implementation. The TSRC is a standing committee at Georgia Tech composed of six senior people, including the Associate Dean of Engineering, the Chair of the School of Mechanical Engineering, a past President of the Health Physics Society, the Georgia Tech Radiation Safety Officer, the Director of the Neely Nuclear Research Center, and the Manager of Capital Projects for Georgia Tech. The DC mobilized on the site in December 1999.

The first activities were the initial confirmatory survey and the packaging and transportation of some miscellaneous radioactive and mixed waste from within the containment building. The DC included GTS Duratek on the team for waste handling and management. Except for the high-activity waste, the waste was packaged and sent to GTS Duratek's facilities in Oak Ridge, TN, for subsequent segregation and shipment to either Envirocare of Utah or to the Barnwell Waste Disposal Facility. The high-activity waste was sent directly from Georgia Tech to the Barnwell disposal site in a shielded cask.

21.6.2 Training

The training of all personnel associated with the project involved two areas of responsibility, Georgia Tech and IT, the decommissioning contractor. Georgia Tech, as the NRC licensee required:

General Employee Training for all personnel
Radiation Worker Qualification for all radiation workers
Radiation Control Zone Unescorted Access Training
Emergency Plan Training for all workers based on site

The Decommissioning Contractor provided all other training including:

Site Specific General Employee Training
OSHA
DOT
Procedure
Daily Safety Status Briefings

21.6.3 Schedule and Surprises

The original schedule of the decommissioning contractor's activities called for 18 months, which included 6 months for the NRC final survey and license termination review. The decommissioning initially proceeded fairly closely to the original schedule with no major delays. The first major delay in schedule was associated with the removal of the graphite. Surprisingly, some of the graphite reflector blocks were reading more than 300 mR/hr and a large number had readings between 20 and 80 mR/hr. Closer investigation revealed not only that the expected ^{14}C, which was present in the graphite as an activated constituent, but also that ^{60}Co was present at up to 300,000 pCi/g concentrations. The higher concentrations were associated with steel heli-coils that had been attached to the end of the graphite stringers to use to remove the stringers during the reactor operations so samples could be introduced for irradiation activities. Discussions with other research reactor D&D personnel who had worked on the Argonne CP series of reactors revealed another common problem that resulted from metal shavings from the saw blades that were used to cut the graphite to fit around the reactor vessel. These metal shavings became activated during the reactor operations producing the ^{60}Co source.

As stated, other graphite blocks were reading from 20 to 80 mR/hr. These blocks contained quantities of europium ^{152}Eu and ^{154}Eu. In reviewing the graphite specification during reactor construction, it was noted that the specified reactor graphite grade AGOT was thermally purified to drive off impurities; however, removal of rare earth elements, such as europium, requires a halogenated chemical purification at elevated temperatures. Envirocare's waste acceptance criteria for ^{60}Co are 30,000 pCi/g and for europium is 20,000 pCi/g. Preliminary analysis of the graphite stringers showed varying levels of activation products throughout the 45,000+ pounds of graphite.

Another surprise resulted from incomplete characterization prior to awarding the decommissioning contract. In particular, because Georgia Tech was operating under a "Possession Only License," there was some concern that destructive coring of the reactor was not permitted within the license operations. Therefore, past experience was used to model the neutron activation of the concrete within the biological shield. Activation was reported as extending 3 inches into the concrete or 67 inches from the center of the reactor core. Once the DC had mobilized and initiated the confirmatory survey, core samples were drilled in the biological shield. These samples showed that the concrete was actually activated 20 inches in to the concrete (~85 inches from the core) and contained significant amounts of iron resulting in more than 150,000 pounds of additional radioactive concrete as waste for disposal.

Monitoring the facility for potential tritium contamination presented difficulties. Because the reactor was a heavy-water-moderated and -cooled reactor, significant levels of tritium contamination were expected where heavy-water spills had occurred. The heavy water had been previously drained and 97% of it was recovered and shipped to the Savannah River Site. However, it was still expected that when coolant systems were opened, tritium contaminated water was likely to have pooled in pipe elbows and low spots. This did occur, but when these systems were opened, the tritium became airborne and gave the semblance of a much greater contamination problem in the facility. Increasing airflow through the ventilation systems resolved this concern.
The second major delay was associated with removal of the bioshield. What had been scheduled for 6 days of effort turned out to be four months of effort.

Another unexpected surprise of increased radioactive contaminated concrete and reinforcement rods associated with the concrete chase surrounding the gas expansion pipeline. The vertical pipeline paralleled the vessel chamber and penetrated through the graphite in a straight line (design flaw) thus permitting an unknown neutron-streaming beam. The pipe chase, with time, became activated.

21.6.4 Educational Objectives

One of the agreed chartering missions, was for the decommissioning to "Become a model for learning, as befits one of the premier locations in the country for the study of health physics" (Bollinger and Eby). Georgia Tech prides itself on the quality of its academic health physics program. Thus, no less standard would be acceptable for the modus operandi of the decommissioning.

Four video cameras were installed inside the containment building to record the decommissioning process. Tapes from these cameras are being summarized to produce a documentary that can be used for other decommissioning projects as well as for students at Georgia Tech. Output from the cameras is displayed in a "war room" where students, regulators, and other interested parties can come and observe the activities in real time. The war room also contains copies of the approved policies and procedures, active rad work permits, work packages, and data from the decommissioning project, allowing students to observe project delivery techniques as a learning experience. The educational objectives are clearly being achieved and it is clear that the project will continue to provide valuable lessons for the students.

21.6.5 Environmental Surveillance

As part of the approved Decommissioning Plan, Georgia Tech agreed to maintain its existing environmental surveillance procedures for one-year post license termination.

The State of Georgia, Environmental Protection Division, set up an on site continuous monitoring scintillation detector down wind at the facility fence line to give real time ambient radiation levels. In addition they established an air monitoring station, a facility rainwater from the tarmac sampling station and numerous thermoluminescent dosimeter stations. They continued their routine sampling of all effluents, soil and vegetation sampling. Although some direct radiation exposure from containers being shipped, no off site contamination has been detected.

21.7 CONCLUSIONS

21.7.1 Accomplishments

- The facility is nearing its objective of unrestricted release.
- A proactive ALARA program was funded and implemented.
- No injuries with over 50,000 contractor man-hours of effort.
- Over 600 tons of radioactive waste removed without incident, some recycled.

21.7.2 Lessons Learned

- Chartering sessions provided team focus on vision and mission.
- Characterize, characterize and characterize to prevent surprises.
- Don't use a fix price contract for an inadequately characterized facility.
- Any accumulated water is a haven for tritium.
- A proactive public relations program mitigates potential problems.
- 10CFR 50.59 is an asset.

21.8 ACKNOWLEDGEMENTS

The authors wish to acknowledge IT Corporation for its fine work as the DC. The authors wish to thank the GSFIC for its financial and contractual support of the D&D project. In addition recognition must be given to the many Georgia Tech staff, the CH2M Hill personnel, and the families for the long extra hours that were required to accomplish a project of this size.

21.9 REFERENCES

Bollinger, A.; Eby, R. "Project I-47 Chartering Meetings," Meeting Summary, CH2M HILL, Atlanta, GA, Aug. 20, 2000.

Fort, E.M. A Historical Site Assessment of the Georgia Tech Research Reactor, Georgia Tech M.S. Thesis, March 1999.

Fort, E.M.; Hereto, NE; Ice, RD; Bostrum, A. A Historical Site Assessment of the Georgia Tech Research Reactor, Health Phys. 76:S158; 1999.

Georgia Institute of Technology Research Reactor Decommissioning Project Radiological Characterization Report, NES Document No. 82A9087, NES, Inc., Danbury, CT, May 1998.

Hertel, N.E.; Sweezy, J.; Blaylock, D.; Ice, R.D. Accident Assessment of Activated Cadmium Control Rods for the GTRR Decommissioning. In Proceedings of the ANS Radiation Protection and Shielding Topical Meeting, Radiation Protection for Our National Priorities: Medicine, the Environment and the Legacy, Spokane, WA, Sept. 2000.

Ice, R.D.; Jawdeh, E.; Strydom, J. Research Reactor De-Fueling and Fuel Shipment, 31st Midyear Topical Meeting, Health Physics Society, Feb. 8-11, 1998.

Ice, Rodney D.; Ice, Ronald D.; Jawdeh, E.; Strydom, J. Contractor Radiation Safety, 43rd Annual Meeting, Health Physics Society, July 1998.

Reg. Guide 1.86, Termination of Operating Licenses for Nuclear Reactors, US Atomic Energy Commission, June 1974.

Nureg-575, Multi-Agency Radiation Survey & Site Investigation Manual, (MARSSIM), Dec. 1997.

Title 10, Code of Federal Regulations, Part 20, Subpart E, Radiological Criteria for License Termination, 1998.

Title 10, Code of Federal Regulations, Part 50.59, Changes, Tests and Experiments, 1998.

SUBJECT INDEX